Cyber Terrorism and Information Warfare

II. U.S. Executive and Congressional Perspectives

Yonah Alexander
and
Michael S. Swetnam

Sixth Volume, Second Series
Terrorism: Documents of International & Local Control

Yonah Alexander
Donald J. Musch
General Editors

Oceana Publications, Inc. / Dobbs Ferry, New York

Terrorism: Documents of International and Local Control, Second Series

9780379214154
9780379214178

Library of Congress Catalog Card Number: 99-75945

Manufactured in the United States of America on acid-free paper.

TABLE OF CONTENTS

Attorney General Documents

CIA Documents

Department of Defense Documents

CHAPTER 1

KEYNOTE ADDRESS BY U.S. ATTORNEY GENERAL JANET RENO ON HIGH-TECH AND COMPUTER CRIME

Delivered at the Meeting of the P-8 Senior Experts' Group on Transnational Organized Crime

Tuesday, January 21, 1997
Chantilly, Virginia

PROCEEDINGS

ATTORNEY GENERAL RENO: Thank you, Mark. I'm very touched by that introduction and I hope I can live up to it. I want to welcome you all to the United States for this first Plenary session of 1997. I am very pleased to be with you today.

Not only is this meeting the first P-8 meeting under the U.S. Presidency, it is the first multilateral meeting of President Clinton's second Administration.

As you know, yesterday the President took the oath of office for his second term. His re-election brings with it the opportunity for me to continue to work with international and domestic law enforcement to bring security to the citizens of our countries, and I consider this a very special privilege.

Besides the historic significance of this day, I want to share with you the excitement and the enthusiasm I feel about and toward the P-8. I view this group like no other: The P-8 countries are a special group made up of the world's most powerful democracies. We are global leaders in so many ways—economically, technologically, legally, and politically. Our small number allows us to act quickly, and our unique membership offers an opportunity to lead the world community that is rarely found in our history. And we are often on the cutting edge—for example—in responding to international terrorism, to international money laundering, to precursor chemicals. This group has so much promise. Through your work, giant strides are being made in several critical areas that have significant global implications.

No area of criminal activity is more on the cutting edge or has greater global implications than crime involving technology and computers. The importance of emerging technologies and the significance of global computer networks cannot be overstated. If properly developed and properly protected, they will be used in virtually all personal communications, financial transactions, information sharing, medical care, and a myriad of other applications. It is, indeed, a very exciting time.

But while new technologies allow us to do things that were previously impossible, they can also be misused in creative ways to threaten public safety and national security. The same technologies that facilitate lightning-fast and ultra-reliable transactions between computers can be misused by hackers, that is, by those who access computers without or in excess of authority. They can access confidential information, steal economic data, disrupt telephone networks, and interfere with the delivery of government and other vital services.

So while the information age holds great promise, law enforcement has a responsibility to ensure that the users of networks are not victimized in new ways.

To protect honest, law abiding citizens, law enforcement must keep pace with advances in computer and telecommunications technologies. We must work to ensure that the international law enforcement community can keep pace with the criminals. This is especially true in the case of computer offenses, which differ from traditional crimes in a number of ways and, as a result, create new and very challenging problems:

> First, international computer crimes are easier to commit. Hackers are not hampered by the existence of international boundaries, since information and property can be transmitted covertly via telephone and data networks. A hacker needs no passport and passes no checkpoints. He simply types a command to gain entry. And there is little need for manpower since a sole hacker, working alone, can effectively steal or erase as much information as he can read, or he can cause extensive damage to global networks.
>
> Secondly, until recently, computer crime has not received the emphasis that other international crimes have engendered. Even now, not all affected nations recognize the threat it poses to public safety or the need for international cooperation to effectively respond to the problem. Consequently, many countries have weak laws, or no laws, against computer hacking—a major obstacle to solving and to prosecuting computer crimes.
>
> Thirdly, law enforcement faces new procedural challenges, many of which are impossible to address without international consensus and cooperation. Consider, if you will, merely locating a hacker whose transmission passes from his computer to a local service provider, then through a telephone network, then crosses an ocean via satellite, and then passes through a university computer on its way to a corporate victim. To make matters worse, this hacker could be in his car, using wireless communications. How do we go about finding this individual? How do we collect the evidence and preserve it in a way that will be useful at trial?

Fourth, law enforcement will be faced with significant technical challenges, such as the widespread use of encryption. In such cases, we will have to find innovative and effective ways to preserve government access to the plain text of encrypted data. We can do this, in part, by supporting international efforts and national policies which promote the development of the emerging key management infrastructure and the use of products which allow for data recovery, as well as by assisting each other in this very difficult area.

I think that these threats and these problems call for the particular experience and the expertise of this group. While important work in the high-tech area is being done under the auspices of other organizations, one thing that sets the P-8 apart from other multilateral groups is its common-sense focus on practical solutions.

And the great thing about practical solutions is that they usually produce real results. Since computer crime is so important to all of our interests, there are several areas that I hope P-8 Experts will address. First, we need adequate laws which will allow us to prosecute hackers and other computer criminals. Second, we need the technical ability to find these individuals, wherever located. Third, we must develop legal procedures that permit timely cooperation in the collection of evidence. And fourth, we need to train law enforcement personnel and devote these technically literate experts to the task at hand.

When countries have inadequate legal structures to combat computer crimes, they provide safe havens for computer criminals, and they can create a major obstacle to obtaining international assistance in multijurisdictional cases. As you know, in 1990, the Council of Europe recommended that European nations adopt harmonious computer crime laws. As a result, several P-8 countries have enacted new laws and joined international efforts to encourage other countries to enact or to strengthen their computer crime laws. However, much work remains to be done in this area.

We need to reach a consensus as to which computer and technology-related activities should be criminalized, and then commit to taking appropriate domestic actions. This would also aid in providing the inevitable legal assistance required to investigate and prosecute these cases. I think it is also important to think about a global legal support regime, which could be used to avoid ad hoc approaches to multiple prosecutions. The unique nature of computer crimes and the unusual problems that can result would make such a regime very useful. Further, it would provide practical solutions as countries determine the best place for a prosecution, the order of prosecutions in a case where multiple countries are affected, and the most fair way to vindicate interests when a crime affects a large number of nations.

When a hacker attacks, the first investigative step is to locate the source of the attack. To do so requires tracing the electronic trail from the victim back to the attacker. However, in today's communications environment, one telecommunications carrier does not carry a communication from end to end. As in the example I mentioned before, a hacker's communication will pass through an array of carriers, often in less than a second, and tracing the electronic trail from victim back to attacker may be difficult or impossible unless the hacker is actually on-line.

One practical solution that our technologically advanced countries should pursue is maintaining access to source information for each link in the chain of transmission. Some countries, including the United States, have required that technical standards be adopted which ensure that "call set-up information" for normal telephone calls is accessible, so that the source of the call can be identified. I think it would be productive for P-8 Experts to consider whether all carriers should carry this kind of information, whether other communications technologies should be similarly designed, and what would be required for countries to share this information with one another. This is a critical time for this issue, as all of us are upgrading our telecommunications systems, because it is far easier to build such requirements into new machines rather than to retro-fit existing equipment.

Finding a criminal who plies his craft through an array of carriers becomes much more challenging when wireless communications are used. In the past, when a perpetrator used a phone to commit a crime, law enforcement could easily find out the exact location that the call came from. They could find out the name of the person who was being billed for the phone line, because the caller would be physically attached to a telephone wire. But today, mobile phones can allow an individual to commit crimes while roaming around a city or even a country.

Even identifying the owner of a particular mobile phone may be difficult, because mobile phones can be altered to transmit phony identifying information. Here, as in most of the areas we discuss, governments would be well-served to work on this problem with the help of industry. Our technical experts tell us that there are practical solutions to the problems created by wireless communications, such as encouraging the encryption of cellular electronic identifiers. I hope that P-8 Experts will work to see that law enforcement is not overtaken by technology in this area, but instead uses technology to thwart crime.

As the globalization of computer networks continues, and as computer criminals become more sophisticated, law enforcement increasingly will need timely access to computer or telecommunications information in all our countries. Up until this point, our regime of mutual legal assistance has served our countries well. But in a hacker case, the trail of evidence sometimes ends abruptly and permanently as soon as the hacker goes off-line. We should consider whether mutual legal assistance treaties and letters rogatory need to be supplemented with procedures that will facilitate the immediate collection and review of evidence, or whether other avenues should be explored. As mechanisms are developed, specially trained lawyers within countries' Central Authorities may be necessary to ensure rapid response to requests for assistance, particularly while a hacker is on-line. Again, the experience and the expertise of the P-8 makes it well-suited to tackle these very difficult problems. Practical solutions are out there—we must work together to find them.

One idea I believe worthy of consideration is formalizing international expedited procedures that protect electronic evidence on foreign soil from alteration or destruction. These could be in the form of "preservation of evidence requests," or "protected seizures," whereby an international request freezes a scene until a domestic judicial search mechanism can be used. Just like technological advances are the product of creativity and ingenuity, our legal work in this area must likewise be imaginative and forward-leaning.

Also in the area of evidence collection, I encourage this group to address the issues involved in analyzing electronic evidence—evidence which can be easily altered or destroyed. We must be able to analyze this evidence in ways that preserve its integrity and make its authenticity irrefutable, both for purposes of domestic prosecution and international cooperation. The ease with which digital evidence can be manipulated has already led to the development of scientific protocols for searching computers and for analyzing data. But we now must strive to ensure that such procedures are internationally accepted.

None of the advances I have discussed are possible without ensuring that law enforcement personnel are capable of addressing high-tech crime by understanding two emerging and converging technologies simultaneously: Computers and telecommunications. The complexity of these technologies, and their constant and rapid change, suggest that countries need to designate investigators and prosecutors to receive appropriate and on-going training. They, in turn, need to work these cases on a full-time basis, immersing themselves in computer-related investigations and prosecutions. Efforts along these lines will dramatically expand enforcement capabilities to solve high-tech crimes. I hope that when you return home, each of you will strongly advocate devoting significant resources to this area, and that we can share our expertise through international training and coordination efforts.

The issues confronting us are very, very difficult, but we can solve them. What will make it all come together in a cohesive way is law enforcement's continued willingness to recognize the new challenges that lay ahead in cyberspace. Whether the challenge is protecting trade secret information, defending intellectual property rights, prosecuting an international hacker, if we do our job right, the people of the world will enjoy the benefits of the information age without becoming its victims.

In closing, I pledge to you my full support in this very critical area. I consider high-tech crime to be one of the most serious issues demanding my attention, and I am doing everything in my power to ensure that the United States actively responds to these challenges. I have instructed Mark Richard to keep me apprised of your work, and I would enjoy the opportunity to contact my counterparts in your countries, if and when the need arises. In fact, this past November, I discussed the threat of high-tech crime with the British Home Secretary, Michael Howard, and he enthusiastically pledged his support to P-8 efforts in this area. Likewise, our Deputy Attorney General had a similar meeting with the German State Secretary of the Interior, Professor Doctor Kurt Schelter, in October of last year. It's an old cliché, but united we stand; divided we fall, and we look forward to working with you in every way we can to address this very important and very complex issue.

Thank you.

CHAPTER 2

LETTER BY ATTORNEY GENERAL JANET RENO IN SUPPORT OF MCCAIN/KERREY/HOLLINGS BILL 5.909 ON ENCRYPTION AND NATIONAL SECURITY

Office of the Attorney General
Washington, D.C. 20530

July 18, 1997

Dear Member of Congress:

Congress is considering a variety of legislative proposals concerning encryption. Some of these proposals would, in effect make it impossible for the Federal Bureau of Investigation (FBI), Drug Enforcement Administration (DEA), Secret Service, Customs Service, Bureau of Alcohol, Tobacco and Firearms, and other federal, state, and local law enforcement agencies to lawfully gain access to criminal telephone conversations or electronically stored evidence possessed by terrorists, child pornographers, drug kingpins, spies and other criminals. Since the impact of these proposals would seriously jeopardize public safety and national security, we collectively urge you to support a different, balanced approach that strongly supports commercial and privacy interests but maintains our ability to investigate and prosecute serious crimes.

We fully recognize that encryption is critical to communications security and privacy, and that substantial commercial interests are at stake. Perhaps in recognition of these facts, all the bills being considered allow market forces to shape the development of encryption products. We, too, plane substantial reliance on market forces to promote electronic security and privacy, but believe that we cannot rely solely on market forces to protect the public safety and national security. Obviously, the government cannot abdicate its solemn responsibility to protect public safety and national security.

Currently, of course, encryption is not widely used, and most data is stored, and transmitted, in the clear. As we move from a plain text world to an encrypted one, we have a critical choice to make: we can either (1) choose robust, unbreakable encryption that protects commerce and privacy but gives criminals a powerful new weapon, or (2) choose robust, unbreakable encryption that protects commerce and privacy and gives law enforcement the ability to protect public safety. The choice should be obvious and it would be a mistake of historic proportions to do nothing about the dangers to public safety posed by encryption without adequate safeguards for law enforcement.

Let there be no doubt: without encryption safeguards, all Americans will be endangered. No one disputes this fact; not industry, not encryption users, no one. We need to take definitive actions to protect the safety of the public and security of the nation. That is why law enforcement at all levels of government—including the Justice Department, Treasury Department, the National Association of Attorneys General, International Association of Chiefs of Police, the Major City Chiefs, the National Sheriffs' Association, and the National District Attorneys Association—are so concerned about this issue.

We all agree that without adequate legislation, law enforcement in the United States will be severely limited in its ability to combat the worst criminals and terrorists. Further, law enforcement agrees that the widespread use of robust non-key recovery encryption ultimately will devastate our ability to fight crime and prevent terrorism.

Simply stated, technology is rapidly developing to the point where powerful encryption will become commonplace both for routine telephone communications and for stored computer data. Without legislation that accommodates public safety and national security concerns, society's most dangerous criminals will be able to communicate safety and electronically store data without fear of discovery. Court orders to conduct electronic surveillance and court-authorized search warrants will be ineffectual, and the Fourth Amendment's carefully-struck balance between ensuring privacy and protecting public safety will be forever altered by technology. Technology should not dictate public policy, and it should promote, rather than defeat, public safety.

We are not suggesting the balance of the Fourth Amendment be tipped toward law enforcement either. To the contrary, we only seek the status quo, not the lessening of any legal standard or the expansion of any law enforcement authority. The Fourth Amendment protects the privacy and liberties of our citizens but permits law enforcement to use tightly controlled investigative techniques to obtain evidence of crimes. The result has been the freest country in the world with the strongest economy.

Law enforcement has already confronted encryption in high-profile espionage, terrorist, and criminal cases. For example:

- An international terrorist was plotting to blow up 11 U.S.-owned commercial airliners in the Far East. His laptop computer, which was seized in Manila, contained encrypted files concerning this terrorist plot.
- A subject in a child pornography case used encryption in transmitting obscene and pornographic images of children over the Internet.

- A major international drug trafficking subject recently used a telephone encryption device to frustrate court-approved electronic surveillance.

And this is just the tip of the iceberg. Convicted spy Aldrich Ames, for example, was told by the Russian Intelligence Service to encrypt computer file information that was to be passed to them.

Further, today's international drug trafficking organizations are the most powerful, ruthless and affluent criminal enterprises we have ever faced. We know from numerous past investigations that they have utilized their virtually unlimited wealth to purchase sophisticated electronic equipment to facilitate their illegal activities. This has included state of the art communication and encryption devices. They have used this equipment as part of their command and control process for their international criminal operations. We believe you share our concern that criminals will increasingly take advantage of developing technology to further insulate their violent and destructive activities.

Requests for cryptographic support pertaining to electronic surveillance interceptions from FBI Field Offices and other law enforcement agencies have steadily risen over the past several years. There has been an increase in the number of instances where the FBI's and DEA's court-authorized electronic efforts were frustrated by the use of encryption that did not allow for law enforcement access.

There have also been numerous other cases where law enforcement, through the use of electronic surveillance, has not only solved and successfully prosecuted serious crimes but has also been able to prevent life-threatening criminal acts. For example, terrorists in New York were plotting to bomb the United Nations building, the Lincoln and Holland Tunnels, and 26 Federal Plaza as well as conduct assassinations of political figures. Court-authorized electronic surveillance enabled the FBI to disrupt the plot as explosives were being mixed. Ultimately, the evidence obtained was used to convict the conspirators. In another example, electronic surveillance was used to stop and then convict two man who intended to kidnap, molest, and kill a child. In all of these cases, the use of encryption might have seriously jeopardized public safety and resulted in the loss of life.

To preserve law enforcement's abilities, and to preserve the balance so carefully established by the Constitution, we believe any encryption legislation must accomplish three goals in addition to promoting the widespread use of strong encryption. It must establish:

- A viable key management infrastructure that promotes electronic commerce and enjoys the confidence of encryption users.
- A key management infrastructure that supports a key recovery scheme that will allow encryption users access to their own data should the need arise, and that will permit law enforcement to obtain lawful access to the plain text of encrypted communications and data.

- An enforcement mechanism that criminalizes both improper use of encryption key recovery information and the use of encryption for criminal purposes.

Only one bill, 5,909 (the McCain/Kerrey/Hollings bill), comes close to meeting these core public safety, law enforcement, and national security needs. The other bills being considered by Congress, as currently written, risk great harm to our ability to enforce the laws and protect our citizens. We look forward to working to improve the McCain/Kerrey/Hollings bill.

In sum, while encryption is certainly a commercial interest of great importance to this Nation, it is not solely a commercial or business issue. Those of us charged with the protection of public safety and national security, believe that the misuse of encryption technology will become a matter of life and death in many instances. That is why we urge you to adopt a balanced approach that accomplishes the goals mentioned above. Only this approach will allow police departments, attorneys general, district attorneys, sheriffs, and federal authorities to continue to use their most effective investigative techniques, with court approval, to fight crime and espionage and prevent terrorism.

Sincerely yours,

Janet Reno
Attorney General

Louis Frech
Director
Federal Bureau of Investigation

Thomas A. Constantine
Director
Drug Enforcement Administration

Raymond W. Kelly
Undersecretary for Enforcement
U.S. Department of Treasury

John W. Magaw
Director
Bureau of Alcohol, Tobacco and Firearms

Barry McCaffrey
Director
Office of National Drug Control Policy

Lewis C. Marletti
Director
United States Secret Service

George J. Weise
Commissioner
United States Customs Service

CHAPTER 3

SUMMARY OF TESTIMONY OF ROBERT S. LITT, DEPUTY ASSISTANT ATTORNEY GENERAL, CONCERNING ENCRYPTION AND H.R. 695, BEFORE THE SUBCOMMITTEE ON TELECOMMUNICATIONS, TRADE AND CONSUMER PROTECTION, OF THE HOUSE COMMERCE COMMITTEE

September 4, 1997

The nation's policy on encryption must carefully balance important competing interests. The Department of Justice has a vital stake in the country's encryption policy because encryption may be used not only to protect lawful data against unauthorized intruders, it may also be used to conceal illegitimate materials from law enforcement. While we support the spread of strong encryption, we believe that the widespread dissemination of unbreakable encryption without any accommodation for law enforcement access is a serious threat to public safety and to the integrity of America's commercial infrastructure.

Public safety and national security must be protected against the threats posed by terrorists, organized crime, foreign intelligence agents, and others. If unbreakable encryption proliferates without accommodations for law enforcement, critical law enforcement tools, including wiretapping and execution of search warrants, would be nullified, and the potential harm to public safety could be devastating. U.S. law enforcement and intelligence agencies do not possess and cannot obtain the resources necessary to decrypt large numbers of encrypted communications and stored data. Our experiences demonstrate that this concern is not theoretical and not exaggerated.

Our goal is to encourage the use of strong encryption to protect privacy and commerce, but in a way that *preserves* (without extending) law enforcement's ability to protect public safety and national security. Accordingly, the Administration has promoted the manufacture and use of key recovery products, aided the development of a global key management infrastructure ("KMI"), and liberalized United States restrictions on the export of robust cryptographic products. We anticipate that market forces will make key recovery products a *de facto* industry standard and thus preserve the balance of privacy and public safety that our Constitution embodies.

Because of its support for key recovery, the Department of Justice cannot support H.R. 695 as it is presently drafted. The bill would discourage the development of a KMI. The bill would also eliminate all export controls on strong encryption and thus would undermine public safety and national security by encouraging the proliferation of unbreakable encryption. We believe it would be unwise simply to lift export controls on encryption for the sake of uncertain commercial benefits. This action would be particularly imprudent when there is the possibility of balancing individual privacy, public safety, and commercial needs through global adoption of a key recovery system. As we have learned through extensive international discussions in the last year, a consensus is now emerging throughout much of the world that the most suitable approach is the use of a "key recovery" or "trusted third party" system.

We look forward to working with this Subcommittee as we continue to develop and implement the Administration's approach.

Testimony of Robert S. Litt Deputy Assistant Attorney General concerning Encryption And H.R. 695 before the Subcommittee on Telecommunications, Trade, and Consumer Protection of the House Commerce Committee

September 4, 1997

Thank you, Mr. Chairman and members of the Subcommittee, for providing me this opportunity to discuss with you the important and complex issue of encryption. The Nation's policy on this issue must carefully balance important competing interests, and it is essential for all interested parties to recognize the validity and importance of all of these interests. The Department of Justice, whose interests I represent, has a vital stake in the country's encryption policy. Encryption will provide all of us the ability to protect lawful data against unauthorized intruders. But encryption can also be used to conceal criminal activity from law enforcement. Although the Department of Justice does not support H.R. 695 in its present form, we look forward to continuing the productive discussions we have had with Congress on this issue, with the goal of arriving at a policy that accommodates all of these interests.

In recent years, the issue of encryption has been vociferously debated in the United States. Having participated actively in these discussions, the Department of Justice believes today, as strongly as ever, that the widespread dissemination of unbreakable encryption without any accommodation for law enforcement access is a serious threat to public safety and to the integrity of America's commercial infrastructure. Our recent experiences only buttress this conclusion.

For example, just last week, in San Francisco, a man named Carlos Salgado, Jr. pleaded guilty to federal computer fraud and stolen credit card trafficking charges for crimes that he tried to obscure from law enforcement by his use of unbreakable encryption. Specifically, Salgado had stolen over 80,000 credit card numbers and intended to sell them for criminal purposes. These credit card accounts had a combined credit limit (and a potential loss to the 1,214 issuing financial institutions) of about one billion dollars. Salgado explicitly insisted on encrypting the stolen credit card numbers before delivering them on a CD-ROM to his purchaser. We were lucky in this case, because Salgado's purchaser was cooperating with the FBI. But if we had discovered this case another way, law enforcement could not have penetrated the information on Salgado's CD-ROM. Crimes like this one have serious implications for law enforcement's ability to protect commercial data as well as personal privacy.

Let me be clear: The Department of Justice supports the spread of strong encryption. Law enforcement's responsibilities and concerns include protecting privacy and promoting commerce over our nation's communications networks. For example, we prosecute under existing laws those who violate the privacy of others by illegal eavesdropping, hacking or theft of confidential information. Indeed, last year the Administration sought, and Congress passed, the National Information Infrastructure Protection Act of 1996, to provide further protection to the confidentiality of stored data. And we help promote commerce by enforcing the laws, including those that protect intellectual property rights, and that combat computer and communications fraud. (In particular, we help to protect the confidentiality of business data through enforcement of the recently enacted Economic Espionage Act.) Our support for robust encryption is a natural outgrowth of our commitment to protecting privacy for personal and commercial interests.

But the Department of Justice protects more than just privacy. We also protect public safety and national security against the threats posed by terrorists, organized crime, foreign intelligence agents, and others. Moreover, we have the responsibility to prosecute serious crime when it does occur. We are gravely concerned that the proliferation and use of *unbreakable* encryption would seriously undermine these duties to protect the American people, even while we favor the spread of *strong* encryption products that permit timely and legal law enforcement access and decryption.

The most easily understood example is electronic surveillance. Court authorized wiretaps have proven to be one of the most successful law enforcement tools in preventing and prosecuting serious crimes, including drug trafficking and terrorism. We have used legal wiretaps to bring down entire narcotics trafficking organizations, to rescue young children kidnapped and held hostage, and to assist in a variety of matters affecting our national security. In addition, as society becomes more dependent on computers, evidence of crimes is increasingly found in stored computer data, which can be searched and seized pursuant to court authorized warrants.

But if unbreakable encryption proliferates, these critical law enforcement tools would be nullified. Thus, for example, even if the government satisfies the rigorous legal and procedural requirements for obtaining a wiretap order, the wiretap would be worthless if the intercepted communications of the targeted criminals amount to an unintelligible jumble of noises or symbols. Or we might legally seize the computer of a terrorist and be unable to read the data identifying his or her targets, plans and co-conspirators. The potential harm to law enforcement and to the nation's domestic security could be devastating.

I want to emphasize that this concern is not theoretical, nor is it exaggerated. Although use of encryption is only in its infancy, we have already begun to encounter its harmful effects in recent investigations, in addition to the Salgado case described above.

- In the Aldrich Ames spy case, Soviet intelligence operatives directed Ames to encrypt computer files that he transmitted to them.
- Ramzi Yousef, recently convicted of conspiring to blow up 10 U.S. owned airliners in Asia, and his coconspirators apparently stored information about their terrorist plot in an encrypted computer file. (Yousef is also one of the alleged masterminds of the World Trade Center bombing.)
- One of the subjects in a child pornography case encrypted pornographic images of children before sending the pictures out on the Internet.
- The subject of a major international drug trafficking case used a telephone encryption device to seriously reduce the effectiveness of a court-ordered wiretap.
- In several major hacker cases, the subjects have encrypted computer files, thereby concealing evidence of serious crimes. In one such case, the government was unable to determine the full scope of the hacker's activity because of the use of encryption.

These are just a few examples of recent cases involving encryption. As encryption proliferates and becomes an ordinary component of mass market items, and as the strength of encryption products increases, the threat to public safety will increase proportionately. It is for this reason that the Attorney General and the leaders of many law enforcement organizations have written to the Congress urging them to support an encryption bill that preserves law enforcement's abilities to protect the public safety and our national security. I have attached a copy of that letter to my statement and would ask that it be made a part of the record.

Some have argued that people have a right to absolute immunity from governmental intrusion, regardless of the costs to public order and safety, and that any new technology that enhances absolute privacy should go unrestricted. But the Founding Fathers recognized that an absolute right to privacy was incompatible with an ordered society, and so our Nation has never recognized such an absolute right. Rather, the Fourth Amendment strikes a careful balance between an individual's right to privacy and society's need, on appropriate occasions, to intrude into that privacy. Our government has always been permitted to invade a person's privacy, for example by searching for and seizing personal communications and papers, when it is necessary to prevent, solve, and prosecute crimes, but, for the most part, we allow this only when the government demonstrates "probable cause" and obtains a warrant from the court.

Unbreakable encryption would upset our delicate constitutional balance, which is one of the bedrock principles of our legal system, by effectively nullifying a court's issuance of a search warrant or wiretap order. The notion that advances in technology should dictate public policy is backwards. Technology should serve society, not rule it. Technology should promote public safety, not defeat it.

Others claim that the fears of law enforcement are overstated. They argue that U.S. law enforcement and intelligence agencies can be given the resources necessary to decrypt encrypted communications. Essentially, they argue that expensive, fast computers can be used to decipher encrypted communications by "brute force" which essentially means trying every possible "key" (a sequence of symbols that determines the transformation from plain text to ciphertext, and vice versa) until the right one is found. They point to one highly publicized success of a group that deciphered a message encrypted with a 56 bit key and argue that law enforcement can surely do the same.

Yet that example underscores the problems that accompany a "brute force" approach. The successful group actually used over 14,000 computers and took over four months—over ten million hours of computer time—to decrypt one single message. That's really not practical for law enforcement if, for example, we're trying to prevent a terrorist attack or find a kidnap victim. And I hope you understand that law enforcement does not have the resources to better that result in any meaningful way. Significantly, the time needed to decrypt a message rises exponentially as the length of the encryption key increases. If the message had been encrypted with a 64-bit key, it would take 10,000 Pentium computers on average *58 years* to crack a single message.

And a new message would require law enforcement to start again from scratch because each message may be encrypted with a different key. During 1995, for example, federal and state courts authorized more than a thousand electronic surveillance court orders, resulting in over two million intercepted communications, each of which could require separate decryption. Given such numbers, brute force attacks are not a feasible solution. This commitment of time and resources is unavailable for every wiretap and every search and seizure executed at federal, state, and local levels.

Additionally, law enforcement agencies at the federal, state, and local level are finding that searches in routine, nonwiretap cases now commonly result in the seizure of electronically stored information. Because storage devices have increased in capacity and decreased in price, the quantity of data seized in "ordinary" cases continues to increase dramatically. If all of these communications and stored files were DES encrypted, brute force attacks would not provide a meaningful and timely solution. Thus, even if tens of thousands of computers were obtained and coordinated (an expensive undertaking, to say the least), the approximately 17,000 federal, state, and local law enforcement agencies could not be given timely access to the evidence we need to prevent and solve crimes.

Finally, many proponents of strong encryption advocate its proliferation precisely *because* it cannot be decrypted by the government. Thus, even if the government could acquire the ability to quickly decrypt DES encrypted communications and information, many of the advocates of absolute privacy would push for even greater key lengths, on the ground that 56 bit DES no longer provided acceptable security. But greater key lengths would, of course, increase the difficulty and cost of decrypting encrypted data even more. We must recognize that it will always be easier and cheaper to devise stronger cryptographic methods than to build computers powerful enough to break them in a reasonable period of time.

Our goal, then, is to encourage the use of strong encryption to protect privacy and commerce, but in a way that preserves law enforcement's ability to protect public safety and national security against terrorism and other criminal threats. We have engaged in extensive international discussions on this topic over the last two years, and a consensus is now emerging throughout much of the world that the way to achieve this balance is through the use of a "key recovery" or "trusted third party" system. Under this system, a key for a given encryption product would be deposited with a trusted third party or "recovery" agent. (Some entities, such as large corporations, might be able to hold their own keys, provided that certain procedural protections were established to preserve the integrity of a law enforcement investigation.) If the government had lawful authority to obtain the encrypted information, for example by a search warrant or a courtordered wiretap, it could likewise obtain the key from the recovery agent in order to decrypt the information it was entitled to get.

I want to emphasize particularly, because our position has often been misrepresented, that a key recovery system would create no new authority to obtain data, to examine personal records, or to eavesdrop. Access to encrypted data could be obtained only as part of a legally authorized investigation, and under the same circumstances that today would authorize access to unencrypted data. The same constitutional and statutory protections that preserve every American's privacy interests today would prevent unauthorized intrusions in a key recovery regime. All we would be doing would be preserving law enforcement's ability to do what it is legally and constitutionally entitled to do today. At the same time, though, individuals and companies would gain the benefit of strong cryptography to protect the confidentiality of their data, whether in storage or in transmission.

Effective law enforcement is not, however, the only reason to support a key recovery system. Business, as well, needs a routinely available method of recovering encrypted information. For example, a company might find that one of its employees had encrypted confidential information in the company's files and then absconded with the key, or just lost it. Without a key recovery system, the company would be out of luck. Key recovery thus serves important private interests as well.

In short, key recovery holds great promise for providing the security and confidentiality that businesses and individuals want and need, while preserving the government's ability to protect public safety and national security. Thus, Administration policy is to promote the manufacture and use of key recovery products, to develop a global key management infrastructure ("KMI"), and to liberalize United States restrictions on the export of robust cryptographic products in the hope that market forces will make such products a de facto industry standard.

For many months, we also have been engaged in serious discussions on this subject with foreign governments, which are now anxious to join us in developing international standards to address this issue on a global scale. In fact, an experts working group of the Organization for Economic Cooperation and Development has issued a statement of principles that acknowledges the need to consider public safety when establishing national cryptographic policies. We believe that key recovery encryption will become the worldwide standard for users of the GII. The United States can be a leader in this process.

If key recovery encryption does become the worldwide standard, U.S. businesses will be able to compete abroad effectively, retaining and even expanding their market share. At the same time, law enforcement agencies will have a legally authorized means of decrypting encoded data. This approach would therefore effectively serve the interests of *all* Americans.

The argument is sometimes made that key recovery encryption is not the solution, because criminals will simply use nonkey recovery encryption to communicate among themselves and to hide evidence of their crimes. But we believe that if our companies develop and market strong key recovery encryption products that will not interoperate with nonkey recovery products and a global KMI arises, key recovery products will become the worldwide standard. Under those circumstances, many criminals will use key recovery products, because products will be easily available from the mass market. And even criminals need to communicate with legitimate organizations such as banks, both nationally and internationally.

The cornerstone of our policy is encouraging the development of key recovery products and a KMI to preserve the balance of privacy and law enforcement that our Constitution embodies. For this reason we cannot support H.R. 695 as it is presently drafted. We believe that the bill would discourage the development of a key management infrastructure. Moreover, we believe that the central provision of the bill, Section 3 which would effectively eliminate all export controls on strong encryption would undermine public safety and national security by encouraging the proliferation of unbreakable encryption.

The first problem that we see with H.R. 695 is its failure to promote development of a key management infrastructure. The Administration believes that the development of a key management infrastructure is critically important for a safe society. H.R. 695 prohibits laws that would require a keyholder to relinquish keys to third parties under certain circumstances. Unfortunately, to the extent that this provision would actually prohibit government from encouraging KMI development, the provision would put public safety and national security at risk and is inadvisable. For example, it might preclude the United States government from utilizing useful and appropriate incentives to use key recovery. The government might not be able to require its own contractors to use key recovery or demand its use in the legally required storage of records regarding such matters as sales of controlled substances or firearms.

We also believe that export controls continue to play an important role in the Administration policy of promoting development of the KMI. We have heard, of course, the argument that the "genie is already out of the bottle" that unbreakable cryptography is already widely available overseas and over the Internet, that its dissemination cannot be halted, and that regulation serves only to handicap U.S. manufacturers seeking to sell their encryption products overseas. We disagree vigorously for a number of important reasons that I would like to explain to you today.

First of all, in recognition of the legitimate interests of U.S. software manufacturers, the Administration, as this Subcommittee is of course aware, has considerably liberalized export controls for certain commercial encryption products. The Administration transferred jurisdiction over commercial encryption products from the Department of State to the Department of Commerce at the end of December, a step that we expect will ease the burden on industry by providing for faster and more transparent decisions on applications for export licenses.

Most significantly, we have allowed unlimited export of key recovery products as well as export of nonkey recovery 56 bit encryption during a two-year transitional period by those companies that commit to the development of key recovery products. This willingness to permit unlimited export of products that incorporate key recovery clearly demonstrates that the Administration is in favor of the spread of strong encryption products, as long as they have accommodations for law enforcement access.

Second, although unbreakable encryption products can be found overseas, these products are not ubiquitous, in part because the export of strong cryptography is controlled today by both the U.S. and other countries. It is worth noting in this regard that export of encryption over the Internet, like any other means of export, is restricted under U.S. law. Although it is difficult to completely prevent encryption products from being sent abroad over the Internet, we believe that the present legal restrictions have significantly limited the use of the Internet as a means of evading export controls.

Third, the products that are available overseas are not widely used because there is not yet an infrastructure to support the distribution of keys among users and to provide interoperability among the different products. Such an infrastructure will have to be created in order to realize the full benefits of encryption, and we should strive to ensure that it is created in a way that preserves public safety.

Fourth, the quality of encryption products offered abroad varies greatly, with some encryption products not providing the level of protection advertised.

Finally, the vast majority of businesses and individuals with a serious need for strong encryption do not and will not rely on encryption downloaded from the Internet from untested sources, but prefer to deal with known and reliable suppliers. For these reasons, export controls continue to serve an important function.

It is also important to consider that our allies strongly concur that unrestricted export of encryption would severely hamper law enforcement objectives. Indeed, when the U.S. let it be known at a December 1995 meeting of the OECD that it was considering allowing the export of some stronger, nonrecoverable encryption, many of our allies expressed dismay at the prospect of such an action. They feared that unbreakable encryption would become so internationally pervasive that criminal organizations and terrorists would be able to use it freely. It follows that the elimination of U.S. export controls, as provided by H.R. 695, would have an even more devastating impact on international law enforcement. It would be a terrible irony if this government which prides itself on its leadership in fighting international crime were to enact a law that would jeopardize public safety and weaken law enforcement agencies worldwide.

In addition, it would be a mistake to assume that if the U.S. were to lift export controls, U.S. companies would have unrestricted access to foreign markets. This assumption ignores the likely reaction of foreign governments to the elimination of U.S. export controls. Up to now, most other countries have not needed to restrict imports or the domestic use of encryption, largely because export controls in the U.S. the world leader in computer technology and other countries have made such restrictions unnecessary. But given other countries' legitimate concerns about the potential worldwide proliferation of unbreakable cryptography, we believe that many of those countries would respond to any lifting of U.S. export controls by imposing import controls, or by restricting use of strong encryption by their citizens. For example, the import and domestic manufacture, sale and use of encryption products have already been restricted in France, Russia and Israel. And the European Union is moving towards the adoption of a key recovery based key management infrastructure similar to that proposed by the Administration. In the long run, then, U.S. companies might not be any better off if U.S. export controls were lifted, but we would have undermined our leadership role in fighting international crime and damaged our own national security interests in the meantime.

In light of these factors, we believe it would be profoundly unwise simply to lift export controls on encryption. The ability of law enforcement to protect of national security, personal privacy, and sensitive commercial data should not be sacrificed for the sake of uncertain commercial benefits, especially when there is the possibility of satisfying both security and commercial needs simultaneously through global adoption of a key recovery system.

We as government leaders should embark upon the course of action that best preserves the balance long ago set by the Framers of the Constitution, preserving both individual privacy and society's interest in effective law enforcement. We should promote encryption products which contain robust cryptography but that also provide for timely and legal law enforcement access and decryption. This is the Administration's policy. We look forward to working with this Subcommittee as we continue to develop and implement our approach.

I would now be pleased to answer any questions you may have.

CHAPTER 4

CONFERENCE ON CRITICAL INFRASTRUCTURE PROTECTION ADDRESS BY ATTORNEY GENERAL JANET RENO

Friday, February 27, 1998

Lawrence Livermore National Laboratory
7000 East Street, Building 123
Auditorium
Livermore, California
(12:00 noon)

MS. RENO: Thank you very much. It is a very special privilege for me to be here today.

This laboratory is such a great institution and it has such a distinguished history. I çan see why, after a half-hour with Bruce, I was the student since I had forgotten most of my chemistry. And it has been one of the most extraordinarily helpful and constructive half-hours I have spent in the almost five years I have been Attorney General.

This is an issue that is critically important to me: How we protect the systems and the networks of this nation that make its businesses run; how we create a system that can provide for the protection of our nation's defenses; how you get to the hospital emergency room on time; how do we protect those whom we hold dear from a threat of chemical weapons in a subway.

Our energy production and distribution channels, our transportation networks and our telecommunication systems are more vulnerable than ever before as we come to rely on technology more than ever.

This generation faces extraordinary challenges as we face the problems associated with weapons of mass destruction. This technology brings us a new century and a new world of incredible opportunities and of daunting challenges which, as Adlai Stevenson would say, stagger the imagination and convert vanity to prayer.

The government, including the Department of Justice, is facing these challenges head on and taking steps to ensure the protection of our critical infrastructures, but we know full well we cannot do it alone. To ensure the protection of our critical networks and systems, we must work as partners, true partners, with the private sector, with the academic world, with great institutions such as this, in this vitally critical effort for this nation.

I am here today to discuss what the Department of Justice, including the FBI, is doing to face the challenges. And I am here to hear from some of you what steps we can take to build a stronger, better, two-way, respectful, trusting partnership with everybody who has been so significantly involved in this effort, some for far longer than we have.

I want a partnership truly based on trust. As Bruce has indicated, in 1995 the President asked me to chair a cabinet committee that would assess the vulnerability of our nation's infrastructures and make recommendations as to how to protect them. The process we started led to the creation of the President's Commission on Critical Infrastructure Protection.

I would like to pay very special tribute to Tom Marsh, who did an extraordinary job. He did not just sit in Washington and listen to people. He went out to communities. He went to so many different places and listened to people because he knew full well how important it was to build a true line of communication in this very sensitive and significant area. And so thank you, Tom, for just some great and wonderful public service. (Applause.)

MS. RENO: As you know, the Administration is presently engaged in determining how to implement this report, so this conference could not be more timely. But one thing is certain, and the commission made sure of that: it is vitally important to the success of any effort that, it be based on the idea that infrastructure protection requires that we work together as never before.

It demands a partnership among all federal agencies with responsibilities for different sectors of the economy or for certain special functions, like law enforcement, intelligence and defense. It also requires a partnership with private industry which owns and operates most of the infrastructures. It calls for a partnership with academia and labs like the one hosting us today.

You have the scientific knowledge to develop technical solutions. I have already been through some of the process that you have been involved in, some of the processes that are actually critical to solving and protecting some of the very critical infrastructures that we have talked about today.

It also requires a partnership with state and local law enforcement. They are used to robbers with guns, but there are new criminals out there who do not have guns. They have computers, and they may have other weapons of mass destruction.

The use of weapons of mass destruction or cyber attacks on infrastructures that could lead to events like power outages or telecommunications breakdowns are not hypothetical. They are not speculative. They can happen. And it requires, in the end, a partnership with the American people who have the right to expect that all of us, whether we are an attorney general or a general, whether we are a scientist or a business person, that all of us are going to work together to protect this nation.

The Department of Justice and the FBI, as I have indicated, want to be strong, good partners. Let me face up to an issue. Some people get suspicious of law enforcement. They say, "I do not want to cooperate. I do not want people to recognize my vulnerability. I do not understand the criminal justice system."

We have a responsibility to work through the concerns that people may have so that they trust us. And I am here today and have been involved in trying to do outreach to those responsible for critical infrastructures to make sure that we hear from you as to how we can be a better, stronger partner in the process. And I have learned today, just from this lab, so much that can be done.

There are other concerns. For example, private business may be concerned about confidentiality. Business does not want to have proprietary information made public. The FBI, on the other hand, has a duty to provide an early warning to the community to prevent further attacks. We must work together to see how we can walk that narrow line and ensure that we do our duty in terms of preventing further attacks while at the same time maintaining the confidentiality of the person or institution or business involved.

The Department of Justice and the FBI have a duty to investigate and prosecute most attacks on the infrastructure, but there are constitutional and other legal limitations on what law enforcement can and cannot do. Fourth Amendment protections against unreasonable search and seizures is one of our citizens most sacred protections.

We must work with scientists as partners to develop technologies and processes that enable us to obtain evidence in strict adherence to the fundamental protections guaranteed our citizens by the Constitution. The private company that is the victim of a cyber attack must likewise understand law enforcement's responsibility to the Constitution.

Some dare to suggest that the Constitution, the most remarkable document that humankind ever put to paper, cannot keep up with modern technology. I say we must not and we will not sacrifice any constitutional protection in order to adapt to new technology.

We must and we will work with you to ensure that we will master the technologies and together, that law enforcement working with the private sector, working with the scientist, will make sure that technology can be adapted to meet the constitutional protections that are so critically important. But to do this, it is going to require that we talk together, that we work together and that we understand the problem. It may be a problem that a scientist can solve, but we need the Fourth Amendment expert working with the scientist to understand.

The FBI works daily to prevent attacks on the infrastructure. And it is prepared to immediately investigate if the attack occurs. United States attorneys and other Justice Department attorneys are available with technical expertise on a 24-hour basis to respond.

And if the plan is carried out, a cyber attack, if it is carried out by agents of a foreign state or international terrorist group, we have the responsibility as well under our foreign counter-intelligence authorities.

In the early stages of a cyber attack on an infrastructure or a power grid, we often have no way of knowing who was behind it, what their motive was or where they attacked from.

It is impossible to determine whether the attack is part of a terrorist plot, a probe by a foreign intelligence service, or a part of a national level military assault by a hostile nation state; or is it simply the work of a disgruntled insider bent on revenge against a supervisor; or is it a young juvenile hacker out to test his skills against the latest firewalls.

At the outset then, it may be premature to mobilize the military or redirect national intelligence assets.

What we do know, however, is that regardless of the perpetrator, his intent or his whereabouts, the intrusion in most cases constitutes a federal crime. This means the Department of Justice and the FBI have the authority and responsibility to investigate it.

Whether the crime is physical or cyber, we need to ensure that as we investigate we are coordinating with other agencies as appropriate. If the attack appears to come from non-U.S. persons located abroad, we would want to call on the intelligence community to assist in gathering information about the perpetrator's intentions; or if the attack seems to be part of a hostile nation's war plan or involves an attack on the Defense Department's own critical infrastructures, DOD obviously has a critical role to play.

Our challenge, our extraordinary challenge, is to identify the attack we need to know: When is it a straight law enforcement investigation that the FBI and the Assistant United States Attorney or Criminal Division lawyer control? When is it something that the National Security Council takes over? When is it something that clearly becomes international as opposed to domestic, and therefore the State Department controls?

What this means is that you do not have any ready answers, but you do have to develop a process—and we are in the process of doing that—to determine when we hand it off from one agency to the next, how we work together to make sure that we adhere to constitutional protections, how we adhere to Fourth Amendment issues, how we continue to adhere to the Constitution.

Now Bruce said you had been talking about that this morning. We have been talking about it constantly in Washington, and it is an extraordinary challenge. And civilian agencies also have important responsibilities and capabilities. Whether it is the Department of Energy in the event of an attack on a nuclear power plant or an electrical power grid, or the Department of Transportation in an attack on our air traffic control or rail systems, all these agencies have crucial roles in the event of a crisis. But the fact remains that law enforcement initially will have the lead responsibility for responding to an imminent or ongoing infrastructure incident.

One example of the partnerships that we need to foster can be found in a major New York hacker case. The FBI, Secret Service, NYNEX and Southwest Bell and a number of private companies and universities worked together to identify and prosecute successfully individuals who had hacked into a telecommunications network, a credit reporting company and other systems.

Meeting our responsibility to protect critical infrastructures, in my view, is one of the central challenges for law enforcement as we face the twenty-first century. As our reliance on the Internet, on automated systems and on other technological advances increases exponentially with every passing month so do our vulnerabilities to infrastructure attacks. Law enforcement must be prepared to confront this challenge and be prepared to do so in partnership with other federal agencies, with the private sector, with academia and with state and local agencies.

And thus today I am announcing the creation of the National Infrastructure Protection Center at the FBI. The NIPC's mission is to detect, to prevent and to respond to cyber and physical attacks on our nation's critical infrastructures and to oversee FBI computer crime investigations conducted in the field.

The center will build on the important foundation laid down by the FBI's Computer Investigations and Infrastructure Threat Assessment Center, which has been subsumed into the NIPC.

To ensure the strong partnerships that I consider vital, the NIPC will include representatives from the Defense Department, the intelligence community and other government agencies. We also very much want to and hope that the private sector will be a participant in this center, very much like it participated in the President's commission.

This is the surest, best, quickest way to build understanding, to learn from each other, to understand the responsibilities, the duties, the processes and the authorities that each agency or institution possesses. But let me be frank again. I know sometimes of the distrust that exists between agencies.

I want to hear from all concerned, all who are dedicated and vitally involved in the protection of our infrastructure; I want to know what we can do to build bridges of trust and understanding and communication, what we can do to better explain the role of law enforcement so that people can understand, what we can do to sit down with scientists and say, "Here is our law enforcement. How do we solve it?" We can do so much through this center if we work together.

To augment our partnership, we want to establish direct electronic connectivity with private industry and the Computer Emergency Response Team, or CERTs, which is located across the country. This is a significant departure from the way law enforcement has traditionally operated. But the challenges of infrastructure protection require imaginative solutions. And I consider our liaison and outreach to the private sector to be absolutely indispensable to our success.

One of the issues the private sector will raise is, "Why should we work with you in developing technology? How do we know that you will maintain confidentiality. What can we do?"

And in the last half-hour I have learned that I might find some examples here at the lab in the partnerships that you have built with the private sector in terms of determining solutions. It is fascinating what we can do if we will only sit down and talk together and build trust, recognizing that we all have one common objective which is the protection of this nation that we hold dear.

The partnerships that we envision will allow the NIPC to fulfill its responsibility as the government's lead mechanism for responding to an infrastructure attack. But the NIPC cannot just react from one crisis to the next. To do our job we will have to be able to prevent crises before they happen, and that requires analysis of information from all relevant sources including law enforcement investigations, intelligence gathering and data provided by industry.

Through partnerships between federal agencies and private industry and with interagency and private sector representation in electronic connectivity to all of our partners, the NIPC will be able to achieve the broadest possible sharing of information and comprehensive analysis of potential threats and vulnerabilities. And through its Watch and Warning Unit, the NIPC will be able to disseminate its analysis and warnings of any imminent threats to a broad audience in and out of government.

This will enable private industry and government agencies to take protective steps before an attack. But, at the same time, we can take steps together to protect the interests of all concerned and balance the responsibilities of everyone involved.

As we build our partnerships, we must ensure that whenever possible we share equipment, technology and know-how with each other and especially with state and local law enforcement who are on the front lines. Local police respond with guns now, but soon they will have to respond with cyber tools to detect an intrusion, to follow through, to find the person, to hold him accountable; and we must be there working with them.

This equipment will be expensive. And you scientists will create so much new equipment so fast that it will be vital that we all work together in every forum possible to make sure that we avoid costly duplication, that we develop research according to sound plans that look both to the defense and the law enforcement and the scientific interest, and that we do as much as we can working together, sharing.

We have established a track record in this area, but we have much to learn, too. One of the most important technological partnerships is the one we have established with the Department of Defense. In 1994 Defense and Justice created a Special Joint Steering Program group and staffed it with both Justice and Defense personnel.

We developed products such as the prototype see-through-the-wall radar; more affordable night vision devices, which have been instrumental in supporting and helping the Border Patrol; concealed weapons and contraband detection systems; and improved lightweight soft-body armor.

In addition to working with DOD, we have developed partnerships with the Department of Energy and with the National Aeronautics and Space Administration.

We point out those as if they are unusual. We should come to accept such partnerships as a way of doing business in everything that those of us involved in the protection of the infrastructure do.

But all of this only begins to touch on the range of things under development and the technologies needed by federal, state and local law enforcement. As technology becomes more essential to the mission of the U.S. criminal justice system, it has become more important that we better organize ourselves to fulfill these new requirements, because neither federal nor local law enforcement can afford to be isolated from scientific and technological developments.

Accordingly, I have directed the creation of a special working group to streamline the Department's management of research and technology development.

Finally, as many of you can sympathize, the information revolution has happened so quickly that kids in junior high school are often more familiar with the new technologies than your local sheriff or the FBI agent. We need to build a law enforcement work force that is educated and equipped to deal with the new technologies and knowledgeable and imaginative enough to think ahead to the next generation of problems.

The NIPC will help us do this by working closely with other interagency groups that are developing training for federal, state and local law enforcement personnel on cyber investigations and weapons of mass destruction.

By creating the NIPC, the Department of Justice is taking an important step: We are creating new partnerships with the private sector and with other government agencies to combat threats to the critical infrastructure.

I also have asked Congress to provide us with $64 million in increased funding to support our expanded efforts to protect the nation's infrastructure in fiscal year 1999.

These additional resources will be critical to support the NIPC and will also allow the FBI to create six additional computer investigation and infrastructure threat assessment squads to be deployed in cities across the country. And it will allow us to hire additional prosecutors to target cyber criminals.

As I mentioned earlier, however, not every attack on a computer network or infrastructure that is used in the United States constitutes an attack on our national security and, in fact, most do not. An unauthorized cyber intrusion could very well be, as I indicated previously, from a little hacker or a disgruntled insider. We will pursue those investigations as part of our law enforcement authority. But, nonetheless, part of protecting our critical infrastructure means working closely with the national security community to fight cyber attacks.

Cyber attacks pose unique challenges. Because of the technological advancements, today's criminals can be more nimble and more elusive than ever before. If you can sit in a kitchen in St. Petersburg, Russia, and steal from a bank in New York, you understand the dimensions of the problem.

Cyber attacks create a special problem, because the evidence is fleeting. You may have gone through this computer 1,500 miles away to break through another computer 5,000 miles away. Simply put, cyber criminals can cross borders faster than law enforcement agents can, as hackers need not respect national sovereignty, nor rely upon judicial process to get information from another country.

If we are to protect our infrastructure we must reach beyond our borders. Cyber threats ignore the borders. The attack can come from anywhere in the world. We must work with our allies around the world to build the same partnerships that we talk about here at home.

And to that end, a little over a year ago, I raised with my colleagues, the ministers of justice of the P8 countries, the eight predominant, largest industrial countries—Canada, France, Germany, the United Kingdom, Italy, Japan, Russia and our government—the issue of cyber crime and urged that we join together in developing a common response. Experts from all our countries and departments worked together in the interim. And last December the ministers came to Washington to meet in a day-long meeting that produced agreement as to the dimension of the problem and produced an action plan that I hope can bring real results in the year to come.

We must join forces around the world if we are to begin to deal with the cyber crime that may affect one person or the cyber threat to our infrastructure that may affect the entire nation.

To do this we must work very closely with our colleagues in the defense and intelligence communities both here and among our allies. And this presents the new partnership. While I am building partnerships with the Department of Defense, I am getting to know the minister of justice and the minister of defense in another country. Sometimes the problem seems so big, but it is so critical that we address it and understand that this great, wide world is now one that can be traveled in seconds.

Together we will determine whether emerging developments are a national security problem, a law enforcement problem, how to attack it, how to proceed. But until evidence is obtained that an incident is a national security matter, it is important that we not jump to conclusions, that we not conclude that we must use extraordinary measures that defy our Constitution.

If it has been determined that an incident is an attack on national security, then the Justice Department has three distinct roles.

First, we can conduct a criminal investigation that runs on a parallel track with the national security elements of the case. Indeed, criminal investigations often yield vital information and leads for the President's national security advisors.

Secondly, we can utilize the FBI's counter-intelligence authorities and techniques when our national security is under cyber attack from a foreign power.

And, third, we will ensure that any national strategy for dealing with a cyber attack is drawn up, executed and assessed with strict fidelity to our Constitution and to our laws.

I think this is the most extraordinarily challenging time that law enforcement has ever faced. Boundaries in this world have shrunk. Technology has burgeoned beyond man's wildest imaginations. It is a time for us to come together and realize that if we work together, if we talk together, if we trust each other and understand that we have one common goal which is the defense of this nation, we can make all the difference. If each discipline goes its own way, ignoring the other, we will not solve the problem, and this nation will be at peril.

This has been, in this one visit and about a brief half-hour, extraordinarily enlightening to me. And I go back to Washington confirmed in the belief that, based on the example of what you do here, we can make a difference and we can translate what you do here to so many other arenas and forums around this country where law enforcement, the private sector, the scientists are going to work together.

Thank you so very much for setting an example.
(Applause.)

(Whereupon, the address by U.S. Attorney General Janet Reno concluded at 12:35 p.m.)

STATE OF CALIFORNIA) County of San Joaquin) ss.

I, Susan Palmer, a Certified Electronic Reporter and Transcriber by the American Association of Electronic Reporters and Transcribers hereby certify that I reported, using the electronic reporting method, the proceedings had of this matter previously captioned herein; that I thereafter transcribed my audio recording to transcription by way of word processing; and that the foregoing transcript, pages 1 to 26, both inclusive, constitutes a full, true and accurate record of all proceedings had upon the said matter, and of the whole thereof. Witness my hand as a Certified Electronic Reporter and Transcriber this 28th day of February 1998.

Susan Palmer, CERT 00124 Palmer Reporting Services

CHAPTER 5

TESTIMONY OF ROBERT S. LITT PRINCIPAL ASSOCIATE DEPUTY ATTORNEY GENERAL BEFORE THE SUBCOMMITTEE ON THE CONSTITUTION, FEDERALISM, AND PROPERTY RIGHTS COMMITTEE ON THE JUDICIARY UNITED STATES SENATE CONCERNING PRIVACY IN A DIGITAL AGE: ENCRYPTION AND MANDATORY ACCESS

Presented on March 17, 1998

Thank you, Mr. Chairman and members of the Committee, for this opportunity to discuss with you the important and complex issue of encryption. Encryption holds the promise of providing all of us with the ability to protect data and communications from unlawful and unauthorized access, disclosure, and alteration. Moreover, encryption can help prevent crime by protecting a wide range of data as we and our valued information become more and more connected to each other and to potential adversaries through the spread of information networks. As a result, the law enforcement community supports the development and widespread use of strong encryption products and services.

At the same time, however, the widespread use of unbreakable encryption presents a tremendous potential threat to public safety and national security. Criminals and terrorists have already begun using encryption to conceal their illegal activities and to defeat important law enforcement and national security objectives. In developing our Nation's encryption policy, we must carefully balance the many different interests that the policy will affect. In seeking that balance, it is essential to understand both the promise and the peril that this technology holds, and to identify responsible ways forward that advance all of the competing interests.

I want to begin, Mr. Chairman, by clarifying the Clinton Administration's recent initiatives regarding encryption. For some time, the Administration's position has been to encourage the design, manufacture, and use of encryption products and services that allow for the plaintext of encrypted data to be recovered. The Administration's approach has in fact found support in the marketplace, in part because businesses and individuals need a routinely available method to recover encrypted information. For example, a company might find that one of its employees lost his encryption key, thus accidentally depriving the business of critically important and time-sensitive data. Or a business may find that a disgruntled employee has encrypted confidential information and then absconded with the key. In this type of case, a data recovery system promotes important private sector interests. Indeed, as the Government implements encryption in our own information technology systems, it also has a business need for plaintext recovery to assure that data and information that we are statutorily required to maintain are in fact available at all times. For these reasons, as well as to protect public safety, the Administration has been affirmatively encouraging the development of data recovery products, recognizing that only their widespread, ubiquitous use will both provide greater protection for data and protect public safety.

In further support of this goal, two weeks ago we set in motion a process of pursuing an intensive dialogue between industry and law enforcement. Our goal in this process is to bring the creative genius of America's technology leaders to bear in developing technical, market-savvy solutions that will enable Americans to realize the benefits of strong encryption while continuing to protect public safety and national security. We do not harbor any illusions that there is one magic technology, a silver bullet that addresses all the needs of the marketplace. But we think constructive dialogue in a variety of areas and fora is far preferable to a stalemate that arises from a battle of wills and rhetoric; working together is better than fighting legislative battles.

The Administration is not advocating any single product, technology, or even technical approach. Rather, we are flexible—provided that the resulting solutions and arrangements preserve the Nation's ability to protect the public safety and defend our national security. These are public interests of the highest order, shared by the Congress and by all of our law-abiding citizens. Industry has the technical know-how to develop commercially viable mechanisms that maintain the government's ability to safeguard its citizens, while protecting our citizens from unwarranted intrusions from any source.

Now let me describe in a little more detail the important law enforcement and national security interests that are at stake in the encryption debate. First, I want to reiterate that the Department of Justice supports the use of strong encryption. Law enforcement's responsibilities and concerns include protecting privacy and promoting secure commerce over our nation's information infrastructure. For example, we prosecute those who violate the privacy of others by illegal eavesdropping, hacking, or stealing confidential information. In the National Information Infrastructure Protection Act of 1996, at the request of the Administration, Congress provided further protection to the confidentiality of stored data. And the Department of Justice helps promote the growth of electronic commerce by enforcing the laws, including those that protect intellectual property rights and that combat computer and communications fraud.

Moreover, the Department of Justice, like other government agencies, realizes that our own information technology systems will increasingly require the use of strong encryption to provide appropriate security for the valuable and sensitive information that we hold on behalf of the American people. The Department, both as an enforcer of the law and as a consumer of encryption technologies, thus has a keen interest in the success of American industry in this area.

However, I don't think that it can reasonably be disputed that the unchecked spread of non-recoverable encryption will also endanger the public safety and our national security. People think of encryption primarily in the context of transmitted communications such as phone calls, and its effect on wiretaps. Indeed, it is absolutely essential that law enforcement preserve the ability to obtain the plaintext of information from lawfully authorized wiretaps and to authenticate this information in court. Court-ordered wiretaps are an essential tool for law enforcement in investigating and prosecuting some of our most important matters involving narcotics dealing, terrorism and organized crime.

But I'd like to focus for a moment on a slightly different aspect here: data stored on computers. It's very common, for example, for drug dealers or terrorists, or any other criminals for that matter, to keep records of their activities in notebooks or other written form. When I was an Assistant United States Attorney, I prosecuted several cases in which we arrested drug dealers and seized their "little black books" pursuant to search warrants or other valid legal authority. These notebooks provided invaluable evidence against the defendant and helped us identify and prosecute other members of the drug ring.

Today, however, we might find that the defendant is using one of the increasingly popular electronic organizers or personal information manager software programs to keep his records instead of a notebook. Or we might find that a swindler running a telemarketing scam has his records on a computer instead of in file cabinets. The switch from written to digital records does not undermine law enforcement interests—as long as the defendant hasn't encrypted the data. But if strong encryption becomes a standard feature, law enforcement will lose its ability to obtain and use this evidence. In fact, commonly available encryption products are already so strong that we cannot break them.

The same problem exists with respect to other types of criminals also. Ramzi Yousef, the mastermind of the World Trade Center bombing, used a laptop computer. Pedophiles who exchange child pornography via computer are already actively using encryption. White collar criminals and economic spies often use computers to steal our businesses' valuable intellectual property. I can't emphasize too strongly the danger that unbreakable, non-recoverable encryption poses: as we move further into the digital age, as more and more data is stored electronically rather than on paper, as very strong encryption becomes built into more and more applications, and as it becomes easier and easier to use this encryption as a matter of routine, our national security and public safety will be endangered—unless we act responsibly.

Some people have suggested that this is a mere resource problem for law enforcement. They believe that law enforcement agencies should simply focus their resources on cracking strong encryption codes, using high-speed computers to try every possible key when we need lawful access to the plaintext of data or communications that is evidence of a crime. But that idea is simply unworkable, because this kind of brute force decryption takes too long to be useful to protect the public safety. For example, decrypting one single message that had been encrypted with a 56-bit key took 14,000 Pentium-level computers over four months; obviously, these kinds of resources are not available to the FBI, let alone the Jefferson City Police Department. Moreover, it is far easier to extend key lengths than to increase computer power. Indeed, 128-bit encryption is already becoming commonplace. In this environment, no one has been able to explain how brute force decryption will permit law enforcement to fulfill its public safety responsibilities.

We believe that the most responsible solution is the development and widespread use of encryption systems that, through a variety of technologies, permit timely access to plaintext by law enforcement authorities acting under lawful authority. I will refer to these systems, collectively, as plaintext recovery systems, although they can encompass a variety of technical approaches. The concept of key recovery, where the key to encryption is held by a trusted third party, is one such approach, but it is by no means the only one that would meet law enforcement's goals.

Some have suggested that law enforcement's access to the plaintext of encrypted data and communications that is evidence of a crime would violate constitutional rights. Although I will discuss in a moment the constitutionality of a mandatory recovery regime, let me begin by reiterating that no such mandatory regime exists, nor does the Administration seek one. Rather, the Administration's efforts have been to encourage the voluntary use of data recovery products. In this context, there is no doubt that the government's efforts are constitutional.

It is certainly difficult to understand how a voluntary regime might violate the Fourth Amendment. As with any kind of stored and transmitted data, it is axiomatic that the government may obtain both encrypted text and decryption keys pursuant to lawful process, which may include a wiretap order, a search warrant issued upon probable cause, a subpoena, or the consent of the party possessing the particular item. Each of these comports with the Fourth Amendment, and voluntary data recovery products do not change this analysis. Additionally, if an individual's encryption key were stored with a third party, Congress could require by legislation that, to compel production of the key, law enforcement would have to meet a standard higher than that required by the Fourth Amendment, much as the Electronic Communications Privacy Act requires a court order to obtain transactional data. If Congress were to address this issue, we would be pleased to work with you to determine the appropriate standard and mechanisms for obtaining keys.

The Committee has requested that I address the legal issues that might be associated with a mandatory plaintext recovery regime. Again, let me restate that the Administration does not advocate such an approach, and believes that a voluntary solution is preferable. Nonetheless, I am prepared to discuss hypothetical legislation prohibiting the manufacture, distribution and import of encryption products that do not contain plaintext recovery technologies, so that the capability to decrypt encrypted data and communications is available to law enforcement upon presentation of valid legal authority.

In considering the Department's views on these issues, I would urge you to keep several caveats in mind. First, the constitutional issues that such a regime would present are undoubtedly novel ones. Indeed, the spectacular growth of the digital world has created many confounding legal issues that the Congress, the courts, the Administration, and our society at large are wrestling with. If history is any guide, changes in technology can lead to changes in our understanding of applicable constitutional doctrine. Moreover, these issues are particularly difficult to address in the abstract, because mandatory plaintext recovery could take a variety of forms. Nonetheless, and with these caveats, it is the best judgment of the Department of Justice that a mandatory plaintext recovery regime, if appropriately structured, could comport with constitutional doctrine.

Let me turn first to the Fourth Amendment. It should be remembered at the outset that the Fourth Amendment does not provide an absolute right of privacy, but protects reasonable expectations of privacy by prohibiting unreasonable searches and requiring that a warrant issue only upon a finding of probable cause by a neutral and detached magistrate. A well-designed plaintext recovery regime would ensure that users' reasonable expectations of privacy were preserved. Any legislation in this area, whether or not it imposed plaintext recovery requirements, should not lessen the showing the government must make to obtain access to plaintext. If a search warrant for data was required before, it should be required under any new regime. By requiring the government to meet current constitutional thresholds to obtain plaintext, such a regime would, in our view, comply with the Fourth Amendment. Moreover, Congress could require under such a regime that even if law enforcement obtains a search warrant for data or communications, it would need additional authority, such as a court order, to obtain the key or other information necessary to perform any decryption if the information is encrypted.

Some have also argued that mandatory plaintext recovery regime would violate the Fifth Amendment's prohibition against compulsory self-incrimination. However, the Fifth Amendment generally prohibits only disclosures that are compelled, testimonial, and incriminating. If a manufacturer of an encryption product were required to maintain information sufficient to allow law enforcement access to plaintext, we believe that there would be no violation of the Fifth Amendment because no disclosure at all would be compelled from the user of the encryption product. If, on the other hand, a mandatory plaintext recovery regime required the user of an encryption product to store his key (or other information needed for recovery) with a third party in advance of using the product, we do not believe that such an arguably compelled disclosure would be testimonial as that term has been interpreted by the Supreme Court. In *Doe v. United States*, 489 U.S. 201 (1988), the Court held that an order compelling a person to execute a form consenting to disclosure of foreign bank accounts did not violate the Fifth Amendment because the form was not testimonial. The compelled disclosure of decryption information to a third party would not seem to be any more testimonial. Moreover, we doubt whether such a disclosure would be incriminating, because unless and until the encryption product is used in the commission of a crime, the key would pose no threat of incrimination against the user.

Finally, it has been suggested that a statutory restriction on the manufacture, import, and distribution of certain types of encryption products would violate the First Amendment. Opponents of encryption restrictions sometimes argue that the First Amendment protects the right of persons to speak in "code"—*i.e.*, to speak in ciphertext—and that a restriction on the distribution of products that make a particular coded communication possible would be analogous to placing a restriction on the use of a foreign language. This First Amendment argument rests on the faulty premise that the creation or dissemination of ciphertext itself is constitutionally protected. But, unlike a foreign language, the ciphertext that is created by strong encryption products cannot be understood by the viewer or listener. When it is heard, such as on a wiretap of a telephone, ciphertext simply takes the form of unintelligible static. In written form, ciphertext may be in the form of letters, numerals and symbols, but no human being can read or "understand" it: it does not contain characters or words or symbols that represent or correspond to any other characters, words or symbols. Accordingly, ciphertext is not like a foreign language, the use of which can convey unique meaning and nuance to the listener or reader. Thus, ciphertext itself—as opposed to the underlying plaintext—has none of the properties of protected "speech" that the Supreme Court has traditionally identified, and, accordingly, the dissemination of ciphertext should not be entitled to First Amendment protection.

A second form of First Amendment argument focuses not on the ciphertext, but on the underlying plaintext. Under this theory, a prohibition on the manufacture or distribution of nonrecoverable encryption products would inhibit an alleged constitutional right of persons to *obscure* their communications in any manner they see fit. Even if legislation would impose such a practical limitation on the manner in which speakers may obscure their underlying communications, it could be drafted so as to pass muster as a permissible time, place and manner restriction—particularly since any such restriction on the "tools" of speech would be unrelated to any communicative impact of the underlying plaintext and the controls would leave open ample and robust alternative channels or methods for obscuring the underlying plaintext.

A related argument is that a communications infrastructure in which recoverable encryption is the *de facto* standard will impermissibly chill a significant quantum of speech because individuals' knowledge of law enforcement's ability to overhear and decipher communications and data will unduly deter them from communicating. But under such a system, the government would have no greater access to the content of private parties' communications than it currently has, and it is well-settled that the government's exercise of its established statutory powers to intercept and seize communications does not create such a "chilling" effect on speech as to transgress the First Amendment, so long as that power is exercised consistent with the Fourth Amendment, and for valid reasons authorized by statute, such as to discover evidence of criminal wrongdoing. *See, e.g., United States v. Ramsey*, 503 F.2d 524, 526 n.5 (7th Cir. 1974) (Stevens, J.) (rejecting argument that "the very existence of wiretapping authority has a chilling effect on free speech and, therefore, . . . violates the First Amendment"); *accord United States v. Moody*, 977 F.2d 1425, 1432 (11th Cir. 1992).

A final type of First Amendment argument often heard is that a restriction on the manufacture and distribution of certain types of encryption products would impermissibly restrict the ability of cryptographers, and others, to disseminate the computer code that is used by computers to transform plaintext into ciphertext. But that argument is based on the mistaken premise that dissemination of the code embedded in encryption products itself is necessarily a form of expression protected by the First Amendment. Most such code is in the form of "object code." Object code is simply an immense string of "0"s and "1"s, representing a bewildering concatenation of thousands or millions of high and low voltage electrical impulses. As such, machine-"readable" cryptographic object codes can reveal to possible "readers" neither the ideas they embody, nor the manner in which the ideas are expressed. And this is especially true where such object code is embedded in a product such as a semiconductor chip, so that even the "0"s and "1"s cannot be discerned. Therefore, a restriction on the dissemination of encryption products containing object code would not violate the First Amendment.

The question would be somewhat more complicated with respect to source code—*i.e.*, the instructions to the computer that human beings write and revise. Some persons do disseminate source code for communicative purposes. Nevertheless, we believe that a restriction on the dissemination of certain encryption products could be constitutional even as applied to those relatively infrequent cases in which such products are in the form of software that is disseminated for communicative reasons, because such a restriction could satisfy the "intermediate" scrutiny that the First Amendment provides for incidental restrictions on communicative conduct. As we have argued in litigation in the export-control context, such intermediate scrutiny would be appropriate because the government's reason for regulating source-code software would not be based on any informational value that its dissemination might have. Instead, regulation would be premised on the fact that such software—like all of the "encryption products" that would be regulated—has *physical*, functional properties that can cause a computer to encrypt information and thereby place plaintext beyond the technical capabilities of law enforcement to recover.

Once again, I would like to emphasize that I have presented our constitutional analysis of a mandatory plaintext recovery system to respond to the Committee's request for our views on the legal issues associated with such systems. As I noted above, this constitutional analysis would depend significantly on the nature of the particular system Congress mandated and the findings which supported it; our analysis is entirely generic. Moreover, I would emphasize again here that it is not the policy of the Administration to seek mandatory plaintext recovery legislation; it is the Department of Justice's hope and expectation that the dialogue with industry that I spoke of earlier will yield outcomes that make sense from both a business and a public policy perspective.

Those who argue against preserving lawful government access to encrypted communications often say that the government should bow to the inevitable and accept, even embrace, the spread of unbreakable encryption, rather than trying to fight it. For example, one of my colleagues recently met with a representative of a large computer company which is critical of the Administration's encryption policy. This industry representative said that he recognized that encryption poses a problem for law enforcement, but that we should recognize that other technologies, such as cars, also create problems for law enforcement, yet we have managed. He said, "We don't ban cars, do we? Then why are you trying to ban encryption?"

Of course, I hope it is clear by now that the Government is not trying to ban encryption. Law enforcement supports the responsible spread of strong encryption. Use of strong encryption will help deter crime and promote a safe national information infrastructure.

But the more fundamental point raised by the analogy to the rise of the automobile is that society "managed" the automobile, not by letting it develop completely unfettered and without regard to public safety concerns, but first by recognizing that cars could cause substantial damage to the public safety, and then by regulating the design, manufacture, and use of cars to protect the public safety. Cars must be inspected for safety on a regular basis. Cars are subject to minimum gasoline mileage requirements and maximum pollutant emission requirements. Cars built today must include seat belts and air bags. Perhaps most closely analogous, the laws of every jurisdiction in the United States closely regulate every aspect of driving cars on the public streets and highways, from driver's licenses to regulation of speed to direction and flow of traffic. Congress and the state legislatures recognized the public safety and health threats posed by the technology of automotive transportation, even as they recognized the dramatic benefits of mobility, productivity, and industrialization that the automobile brought with it. Elected government representatives of the people have consistently acknowledged and acted on their sworn responsibilities by assessing the public safety issues at stake and then regulating the technology accordingly.

Perhaps most relevant to the policy issues posed by encryption is the practice, begun by most states about a hundred years ago, of requiring cars to be registered and to bear license plates. More recently, federal law has required all vehicles to bear a vehicle identification number, or VIN. As you may recall, it was the VIN in the Oklahoma City bombing case that led the FBI to the truck rental office at which Timothy McVeigh rented the truck he used. We now recognize that license plates and VIN's afford victims of accidents, victims of car theft, and law enforcement officials with an essential means of identifying vehicles and obtaining information on the movements of criminals. Just as legislatures in the early 1900's acted to manage the risks posed by automotive technology, government leaders today, as the 21st century approaches, must bring the same sensitivity to the need to preserve and advance public safety in the face of encryption in the information age. And such a regulatory scheme, if constructed properly, will, like license plates, have benefits for businesses and consumers as well.

Of course, no analogy is perfect. Computers are not cars, and plaintext recovery is not a speed limit. But my broader point is an important one. The Framers of our Constitution determined that individuals would not have an absolute right of privacy. The Constitution recognizes that there are certain circumstances in which it is appropriate for law enforcement to obtain information that the individual wants to keep private: for example, when a judge finds probable cause to believe that information is evidence of a crime. Decisions as to where that line should be drawn are political and legal ones, not scientific or business ones; they should be made by this Congress and the courts, not by programmers or marketers. Policy should regulate technology; technology should not regulate policy. Just as in the first part of the twentieth century, the law had to take account of the changes in society brought about by the automobile, the law will have to take account of the changes brought about by encryption.

We at the Department of Justice look forward to continuing the productive discussions we have had with this Committee and the Congress on encryption issues. We share the goal of arriving at a policy and marketplace that appropriately balance the competing public and private interests in the spread of strong encryption.

I would be pleased to answer any questions you may have.

CHAPTER 6

DEPARTMENT OF JUSTICE FAQ ON ENCRYPTION POLICY

April 24, 1998

A. GENERAL POLICY ISSUES

1. What is the Administration's Policy on Encryption?

2. Aren't you really trying to force the market where it won't go?

3. Some bills currently before Congress, such as H.R. 695, would make the criminal use of encryption a crime. Isn't this sufficient?

4. Does the government want to hold everyone's private keys?

B. LAW ENFORCEMENT ISSUES

5. Why does law enforcement oppose the use of encryption? Don't you realize that it will make your job easier by stopping crime?

6. We don't ban cars, do we? Then why are you trying to ban encryption?

7 We lived without wiretaps for centuries—couldn't we do so again?

8. Aren't you overstating the threat?

9. Isn't the government's policy unworkable because strong unrecoverable encryption is widely available and, therefore, criminals will not use data recovery products even if they are widely available and commonly used?

10. Shouldn't we solve this problem by equipping law enforcement with the resources necessary to break encryption in particularly important cases?

11. I heard about one group of Internet users that worked together to crack a 56-bit encrypted message. If they did it, why can't the federal government?

12. If even 56-bit DES encryption stops law enforcement, why did the Administration decide to allow export of such products?

13. Don't Americans have a right to privacy?

14. Why does law enforcement want to be able to snoop on everyone's private communications?

C. CONSTITUTIONAL ISSUES

15. Wouldn't the use by law enforcement of recovery systems in encryption products violate the Fourth Amendment?

16. What about a mandatory plaintext recovery regime? Wouldn't that violate the Fourth Amendment?

17. Would such a hypothetical mandatory plaintext recovery regime violate the Fifth Amendment's prohibition against compulsory self-incrimination?

18. What about the First Amendment? Doesn't the First Amendment protect the right of persons to speak in "code"? Wouldn't a restriction on encryption products be analogous to placing a restriction on the use of a foreign language? Wouldn't restriction of available encryption products "chill" free speech?

A. GENERAL POLICY ISSUES

1. What is the Administration's Policy on Encryption?

The Administration's policy is to promote the development and use of strong encryption which enhances the privacy of communications and stored data while preserving law enforcement's ability to gain access to evidence as part of a legally authorized search or surveillance. We are willing to look at any options that advance these goals, as well as protecting national security, securing electronic commerce and preserving U.S. competitiveness. The Administration has identified one method to achieve the necessary balance—the use of encryption products that incorporate recovery systems. With such products, law enforcement agents can, pursuant to lawful process, obtain recovered "plaintext." The Administration is open to other approaches.

2. Aren't you really trying to force the market where it won't go?

Not at all. Indeed, we know that hardware and software companies have begun to develop data recovery products in response to the needs of businesses and individuals. For example, dozens of companies are now members of the "Key Recovery Alliance," including some of the largest computer companies in the United States. That such products are economically viable is not surprising. For example, companies may need access to encrypted data when employees are ill or otherwise unavailable, and individuals may forget their keys.

However, while industry must take the lead in developing cryptographic products and services, it is also true that market forces alone will not adequately protect public safety and national security. For example, market forces alone are not permitted to determine whether an airline should be allowed to fly—the government grounds unsafe airlines to protect public safety. Market forces alone do not determine whether meat is safe to sell in supermarkets and restaurants. Likewise, when the government pursues a cryptographic policy, it must take into account the need to protect public safety and national security.

3. Some bills currently before Congress, such as H.R. 695, would make the criminal use of encryption a crime. Isn't this sufficient?

Making the use of encryption a crime if in furtherance of the commission of another crime makes an important statement, because encryption can pose a significant obstacle to the investigation and prosecution of criminal offenses.

That said, the mere fact that the criminal use of encryption would itself be a crime would be unlikely, standing alone, to prevent most criminals from using encryption. Moreover, since the encrypted data cannot be decrypted without recovery systems—and the plaintext cannot be ascertained—it would be difficult, if not impossible, to prove in most cases that the encryption was used in the furtherance of a crime. Finally, such a prohibition would fail to address the true public safety threat: that terrorists, child pornographers, drug dealers, and any other criminals could render useless court-authorized searches and wiretaps.

4. Does the government want to hold everyone's private keys?

No, the government does not want to hold the keys of private citizens or commercial enterprises.

Actually, the Administration encourages the design, manufacture, and use of encryption products and services that allow for recovery of the plaintext of encrypted data, including the development of plaintext recovery systems, which permit through a variety of technical approaches timely access to plaintext either by the owners of data or by law enforcement authorities acting under lawful authority. Only the widespread use of such systems will both provide greater protection for data and protect public safety.

The Administration is not advocating any single product, technology, or even technical approach, and is certainly not insisting upon "escrow" of keys with the government. Key recovery, for example, where the encryption key is held by a trusted third party, is merely one possible approach, and is by no means the only one that would meet law enforcement's goals. Rather, we are flexible—provided that the resulting solutions and arrangements preserve the Nation's critical abilities to protect the public safety and defend our national security.

B. LAW ENFORCEMENT ISSUES

5. Why does law enforcement oppose the use of encryption? Don't you realize that it will make your job easier by stopping crime?

We do not oppose the use of encryption—just the opposite, because strong encryption can be an extraordinary tool to prevent crime. We believe that the use of strong cryptography is critical to the development of the "Global Information Infrastructure," or the GII. We agree that communications and data must be protected—both in transit and in storage—if the GII is to be used for personal communications, financial transactions, medical care, the development of new intellectual property, and other applications.

The widespread use of unrecoverable encryption by criminals, however, poses a serious risk to public safety. Encryption may be used by terrorist groups, drug cartels, foreign intelligence agents, and other criminals to secure their data and communications, thus nullifying the effectiveness of search warrants and wiretap orders. The Department's goal—and the Administration's policy—is to promote the development and use of strong encryption that enhances the privacy of communications and stored data while also preserving law enforcement's current ability to gain access to evidence as part of a legally authorized search or surveillance.

At bottom, it is important to recognize that society has an important choice to make. On the one hand, it can promote the use of unrecoverable encryption, and give a powerful tool to the most dangerous elements of our global society. On the other hand, it can promote the use of recoverable encryption and other techniques, achieve all of the benefits, and help protect society from these criminals. Faced with this choice, there is only one responsible solution.

6. We don't ban cars, do we? Then why are you trying to ban encryption?

The administration generally, and law enforcement particularly, are not trying to ban encryption. Law enforcement supports the responsible spread of strong encryption. Use of strong encryption will help deter crime and promote a safe national information infrastructure.

The more fundamental point raised by the analogy to the rise of the automobile is that society "managed" the automobile, not by letting it develop completely unfettered and without regard to public safety concerns, but first by recognizing that cars could cause substantial damage to the public safety, and then by regulating the design, manufacture, and use of cars to protect the public safety. Cars must be inspected for safety on a regular basis. Cars are subject to minimum gasoline mileage requirements and maximum pollutant emission requirements. Cars built today must include seat belts and air bags. Indeed, the laws of every jurisdiction in the United States closely regulate every aspect of driving cars on the public streets and highways, from driver's licenses to regulation of speed to direction and flow of traffic. Congress and the state legislatures recognized the public safety and health threats posed by the technology of automotive transportation, even as they recognized the dramatic benefits of mobility, productivity, and industrialization that the automobile brought with it. Elected government representatives of the people have consistently acknowledged and acted on their sworn responsibilities by assessing the public safety issues at stake and then regulating the technology accordingly.

Perhaps most analogous to the policy issues posed by encryption is the practice, begun by most states about a hundred years ago, of requiring cars to be registered and to bear license plates. More recently, federal law has required all vehicles to bear a vehicle identification number, or VIN. As you may recall, it was the VIN in the Oklahoma City bombing case that led the FBI to the truck rental office at which Timothy McVeigh rented the truck he used. We now recognize that license plates and VIN's afford victims of accidents, victims of car theft, and law enforcement officials with an essential means of identifying vehicles and obtaining information on the movements of criminals. Just as legislatures in the early 1900's acted to manage the risks posed by automotive technology, government leaders today, as the 21st century approaches, must bring the same sensitivity to the need to preserve and advance public safety as encryption use expands in the information age. And such a regulatory scheme, if constructed properly, will, like license plates, have benefits for businesses and consumers as well.

Of course, no analogy is perfect. Computers are not cars, and plaintext recovery is not a speed limit. But the broader point is an important one: The Framers of our Constitution determined that individuals would not have an absolute right of privacy. The Constitution recognizes that there are certain circumstances in which it is appropriate for law enforcement to obtain information that an individual wants to keep private: for example, when a judge finds probable cause to believe that such information is evidence of a crime. Decisions as to where that line should be drawn are political and legal ones, not scientific or business ones; they should be made by the Congress, the Executive, and the courts, not by programmers or marketers. Policy should regulate technology; technology should not regulate policy. Just as in the first part of the twentieth century, when the law had to take account of the changes in society brought about by the automobile, the law will have to take account of the changes brought about by encryption.

7. We lived without wiretaps for centuries—couldn't we do so again?

Court-authorized wiretaps have proven to be one of the most successful law-enforcement tools in preventing and prosecuting serious crimes, including terrorism. The inability of law enforcement to conduct effective wiretaps would have a tremendous impact, especially as the use of "traditional investigative techniques" is no substitute for wiretaps. In fact, under 18 U.S.C. § 2518(1)(c), such techniques must have been tried, be expected to fail, or be too dangerous to use, before a wiretap order may be issued. In other words, wiretaps may only be used when necessary. As society has becoming increasingly reliant on wire communication, law enforcement's need to access the contents of those communications in appropriate circumstances has also increased.

It is also important to recognize that widespread use of unrecoverable encryption will not merely negate wiretaps: the effect of encryption on court-authorized searches and seizures of computer data will also be significant. As society becomes more dependent on computers, evidence (and the fruits) of crime increasingly will be found in stored computer data, which can be searched and seized pursuant to court authorized warrants. But if unbreakable encryption proliferates, this critical law enforcement tool could also be nullified. And this would affect not only our ability to prosecute cases of terrorism and drug trafficking, but any case that relies on documents, such as fraud and child pornography cases.

If American society is to be protected as it rightfully expects and demands, law enforcement agents must have investigative tools that work. To the extent society is unwilling to grant law enforcement such tools, it must be willing to accept fewer successful investigations, fewer successful prosecutions, and, consequently, more crime that goes unprosecuted.

8. Aren't you overstating the threat?

Not at all. Law enforcement has already confronted encryption in high-profile espionage, terrorist, and criminal cases. For example:

- An international terrorist was plotting to blow up 11 U.S.-owned commercial airliners in the Far East. His laptop computer, which was seized during his arrest in Manila, contained encrypted files concerning this terrorist plot.
- A subject in a child pornography case used encryption in transmitting obscene and pornographic images of children over the Internet.
- A major international drug trafficking subject recently used a telephone encryption device to frustrate court-approved electronic surveillance.

And this is just the tip of the iceberg. Convicted spy Aldrich Ames, for example, was told by the Russian Intelligence Service to encrypt computer file information that was to be passed to it.

There have also been numerous other cases where law enforcement, through the use of electronic surveillance, has not only solved and successfully prosecuted serious crimes but has also been able to prevent life-threatening criminal acts. For example, terrorists in New York were plotting to bomb the United Nations building, the Lincoln and Holland Tunnels, and the main federal building in New York City as well as conduct assassinations of political figures. Court-authorized electronic surveillance enabled the FBI to disrupt the plot as explosives were being mixed. Ultimately, the evidence obtained was used to convict the conspirators. In another example, electronic surveillance was used to stop and then convict two men who intended to kidnap, molest, and kill a child. In all of these cases, the use of unrecoverable encryption might have seriously jeopardized public safety and resulted in the loss of life.

As encryption proliferates and becomes an ordinary component of mass market items, and as the prevalence of encryption products increases to the point of regularly denying law enforcement access to intercepted communications or physical evidence, the threat to public safety will increase greatly.

9. Isn't the government's policy unworkable because strong unrecoverable encryption is widely available and, therefore, criminals will not use data recovery products even if they are widely available and commonly used?

No policy will guarantee that, in every case, law enforcement's needs are met—some criminals won't use recoverable encryption under any circumstances. However, many criminals will use encryption that permits access by law enforcement, if that is the type of encryption that is commonly used and included in over-the-counter software. Criminals use telephones today, even though they are aware that telephones can be tapped. What we want to avoid is a situation where common street-corner drug dealers regularly without thinking make their record books and notes utterly unreadable by law enforcement at the click of a mouse button. In this regard, we hope that the availability of highly reliable encryption that provides recovery systems will reduce the demand for other types of encryption, and increase the likelihood that criminals will use recoverable encryption.

10. Shouldn't we solve this problem by equipping law enforcement with the resources necessary to break encryption in particularly important cases?

Additional resources alone will not solve this problem. It is not possible to build machines with any reasonable resources that would permit law enforcement to break even 56-bit DES encryption in the time necessary to be useful in real cases. Obviously, stronger encryption would be even more difficult to crack. In many cases, it might be difficult even to determine the type of encryption used.

This is especially significant in investigations, which can be extremely time-critical. Particularly in the case of wiretaps, decrypting messages weeks or months after interception will not protect the public. Wiretaps are used only in the most critical cases, and often provide crucial information just before a crime is to occur. Near real-time access is necessary, as days or weeks are too long to wait to find out that a terrorist attack is about to occur.

Even if the FBI were able to build a supercomputer that could periodically crack a single message encoded with 56-bit DES, each wiretap or search can result in thousands of messages or files to be decoded. Cracking all of those messages is unrealistic. And, obviously, it would be impossible to supply such a supercomputer to every state and local law enforcement agency around the country. It will always be easier and cheaper to devise stronger cryptographic methods than to build computers powerful enough to break these methods in a reasonable period of time.

11. I heard about one group of Internet users that worked together to crack a 56-bit encrypted message. If they did it, why can't the federal government?

That example actually underscores the problems that accompany a "brute force" approach. The successful group actually used over 14,000 computers and took over four months—over ten million hours of computer time—to decrypt one single message. That is not practical for law enforcement, especially if, for example, we are trying to prevent a terrorist attack or find a kidnap victim. Significantly, the time needed to decrypt a message rises exponentially as the length of the encryption key increases. If the message had been encrypted with a 64-bit key, it would take 10,000 Pentium computers on average *58 years* to crack a single message.

And a new message would require law enforcement to start again from scratch because each message may be encrypted with a different key. During 1995, for example, federal and state courts authorized more than a thousand electronic surveillance court orders, resulting in over two million intercepted communications, each of which could require separate decryption. Given such numbers, brute force attacks are not a feasible solution. This commitment of time and resources is unavailable for every wiretap and every search and seizure executed at federal, state, and local levels.

Additionally, law enforcement agencies at the federal, state, and local level are finding that searches in routine cases now commonly result in the seizure of electronically stored information. Because storage devices have increased in capacity and decreased in price, the quantity of data seized in "ordinary" cases continues to increase dramatically. If all of these communications and stored files were encrypted with unrecoverable cryptographic systems, brute force attacks would not provide a meaningful and timely solution. Thus, even if tens of thousands of computers were obtained and coordinated (an expensive undertaking, to say the least), the approximately 17,000 federal, state, and local law enforcement agencies could not be given timely access to the evidence needed to prevent and solve crimes.

12. If even 56-bit DES encryption stops law enforcement, why did the Administration decide to allow export of such products?

In developing encryption policy, the government must balance the competing needs and desires for individual privacy, international economic competitiveness, law enforcement, national security, and secure electronic commerce. The decision to permit export of 56-bit DES or equivalent strength products during a two-year window reflects this balance. The Administration's goal is to encourage the use of strong encryption to protect privacy and commerce, but in a way that preserves law enforcement's ability to preserve public safety and protect national security against criminal, terrorist, and military threats.

That's why export of 56-bit DES is permitted during that two-year window only when supported by a satisfactory business and marketing plan for recoverable items and services. Renewal of these export licenses is required for every six-month period and will depend on the applicant's adherence to explicit benchmarks and milestones as set forth in the plan approved with the initial classification request and amendments as approved by the Commerce Department's Bureau of Export Administration. Building an infrastructure that supports recovery serves the goals of preserving public safety and national security, while the Administration recognizes that this process can take time, and that privacy and security must be provided in the interim.

13. Don't Americans have a right to privacy?

Privacy is an extremely important value to be protected, and people sometimes lose sight of the fact that law enforcement is responsible, in part, for protecting privacy in a variety of circumstances. For example, we prosecute violations of the wiretap statute, as well as cases where data confidentiality has been breached. *See, e.g.*, 18 U.S.C. § 1030(a)(2). But our society has never recognized an absolute right to privacy. Rather, the Fourth Amendment strikes a careful balance, permitting government invasion of privacy to protect public safety and to prosecute crimes, but only when law enforcement can make the necessary showing, such as demonstrating "probable cause" to a neutral and detached magistrate. For example, most people would think it was justifiable for the police to search a man's bedroom pursuant to a search warrant—normally one of the most private places in one's life—if there were probable cause to believe that he had murdered someone there. In the Information Age, unbreakable encryption would upset this delicate constitutional balance, which is one of the bedrock principles of our legal system, by effectively nullifying a court's issuance of a search warrant or wiretap order.

14. Why does law enforcement want to be able to snoop on everyone's private communications?

Government should not be able to access *arbitrarily* the plaintext of encrypted communications of citizens or businesses. Law enforcement should obtain access pursuant to legal procedures such as those set out by 18 U.S.C. § 2518, *i.e., only as part of a legally authorized investigation*, and only after making the necessary legal showing. The same constitutional protections—such as the requirement that a search warrant or Title III order be obtained from a neutral judicial official, upon determination of probable cause—that preserve every American's privacy interests today will continue to prevent unauthorized intrusions in a key recovery regime.

C. CONSTITUTIONAL ISSUES

15. Wouldn't the use by law enforcement of recovery systems in encryption products violate the Fourth Amendment?

It is difficult to understand how use of recovery systems under the present, voluntary regime might violate the Fourth Amendment. As with any kind of stored and transmitted data, it is axiomatic that the government may obtain both encrypted text and decryption keys pursuant to lawful process, which may include a wiretap order, a search warrant issued upon probable cause, a subpoena, or the consent of the party possessing the particular item. Each of these procedures comports with the Fourth Amendment, and voluntary data recovery products do not change this analysis. Additionally, if an individual's encryption key were stored with a third party, Congress could require by legislation that, to compel production of the key, law enforcement would have to meet a standard higher than that required by the Fourth Amendment, much as the Electronic Communications Privacy Act requires a court order to obtain transactional data.

16. What about a mandatory plaintext recovery regime? Wouldn't that violate the Fourth Amendment?

The Administration does not advocate a mandatory approach, and believes that a voluntary solution is preferable. Nonetheless, many have asked about the constitutionality of hypothetical legislation prohibiting the manufacture, distribution and import of encryption products that do not contain plaintext recovery technologies, so that the capability to decrypt encrypted data and communications is available to law enforcement upon presentation of valid legal authority.

A discussion of the constitutionality of such hypothetical legislation must be prefaced with several caveats. First, the constitutional issues that such a regime would present are undoubtedly novel ones. Indeed, the spectacular growth of the digital world has created many confounding legal issues that the Congress, the courts, the Administration, and our society at large are wrestling with. If history is any guide, changes in technology can lead to changes in our understanding of applicable constitutional doctrine. Moreover, these issues are particularly difficult to address in the abstract, because mandatory plaintext recovery could take a variety of forms. Nonetheless, and with these caveats, it is the best judgment of the Department of Justice that a mandatory plaintext recovery regime, if appropriately structured, could comport with constitutional doctrine.

The Fourth Amendment does not provide an absolute right of privacy, but protects reasonable expectations of privacy by prohibiting unreasonable searches and requiring that a warrant issue only upon a finding of probable cause by a neutral and detached magistrate. A well-designed plaintext recovery regime would ensure that users' reasonable expectations of privacy were preserved. Any legislation in this area, whether or not it imposed plaintext recovery requirements, should not lessen the showing the government must make to obtain access to plaintext. If a search warrant for data was required before, it should be required under any new regime. By requiring the government to meet current constitutional thresholds to obtain plaintext, such a regime would, in our view, comply with the Fourth Amendment. Moreover, Congress could require under such a regime that even if law enforcement obtains a search warrant for data or communications, it would need additional authority, such as a court order, to obtain the key or other information necessary to perform any decryption if the information is encrypted.

17. Would such a hypothetical mandatory plaintext recovery regime violate the Fifth Amendment's prohibition against compulsory self-incrimination?

Again, it must be clearly stated that the Administration does not advocate a mandatory plaintext recovery regime. The Administration believes that a voluntary solution is preferable.

However, in response to questions about the Fifth Amendment, we note that the Fifth Amendment generally prohibits only disclosures that are compelled, testimonial, and incriminating. If a manufacturer of an encryption product were required to maintain information sufficient to allow law enforcement access to plaintext, we believe that there would be no violation of the Fifth Amendment because no disclosure at all would be compelled from the user of the encryption product. If, on the other hand, a mandatory plaintext recovery regime required the user of an encryption product to store his key (or other information needed for recovery) with a third party in advance of using the product, we do not believe that such an arguably compelled disclosure would be testimonial as that term has been interpreted by the Supreme Court. In *Doe v. United States*, 489 U.S. 201 (1988), the Court held that an order compelling a person to execute a form consenting to disclosure of foreign bank accounts did not violate the Fifth

Amendment because the form was not testimonial. The compelled disclosure of decryption information to a third party would not seem to be any more testimonial. Moreover, we doubt whether such a disclosure would be incriminating, because unless and until the encryption product is used in the commission of a crime, the key would pose no threat of incrimination against the user.

18. What about the First Amendment? Doesn't the First Amendment protect the right of persons to speak in "code"? Wouldn't a restriction on encryption products be analogous to placing a restriction on the use of a foreign language? Wouldn't restriction of available encryption products "chill" free speech?

Again, the Administration prefers a voluntary solution. Nevertheless, many ask about whether a mandatory plaintext recovery regime would in some manner violate the First Amendment.

A First Amendment argument that encrypted speech is like a foreign language rests on the faulty premise that the creation or dissemination of ciphertext itself is constitutionally protected. But, unlike a foreign language, the ciphertext that is created by strong encryption products cannot be understood by the viewer or listener. When it is heard, such as on a wiretap of a telephone, ciphertext simply takes the form of unintelligible static. In written form, ciphertext may be in the form of letters, numerals and symbols, but no human being can read or "understand" it: it does not contain characters or words or symbols that represent or correspond to any other characters, words or symbols. Accordingly, ciphertext is not like a foreign language, the use of which can convey unique meaning and nuance to the listener or reader. Thus, ciphertext itself—as opposed to the underlying plaintext—has none of the properties of protected "speech" that the Supreme Court has traditionally identified, and, accordingly, the dissemination of ciphertext should not be entitled to First Amendment protection.

A second form of First Amendment argument focuses not on the ciphertext, but on the underlying plaintext. Under this theory, a prohibition on the manufacture or distribution of nonrecoverable encryption products would inhibit an alleged constitutional right of persons to *obscure* their communications in any manner they see fit. Even if legislation would impose such a practical limitation on the manner in which speakers may obscure their underlying communications, it could be drafted so as to pass muster as a permissible time, place and manner restriction—particularly since any such restriction on the "tools" of speech would be unrelated to any communicative impact of the underlying plaintext and the controls would leave open ample and robust alternative channels or methods for obscuring the underlying plaintext.

A related argument is that a communications infrastructure in which recoverable encryption is the *de facto* standard will impermissibly chill a significant quantum of speech because individuals' knowledge of law enforcement's ability to overhear and decipher communications and data will unduly deter them from communicating. But under such a system, the government would have no greater access to the content of private parties' communications than it currently has, and it is well-settled that the government's exercise of its established statutory powers to intercept and seize communications does not create such a "chilling" effect on speech as to transgress the First Amendment, so long as that power is exercised consistent with the Fourth Amendment, and for valid reasons authorized by statute, such as to discover evidence of criminal wrongdoing. *See, e.g., United States v. Ramsey*, 503 F.2d 524, 526 n.5 (7th Cir. 1974) (Stevens, J.) (rejecting argument that "the very existence of wiretapping authority has a chilling effect on free speech and, therefore, . . . violates the First Amendment"); *accord United States v. Moody*, 977 F.2d 1425, 1432 (11th Cir. 1992).

A final type of First Amendment argument often heard is that a restriction on the manufacture and distribution of certain types of encryption products would impermissibly restrict the ability of cryptographers, and others, to disseminate the computer code that is used by computers to transform plaintext into ciphertext. But that argument is based on the mistaken premise that dissemination of the code embedded in encryption products itself is necessarily a form of expression protected by the First Amendment. Most such code is in the form of "object code." Object code is simply an immense string of "0"s and "1"s, representing a bewildering concatenation of thousands or millions of high and low voltage electrical impulses. As such, machine-"readable" cryptographic object codes can reveal to possible "readers" neither the ideas they embody, nor the manner in which the ideas are expressed. And this is especially true where such object code is embedded in a product such as a semiconductor chip, so that even the "0"s and "1"s cannot be discerned. Therefore, a restriction on the dissemination of encryption products containing object code would not violate the First Amendment.

Somewhat more complicated questions might be raised if such legislation were to reach encryption products in the form of source code—*i.e.*, the instructions to the computer that human beings write and revise. Some persons do disseminate source code for communicative purposes. Nevertheless, we believe that a restriction on the dissemination of certain encryption products could be constitutional even as applied to those relatively infrequent cases in which such products are in the form of software that is disseminated for communicative reasons, because such a restriction could satisfy the "intermediate" scrutiny that the First Amendment provides for incidental restrictions on communicative conduct. As we have argued in litigation in the export-control context, such intermediate scrutiny would be appropriate because the government's reason for regulating source-code software would not be based on any informational value that its dissemination might have. (Indeed, such legislation would not restrict the publication of any ideas reflected in such source code.) Instead, regulation would be premised on the fact that such software—like all of the "encryption products" that would be regulated—has *physical*, functional properties that can cause a computer to encrypt information and thereby place plaintext beyond the technical capabilities of law enforcement to recover.

CHAPTER 7

UNITED STATES DEPARTMENT OF JUSTICE
REMARKS OF
THE HONORABLE JANET RENO,
ATTORNEY GENERAL
8TH ANNUAL EXECUTIVE LEADERSHIP CONFERENCE

Richmond, Virginia
October 13, 1998

Transcribed from the provided tape for:
The U.S. Department of Justice

PROCEEDINGS

ATTORNEY GENERAL RENO: I want to thank the IAC for this wonderful opportunity to talk about how we can built trust, how we can build understanding and how we can work together.

I think this is one of the most exciting times in our history. It is certainly one of the most exciting and challenging times in law enforcement's history. We have an opportunity to build systems and networks that can help us solve crimes in ways that we never dreamed would be possible. But we also have challenges that stagger the imagination and convert vanity to prayer.

How do we face these issues? What do we do?

I am convinced that law enforcement cannot shy away. We must grasp this opportunity. Our national security depends on it. The rule of law depends on it. We cannot turn away. We cannot say we don't understand. But the only way we can meet the challenge is through a partnership between government and industry, between government and the scientists who have made this possible. And we will only be successful if that partnership is built based on certain principles.

First of all, your very existence recognizes that the interconnectivity of the information networks make it essential that law enforcement rely on industry and trust industry. But I think one thing is clear. In all my conversations, at whatever level, with respect to information technology, it is clear that industry and that American business cares deeply about the rule of law, cares deeply about our national security, and in short, we are all in this together. We must trust each other. We must know what we are talking about. And we must be able to communicate with each other in words that people understand.

Now, that includes not just the scientists and the technological experts. That includes lawyers and managers, who sometimes get it confused.

(Laughter.)

ATTORNEY GENERAL RENO: As we talk and as we build trust, one of the first essential principles that we have got to deal with is how we provide security for this tremendous network of information and communications that has been developed. How can we ensure privacy so that I can talk to my friends, thinking that my conversation is not being monitored? How can I ensure privacy, so that the ingredients about my life that have no purpose for anyone else do not end up in one giant database?

How do we achieve the wonderful opportunities that information technology affords us while at the same time adhering to the principles of the Constitution that have been so important to this Nation's history and its very being? How do we work together to solve problems?

Industry should be able to make a profit. And government must realize this. And yet, as it makes a profit, the Nation should be able to develop a cohesive information network, unfettered by fragmentation, duplication and immediate obsolescence that makes the cost of the system so prohibitive?

These are issues that we must confront in law enforcement. We have learned some hard lessons. And, frankly, in some instances, we have learned another. Too often, when I am asked to talk, I am asked to focus on one issue in the whole panoply of issues that we face in information technology. I am going to try to go through the issues that I confront daily, to try to share with you, and hope that I can establish an ongoing dialogue with the IAC about how we solve these problems together.

First is a very simple issue, but very complex. How does law enforcement communicate? We are not talking about fancy technology. We are talking about how do we communicate by radios or by cell phones. In Oklahoma City, it was startling to see the inability of police and State police and Federal officials to communicate with each other.

It becomes imperative that we communicate when we consider the threat of weapons of mass destruction and how essential it will be for first-responders within communities to be able to communicate and understand the dimension of the problem.

We have an opportunity that has been forced on us in law enforcement. It is an opportunity to—[off microphone]—resources. So that I do not look out and see three different radio towers, when one radio tower could do it; that we have an interoperable system, and we build an interoperable system so that we can communicate together; that we develop means of ensuring security; and that we develop the means to talk together.

Now, there are two problems. One, the cost, particularly for State and local first-responders and police officers, particularly in more rural areas and mountainous areas, where the problems of communication become more complex. How do we in law enforcement, in a partnership with State and local communities, develop that system? And how do we do it with limited expertise?

The answer is simple: a partnership with industry, a partnership based on trust, recognizing that we do not have all the answers, but if we sit down together and plan with respect to costs, plan with respect to sharing resources, plan with respect to the interoperability, we can solve the problem.

But it will all be for naught, as we see technology develop, if we cannot preserve some of the fine, old tools of law enforcement. One of the most important tools that has served law enforcement for the last 30 years has been the ability, upon a court order and with very narrow terms, to intercept the communication of another. This tool has been absolutely invaluable to law enforcement in prosecuting drug traffickers, terrorists and kidnappers.

I have personally been involved in a wiretap that probably saved a child's life.

Now, some people say that that is Big Brother watching you. Let me just explain what a wiretap is and what you have to do to get a wiretap. You have to prepare an affidavit, based on solid facts, that you have probable cause to believe that the electronic communication is being used in the commission of a crime and probable cause to believe that you will be able to secure evidence of the commission of that crime.

The prosecutor must present it to the court. In so doing, they must tell the court that there is no other law enforcement method by which that information can be secured, either because it is not reasonably possible or because it is too dangerous to pursue the other law enforcement methods.

We must minimize any innocent conversation overheard on this wiretap, or Title III, as we call it. It is a very strenuous process for law enforcement, but it is critical.

But for five years now, we have grappled with two big issues: how we maintain the capacity to surveil electronically in a digital system, for which Congress has passed what we call CALEA, the Computer Assistance Law Enforcement Act. In this instance, we have tried to work with the industry to see how we can properly implement CALEA.

I think there has been a—[off microphone]—together as we should. We have not put ourselves in the other person's shoes, although we have begun to do that. But we have much, much more to do.

The second issue is how we maintain the ability to surveil electronically in the face of strong encryption, while at the same time still providing for the distribution of strong encryption to ensure secure private communication. Both of these goals are very, very important. And we have made progress this past summer on the issue of encryption.

We made it because we sat down and talked with the industry. We made it because we sat down and listened, and listened with a listening ear, and said, let's work together. We recognized that there were no solutions that would guarantee 100 percent of all the answers.

We realized that different solutions would serve different problems. But we realized that we must work together. And we have done so by agreeing to develop a technical support center, in which we bring in the private sector and industry and science, and work together with law enforcement to solve the problems presented by encryption.

These are hard lessons, with more to learn. But the clear implication of everything is that we are all in this together. Industry cares just as much about the national security and the well-being of this Nation as law enforcement. Law enforcement cares that communication should be secure, that it should be private, because that is the best prevention for crime, computer crimes, that we know.

But we now have an opportunity that you have created for us, that just—[off microphone]—law enforcement solves crime based on information. The more information it has, the better it can solve a crime.

I used to think when I was a prosecutor in Dade County, wouldn't it be wonderful if I had a big database that I could put every convenience store robbery within a five-county area, trace the battered Oldsmobile with the battered right fender, and pick up the pieces of every crime that Oldsmobile was involved in. Well, now, some good detectives do it either just by manually poring over police report after police report, and some have close to photographic memories. But we all know—and now we have the opportunity—to develop in this information age a global information network that presents such opportunities for law enforcement, that we cannot sit still and watch it be fragmented and interoperable.

It is so exciting what you have permitted us to do in terms of detecting at the scene of a crime, taking a DNA sample, immediately transporting it by computer to a databank, immediately making the identification, or excluding an individual from the identification, and immediately giving law enforcement clues that it never had before at such an early time.

We can now take information with respect to terrorists, and combine that into one system, and better understand the situation. And the Oldsmobile is a snap now, thanks to what you have done.

This spring, I met criminal justice practitioners from all levels of government, who were as excited as I am about the possibility and the opportunity of establishing a global criminal justice information network. In their report to me, they identified numerous challenges, including security. If I am going to put my precious information into a system, who do I trust out there?

Well, we can develop means of giving limited access, while at the same time letting people know that this particular police department may have critical information that can be very useful. We can address those problems by what you have developed.

The next issue they presented was privacy. What if the police department decides it is going to develop the greatest database and information sharing possible, and it takes everybody's phone book, bank records, health records? America is going to rise up in arms over that. We have got to address standards of privacy.

They address the issue of the fact that we have incompatible and interoperable systems, the lack of linkages and automation in many areas, the unavailability of information because it is not being captured. Your information network that you are making possible will not be worth very much unless we get the accurate information into the system. And that is one of the problems that we face now.

They spoke of the lack of funding resources at all levels of government, and the lack of sharing information between Federal and State and local entities. Making these problems even harder to solve is that they indicated that there is also a lack of vision regarding future technological possibilities and a lack of understanding, education and resources at policy decisionmaking levels throughout the Federal, State and local systems.

As the Department of Justice continues to work with this group to identify ways of overcoming these challenges, we must all do our part in ensuring that this type of information sharing becomes a reality and not just a dream. Later this year, we will be hearing a report from the IAC, reflecting your assessment of challenges for the global criminal justice information network capability. I am anxious to hear these challenges. And I am more anxious to work together with you to pursue solutions to the problems created by these challenges.

I encourage each of you to see this initiative through. As I have said to some of you before, there is a proliferation of systems, different names, different goals, different objectives, different means of communicating. I have never seen so many information systems as I have seen in the last five and a half years. If we can get them all going in the same direction, if we can get them all going in the same direction while at the same time complying with constitutional safeguards and privacy considerations and security considerations, we are going to have a tremendous tool for law enforcement.

The last issue that I want to deal with is the issue of cybercrime. When a man can sit in a kitchen in St. Petersburg, Russia, and steal from a bank in New York, using his computer, when someone can steal your identity, and then extort from you money because they have stolen your credit card information, when they can extort you in other ways, when they can convey child pornography through the Internet, and they can stalk through the Internet, when they can make boiler rooms more effective on the Internet than ever before, it is imperative that we come together and again look to trust and partnership in determining how we can solve these crimes.

The gun is going to be obsolete in this next millennium. The black book of the drug dealer is now stored on the computer that is encrypted in many instances. Cell phones are tossed away like bubble gum. A tool, again, of the information age.

Law enforcement is sometimes criticized now, again, by the industry, who say, look, I know you all are trying, but your equipment is outmoded and you just do not have the tools necessary to keep up. Or, I know you are trying, but you do not have the expertise that is really necessary. We would like to be able to help you, but we do not know where to start.

We have made some progress. We have developed CART teams, which are forensic teams, across the country, responsible for data recovery and data analysis from the drug dealer's black book, that has become the computer. We have developed the Computer Crime and Intellectual Property Section in the Criminal Division, that has forged new partnerships around the world. And each U.S. Attorney is developing an expertise in their office.

But we have got to do more. We have got to make sure that your clients in the private sectors—the bankers or others—report computer crimes inflicted on them. If there is a theft, if there is an intrusion, a banker does not like to report it, because it indicates that the bank might not be as secure as one might like. Or they have developed a pretty good prevention system, and they just want to improve it. Or they are dubious about law enforcement's capacity to do the job.

We have got to develop sufficient trust and confidence so that the bankers and others will be willing to report crime. That is the only way we are going to deter it. If the casual, 18-year-old hacker thinks nobody is going to do anything about it, he is going to hack away. If we are to develop the information age as we envision it, we are going to have to develop it according to the rule of law. And we are going to have to make sure that the law keeps up with technology.

And so we must work together to solve the technical and legal problems we face. We need to work together in one specific area. We are developing the technological equipment at the Federal level. But police departments and sheriffs offices across the country may not begin to afford some of the sophisticated equipment they need. And we need to work together to develop a comprehensive system in this country of computer forensic systems and equipment that is available on a reasonable and on a national basis for State and local law enforcement, and sharing with the Federal system.

But there will be legal and technical problems. Let me give you an example. Let's talk about the use of digital signatures. Digital signatures allow individuals to electronically sign documents. In a way, this change in the way documents can be signed raises the same issues for us that I talked about in comparing a traditional bank robbery investigation to an investigation of a bank robbery by hacking.

Just like the traditional bank robbery is not like the bank robbery by hacking, the traditional signature is not like the electronic signature. As you all know, the electronic signature is not really even a signature by the conventional definition.

We need to work together in government and industry to be able to determine whether an electronic signature is genuine. If someone falsely files a claim based upon a digital signature, and we seek to prove that the claim was fraudulent, how do we prove in court that the defendant signed the document? After all, in a criminal case, the government cannot ask the defendant if he was the only one to use this encryption key, and whether he kept his encryption key secure. He can simply plead his right against self-incrimination.

The problem is that we cannot conduct a traditional document or handwriting analysis to verify his digital signature the same way we can for a handwritten signature. So how do we evaluate the authenticity of a digital signature? How do we prove that a person's electronic signature was actually made by that person?

Digital signatures have their place. And we have got to ensure their place by working together to ensure that we solve problems just like this. We must challenge ourselves to develop ways that we can apply this technology in a way that meets everybody's needs.

But the problem is greater than one bank theft. Increased reliance on network systems increases our vulnerability to that terrorist, who, with a concentrated attack on our information infrastructure, can bring—[off microphone]. [off microphone]—got to help America understand how we deal with this.

The challenge is clearly defined. How does law enforcement accept and meet our responsibilities when we do not and cannot control the technologies and the systems to which the responsibility relates? The answer is, again, a partnership with those who own and operate the systems, and work with them to fulfill our responsibilities.

But how do we develop this partnership? First of all, we had to do more to get our house in order than just developing forensic teams, finding some experts on cyber issues. We had to begin by establishing a center for expertise that I hope will come to represent the best of the government. This center is needed to serve as the government's lead mechanism for responding to computer crimes and attacks against our critical infrastructures. These attacks can include both cybercrimes and acts of terrorism.

On February 27, 1998, I announced the creation of the National Infrastructure Protection Center at the FBI. The Center is a true interagency organization. It includes representatives from the Department of Defense, Treasury, the intelligence community, State and local law enforcement, and other government agencies. It is charged with detecting, preventing and responding to both cyber and physical attacks, including acts of terrorism, on our Nation's critical infrastructures.

The critical infrastructures to be protected by the NIPC include those services—mostly privately owned—that are vital to our national security and to our economy. This includes everything from technology to telecommunications to transportation.

The NIPC will also serve to coordinate greater technical assistance and other computer-related cases. It will be a critical mechanism for addressing the national security and law enforcement challenges of the 21st century.

But the government cannot go it alone. Because it does not design, employ, own, or maintain most of the critical network components. Thus, we knew from the beginning of this effort that to protect our Nation's infrastructure, we needed the private sector to be an equal partner in such initiatives as the NIPC. Because much of the relevant expertise in the infrastructures and in industry technology resides in the private sector. We need to ensure direct relationships with private industry, with academic institutions, and with entities such as the Computer Emergency Response Team at Carnegie-Mellon University, and other CERT's across the country.

This will facilitate the two-way exchange of information, and enable us to keep up to date on the latest technologies. This is not only the best way to ensure the protection of the Nation and its citizens, it is the only way.

We are also establishing a national program of information sharing between the government and private industry, called the INFOGARD. For those of you who do not know, INFOGARD began as a pilot project in Cleveland, established by private sector companies and a Federal enforcement office. It is a system in which industry and government have joined together to share threat and vulnerability information. INFOGARD operates a secure E-mail system and secure Web site to facilitate the two-way exchange of information among its members.

It also provides a vehicle for training, seminars and policy discussions. This fall, the NIPC will expand INFOGARD into a national program, with local chapters in every State, making up one national partnership. The national program is necessary because what happens to a bank in Chicago may be more relevant to a bank in Cleveland than what happens to the power company in Kansas City.

Since cyberspace knows no boundaries, INFOGARD, too, has to transcend physical borders. This type of proactive and close partnership is a significant departure from the way law enforcement has traditionally operated. But the challenges of infrastructure protection require imaginative solutions. And I consider liaison and outreach to the private sector in the development of this Center to be absolutely indispensable to its success. And I would ask you to share with me suggestions as to how I can improve the outreach, what I can do to truly build a partnership with the private sector and the Center.

The system is not going to work unless government has at least sufficient expertise to begin to look at the problem. It does not. The shortage of technology workers with the skills necessary to enable to government to meet these challenges is one of the big problems we face.

We need to make sure that the government recruits and retains the highest quality talent. This personnel problem is of increasing concern due to the increased vulnerability of government systems and communications infrastructure, as well as the growing number of foreign governments and non-state actors that are exploring development of information warfare capabilities.

The demand for quality, high-tech employees will continue to increase, especially as each government department and agency assumes its full responsibility, as required under Presidential Decision Directive 63, to protect its information systems and its physical—[off microphone].

How can the government recruit the best, even when the talent pool is limited and private sector opportunities abound? First, of course, is that we offer the opportunity for government service. Now, some people these days say, Janet, what are you doing getting cussed at, fussed at and figuratively beaten around the head?

(Laughter.)

ATTORNEY GENERAL RENO: Government service is not that much fun.

(Laugther.)

ATTORNEY GENERAL RENO: These five and a half years have been an extraordinary time for me. I always thought that public service was one of the great callings that anybody could undertake. I have now had the opportunity to visit most of this country, most States. This is such a great Nation. Its people want to be represented and publicly served by good and—

[End of audio.]

[End of transcript.]

CHAPTER 8

TESTIMONY OF RONALD D. LEE
ASSOCIATE DEPUTY ATTORNEY GENERAL
DEPARTMENT OF JUSTICE
BEFORE
THE HOUSE COMMITTEE ON THE JUDICIARY
SUBCOMMITTEE ON COURTS AND INTELLECTUAL PROPERTY
ON THE SECURITY AND FREEDOM THROUGH ENCRYPTION ACT

March 4, 1999

Mr. Chairman, thank you for the opportunity to testify about the Department of Justice's views on export controls on encryption, and particularly the proposed Security and Freedom through Encryption (SAFE) Act, recently introduced by Mr. Goodlatte as H.R. 850. As you are aware, export controls on encryption is a complex and difficult issue that we are attempting to address with our colleagues throughout the Administration. In my testimony, I will first outline the basic perspective and recent initiatives of the Department of Justice on encryption issues, and will then discuss some specific concerns with the SAFE Act.

The Department of Justice supports the spread of strong, recoverable encryption. Law enforcement's responsibilities and concerns include protecting privacy and commerce over our nation's communications networks. For example, we prosecute under existing laws those who violate the privacy of others by illegal eavesdropping, hacking or theft of confidential information. Over the last few years, the Department has continually pressed for the protection of confidential information and the privacy of citizens. Furthermore, we help protect commerce by enforcing the laws, including those that protect intellectual property rights, and that combat computer and communications fraud. (In particular, we help to protect the confidentiality of business data through enforcement of the recently enacted Economic Espionage Act.) Our support for robust encryption is a natural outgrowth of our commitment to protecting privacy for personal and commercial interests.

But the Department of Justice protects more than just privacy. We also protect public safety and national security against the threats posed by terrorists, organized crime, foreign intelligence agents, and others. Moreover, we have the responsibility for preventing, investigating, and prosecuting serious criminal and terrorist acts when they are directed against the United States. We are gravely concerned that the proliferation and use of non-recoverable encryption by criminal elements would seriously undermine these duties to protect the American people, even while we favor the spread of strong encryption products that permit timely and legal law enforcement access and decryption.

The most easily understood example is electronic surveillance. Court-authorized wiretaps have proven to be one of the most successful law enforcement tools in preventing and prosecuting serious crimes, including drug trafficking and terrorism. We have used legal wiretaps to bring down entire narcotics trafficking organizations, to rescue young children kidnapped and held hostage, and to assist in a variety of matters affecting our public safety and national security. In addition, as society becomes more dependent on computers, evidence of crimes is increasingly found in stored computer data, which can be searched and seized pursuant to court-authorized warrants. But if non-recoverable encryption proliferates, these critical law enforcement tools would be nullified. Thus, for example, even if the government satisfies the rigorous legal and procedural requirements for obtaining a wiretap order, the wiretap would be worthless if the intercepted communications of the targeted criminals amount to an unintelligible jumble of noises or symbols. Or we might legally seize the computer of a terrorist and be unable to read the data identifying his or her targets, plans and co-conspirators. The potential harm to public safety, law enforcement, and to the nation's domestic security could be devastating.

I want to emphasize that this concern is not theoretical, nor is it exaggerated. Although use of encryption is still not universal, we have already begun to encounter its harmful effects. For example, in an investigation of a multi-national child pornography ring, investigators discovered sophisticated encryption used to protect thousands of images of child pornography that were exchanged among members. Similarly, in several major hacker cases, the subjects have encrypted computer files, thereby concealing evidence of serious crimes. In one such case, the government was unable to determine the full scope of the hacker's activity because of the use of encryption. The lessons learned from these investigations are clear: criminals are beginning to learn that encryption is a powerful tool for keeping their crimes from coming to light. Moreover, as encryption proliferates and becomes an ordinary component of mass market items, and as the strength of encryption products increases, the threat to public safety will increase proportionately.

Export controls on encryption products have been in place for years and exist primarily to protect national security and foreign policy interests. The nation's intelligence gathering efforts often provide valuable information to law enforcement agencies relating to criminal or terrorist acts, and we believe that this capability cannot be lost. Nonetheless, U.S. law enforcement has much greater concerns about the use of non-recoverable encryption products by criminal elements within the United States that prevent timely law enforcement decryption of lawfully-seized encrypted data and communications relating to criminal or terrorist activity.

The Department of Justice, and the law enforcement community as a whole, supports the use of encryption technology to protect data and communications from unlawful and unauthorized access, disclosure, and alteration. Additionally, encryption helps to prevent crime by protecting a range of valuable information over increasingly widespread and interconnected computer and information networks. At the same time, we believe that the widespread use of unbreakable encryption by criminal elements presents a tremendous potential threat to both public safety and national security. Accordingly, the law enforcement community supports the development and widespread use of strong, recoverable encryption products and services.

The Department believes that encouraging the use of recoverable encryption products is an important part of protecting business and personal data as well as protecting public safety. In addition, this approach continues to find support among businesses and individuals that foresee a need to recover information that has been encrypted. For example, a company might find that one of its employees lost his encryption key, thus accidentally depriving the business of important and time-sensitive business data. Similarly, a business may find that a disgruntled employee has encrypted confidential information and then absconded with the key. In these cases, a plaintext recovery system promotes important private sector interests. Indeed, as the Government implements encryption in our own information technology systems, it also has a business need for plaintext recovery to assure that data and information that we are statutorily required to maintain are in fact available at all times. For these reasons, as well as to protect public safety, the Department has been affirmatively encouraging the voluntary development of data recovery products, recognizing that only their ubiquitous use will both provide both protection for data and protection of public safety.

Because we remain concerned with the impact of encryption on the ability of law enforcement at all levels of government to protect the public safety, the Department and the FBI are engaged in continuing discussions with industry in a number of different fora. These ongoing, productive discussions seek to find creative solutions, in addition to key recovery, to the dual needs for strong encryption to protect privacy and plaintext recovery to protect public safety and business interests. While we still have work to do, these dialogues have been useful because we have discovered areas of agreement and consensus, and have found promising areas for seeking compromise solutions to these difficult issues. While we do not think that there is one magic technology or solution to all the needs of industry, consumers, and law enforcement; we believe that by working with those in industry who create and market encryption products, we can benefit from the accumulated expertise of industry to gain a better understanding of technology trends and develop advanced tools that balance privacy and security.

We believe that a constructive dialogue on these issues is the best way to make progress, rather than seeking export control legislation. Largely as a result of the dialogue the Administration has had with industry, significant progress was made on export controls. Recent updates were announced by Vice President Gore on September 16, 1998, and implemented in an interim rule, which was issued on December 31, 1998. The Department of Justice supports these updates to export controls, which liberalized controls on products that have a bit length of 56-bits or less, and permits the export of unlimited-strength encryption to certain industry sectors, including banks, financial institutions, insurance companies, and medical facilities. These changes allow these sectors, which possess large amounts of highly personal information, to use products that will protect the privacy of their clients. We also expanded our policy to permit recoverable exports, such as systems managed by network administrators, to foreign commercial firms. We learned about these systems through our dialogue with industry, and they are largely consistent with the needs of law enforcement. In addition, the Department, in conjunction with the rest of the Administration, intends to continue our dialogue with industry, and will evaluate the export control process on an ongoing basis in order to ensure that the balance of interests remains fair to all concerned.

At the same time, the Department of Justice is also trying to address the threat to public safety from the widespread use of encryption by enhancing the ability of the Federal Bureau of Investigation and other law enforcement entities to obtain the plaintext of encrypted communications. Among the initiatives is the funding of a centralized technical resource within the FBI. This resource, when fully established, will support federal, state, and local law enforcement in developing a broad range of expertise, technologies, tools, and techniques to respond directly to the threat to public safety posed by the widespread use of encryption by criminals and terrorists. It will also allow law enforcement to stay abreast of rapid changes in technology. Finally, it will enhance the ability of law enforcement to fully execute the wiretap orders, search warrants, and other lawful process issued by courts to obtain evidence in criminal investigations when encryption is encountered.

The proposed Security and Freedom through Encryption Act raises several concerns from the perspective of the Department of Justice. First, we share the deep concern of the National Security Agency that the proposed SAFE Act would harm national security and public safety interests through the liberalization of export controls far beyond our current policy, and contrary to our international export control obligations. We are similarly concerned that a decontrol of unbreakable encryption will cause the further spread encryption products to terrorist organizations and international criminals and frustrate the ability of law enforcement to combat these problems internationally.

The second problem is that the Act may impede the development of products that could assist law enforcement to access plaintext even when also demanded by the marketplace. The Administration believes that the development of such products is important for a safe society. Unfortunately, to the extent that this provision would actually prohibit government from encouraging development of key management infrastructures and other similar technologies, the provision could preclude U.S. government agencies from complying with statutory requirements and would put public safety and national security at risk. For example, it might preclude the United States government from utilizing useful and appropriate incentives to use key recovery techniques. The government might not be able to require its own contractors to use key recovery or demand its use in the legally required storage of records regarding such matters as sales of controlled substances or firearms.

It is also important to consider that our allies concur that unrestricted export of encryption poses significant risk to national security, especially to regions of concern. As recently as December 1998, the thirty-three members of the Wassenaar Arrangement reaffirmed the importance of export controls on encryption for national security and public safety purposes and adopted agreements to enable governments to review exports of hardware and software with a 56-bit key length and above and mass-market products above 64 bits, consistent with national export control procedures. Thus, the elimination of U.S. export controls, as provided by the proposed Act, would severely hamper the international community's efforts to combat such international public safety concerns as terrorism, narcotics trafficking, and organized crime.

In light of these factors, we believe that the Administration's more cautious balanced approach is the best way to protect our national interests, including a strong U.S. industry and promoting electronic commerce, while simultaneously protecting law enforcement and national security interests. We believe that legislation that eliminates all export controls on encryption could upset that delicate balance and is contrary to our national interests.

We as government leaders should embark upon the course of action that best preserves the balance long ago set by the Framers of the Constitution, preserving both individual privacy and society's interest in effective law enforcement. We should promote encryption products which contain robust cryptography but that also provide for timely and legal law enforcement plaintext access to encrypted evidence of criminal activity. We should also find ways to support secure electronic commerce while minimizing risk to national security and public safety. This is the Administration's approach. We look forward to working with this Subcommittee as it enters the markup phase of this bill.

CHAPTER 9

STATEMENT BY ATTORNEY GENERAL JANET RENO PRESS CONFERENCE TO ANNOUNCE CYBERCITIZEN PARTNERSHIP ITAA POLICY SUMMIT

Monday, March 15, 1999

Thank you. I am pleased to be here today to announce a new partnership between law enforcement and the technology community. We are coming together because we recognize that while technology offers our society great benefits, it also creates great challenges that can only be addressed if we work together.

A decade ago "cybercrime" and "cyberterrorism" didn't really exist outside of Hollywood movies; today they are very real threats. At the same time, we have raised a generation of new computer users, whose ability to utilize powerful technology is not necessarily matched by their understanding of the responsibilities that come with such power.

We cannot allow cyberspace to become the wild west of the information age. But if we are to ensure public safety and responsible computer use, then government, industry and the public must all work together.

Today we are announcing the formation of a "Cybercitizen Partnership," a partnership between the Department of Justice and computer-related industries. This partnership has grown out of several years of conversations between the Department and industry, and a recognition that together we can make a far greater impact than could either sector working on its own.

As a first step in our partnership, I am pleased to announce three initiatives that we will pursue together in the coming months.

First, we will begin a new educational campaign targeting the nation's youth on the appropriate use of computers and the Internet. This media campaign will be designed to educate our youth that the privacy and security values they respect in the physical world are just as important on-line.

All children know that it's wrong to break into their neighbor's house or read their best friend's diary. Unfortunately, fewer realize that its wrong to break into their neighbor's computer and snoop through their computer files. We need kids to understand that hacking is the same as breaking and entering—that being a hacker doesn't make them "cool" or show their smarts—it makes them a criminal.

By fostering cyber-ethics and promoting responsible computer use, we can better ensure the privacy and security of all Americans.

Second, we are establishing a personnel exchange program between law enforcement and the computer industry that we believe will help both industry and government serve the public more effectively. Industry candidates will share their expertise in system security with the government, and will learn how the government responds to the growing threats of cyberterrorism and cybercrime.

We in government will send some of our best and brightest to work in industry, to get an inside view of various infrastructures.

These exchanges will also help establish relationships rooted in trust, that will foster a closer and more effective partnership in the future.

Finally, our government-industry partnership will create a directory of computer industry professionals: a "Yellow Pages" of the computer industry. The purpose of this directory will be to provide both law enforcement and industry with a listing of computer professionals where each can find help, whether the help needed is to design a more secure system or locate an expert witness for testimony in a critical case.

For example, agents and prosecutors presently have no easy way to contact experts when they come across difficult technical problems that only the most knowledgeable industry individuals can solve.

With a Yellow Pages in place, law enforcement will know where to turn when they encounter difficult problems.

In the coming months and years I am certain that we will find more ways to work together. As computers play an ever-increasing role in our lives, the threats to our security and privacy posed by computer crime will also grow. I believe that the Cybercitizen Partnership will play a key role in readying us to respond to that threat.

CHAPTER 10

ITAA AND ATTORNEY GENERAL JANET RENO UNVEIL NEW TECH PARTNERSHIP; CAMPAIGN WILL PROMOTE CYBERCITIZENSHIP

March 15, 1999

For More Information Contact:

Bob Cohen ITAA
(703) 284-5333
bcohen@itaa.org

Tinabeth Burton
ITAA
(703) 284-5305
tburton@itaa.org

Carole Florman
Department of Justice
(202) 616-2777

Washington, DC—The Information Technology Association of America (ITAA) and the Department of Justice (DOJ) today announced the Cybercitizen Partnership, a new alliance between the high tech industry and the United States government designed to promote computer ethics and civic responsibility in the cyber age. The partnership also aims to aid law enforcement and industry in the battle against a new breed of criminal—the on-line outlaw. ITAA President Harris Miller, Attorney General Janet Reno and several high tech CEOs who support the plan announced the new program at a press conference today.

DOJ and ITAA said that computer crime is a growing problem worldwide, increasing concern that new information-based assets must be protected and preserved. A recent survey by the Computer Security Institute estimates that more than 60 percent of companies have experienced financial losses due to cyber crime. The proliferation of low cost computers and networks have spread information technology to every quarter of society. Today's partnership will conduct a major national campaign to educate, raise awareness and provide resources to empower concerned users and citizens.

"We cannot allow cyberspace to become the wild west of the information age," said Attorney General Janet Reno. "If we are to ensure public safety and responsible computer use, then government, industry and the public must all work together. This Cybercitizen Partnership is an exciting beginning."

ITAA President Harris Miller agreed: "I predict we will see cyber crime proliferate in the very near future if we as an industry, working with government, do not take every necessary step to educate computer users on the consequences of irresponsible cyber behavior. There is simply no greater priority than protecting what has rapidly become the bricks and mortar of this country's economic well-being—our information technology infrastructure—from falling victim to cyber criminals and delinquents. We are delighted to play a leadership role in critical infrastructure protection by collaborating with the Justice Department on the Cybercitizen Partnership."

One ITAA member company, Computer Sciences Corporation, has taken a leadership role within ITAA to actively address cyber concerns and is representative of many of the members.

"As a provider of I/T solutions for government and industry, we have a first hand view of the growing potential for cybercrime and the need to protect our country's technology infrastructure," said Van Honeycutt, CSC's chairman, president, and chief executive officer. "Government and the private sector can always accomplish more by working together on challenging issues such as this, and CSC is proud to be the first company to contribute to the private sector fund for the Cybercitizen Partnership. We encourage others within the I/T and telecommunications industries to do their part to ensure the future safety of our society."

The partnership as it is currently conceived will consist of three complementary segments. The first segment is a "good cybercitizenship" public awareness campaign designed to engage children, young adults and the wider user community on the basics of critical information protection and security, and the limits of acceptable online behavior. The goal of the campaign is to foster responsible cyber citizenship, educate the public on how to protect cyber resources, and decrease the likelihood of attacks through a heightened level of civic preparedness. A second component is a user-friendly computer and network security directory to help public and private sector organizations quickly find the computer security resources they need to protect information assets. Finally, an Information Security Professional fellowship program between industry and government will raise the awareness levels of participants with respect to the views, perspectives and needs of their respective counterparts. Knowledge and experience will be gained that will help shape the development and enhance the utility of potential information products to be shared within and between government and industry.

ITAA consists of 11,000 direct and affiliate members throughout the U.S. which produce products and services in the IT industry. The association plays a leading role in public policy issues of concern to the IT industry, including taxes and finance policy, intellectual property, critical infrastructure protection, telecommunications law, encryption, securities litigation reform, and human resources policy. ITAA members range from the smallest IT start ups to industry leaders in the software, services, systems integration, telecommunications, Internet, and computer consulting fields. ITAA also serves as one of three Information and Communications (I&C) Sector Coordinators under President Clinton's critical information infrastructure protection (CIIP) Decision Directive 63 (PDD 63). Learn more about ITAA and its positions on the issues by connecting to its web site at http://www.itaa.org.

The computer crimes and intellectual property section of the Department of Justice is responsible for implementing the Department's Computer Crime Initiative, a comprehensive program designed to address the growing global computer crime problem. Section attorneys actively work with other government officials (e.g., the FBI, Department of Defense, NASA) the private sector, academic institutions, and foreign representatives to develop a global response to cyberattacks. For more information about the Department of Justice's computer crimes and intellectual property section, check out its web site http://www.usdoj.gov/criminal/cybercrime/ccips.html

CHAPTER 11

STATEMENT OF THE DIRECTOR OF CENTRAL INTELLIGENCE GEORGE J. TENET AS PREPARED FOR DELIVERY BEFORE THE SENATE ARMED SERVICES COMMITTEE HEARING ON CURRENT AND PROJECTED NATIONAL SECURITY THREATS

February 2, 1999

INTRODUCTORY REMARKS

Good morning, Mr. Chairman. In this last annual threat assessment of the 20th century, I must tell you that US citizens and interests are threatened in many arenas and across a wide spectrum of issues. What is noteworthy is the manner in which so many issues are now intertwined and so many dangers mutually reinforcing.

Why is this so? To some degree it involves historic legacies fueled by the continued crumbling of Cold War constraints. We see this in the ongoing turmoil of the Balkans, the increasing violence in Africa, and the renewed volatility of the Subcontinent. But in today's world, these problems fester amidst new dangers—dangers that flow from new factors, such as the increasing availability of sophisticated technology and the ease and speed with which it can be applied by those hostile to the United States. In a very real sense, we live at a moment when the past and the future are colliding. In other words, today we must still deal with terrorists, insurgents, and others who have hundreds of years of history fueling their causes—but chances are they will be using laptop computers, sophisticated encryption, and weaponry their predecessors could not even have imagined.

Transnational Issues: WMD Proliferation

No issue is more emblematic of these new challenges than the proliferation of weapons of mass destruction. As you know, 1998 saw the nuclear tests in South Asia, continued concerns about Iraq's WMD programs, accelerated missile development in Iran, North Korea, Pakistan and India, and broader availability of BW and CW relevant technologies. Particularly worrisome to the Intelligence Community is the security of Russian WMD materials, increased cooperation among rogue states, and more effective efforts by proliferants to conceal illicit activities. US intelligence is increasing its emphasis and resources on many of these issues, but I must tell you that there is a continued and growing risk of surprise.

Looking at the supply-side first: Russian and Chinese assistance to proliferant countries has merited particular attention for several years. This year, unfortunately, is no exception. I mentioned in my statement last year that Russia had just announced new controls on transfers of missile-related technology. There were some positive signs in Russia's performance early last year but, unfortunately, there has not been a sustained improvement. Especially during the last six months, expertise and materiel from Russia has continued to assist the Iranian missile effort in areas ranging from training, to testing, to components. This assistance is continuing as we speak, and there is no doubt that it will play a crucial role in Iran's ability to develop more sophisticated and longer range missiles.

Making matters worse, societal and economic stress in Russia seems likely to grow, raising even more concerns about the security of nuclear weapons and fissile material. Although we have not had recent reports of weapons usable nuclear material missing in Russia, what we have noticed are reports of strikes, lax discipline, and poor morale, and criminal activity at nuclear facilities. For me, Mr. Chairman, these are alarm bells that warrant our closest attention and concern.

The China story is a mixed picture, Mr. Chairman. China's senior leaders are actively studying membership in the Missile Technology Control Regime and have pledged to prevent the export of materials or technology that could assist missile and nuclear programs in South Asia. Beijing has promulgated controls on dual-use nuclear technology and tightened chemical export controls.

We cannot yet assure you, however, that the new export control mechanisms will be effective. Both the Chinese Government and Chinese firms have long-standing and deep relationships with proliferant countries, and we are not convinced that China's companies fully share the commitments undertaken by senior Chinese leaders. While all aspects of China's proliferation behavior bear continued watching, we see more signs of progress on nuclear matters than on missile assistance. Moreover, the restructuring of China's defense industrial bureaucracy—including entities charged with export oversight—holds the potential to create confusion and incentives that would impede the effectiveness of this system. In short, Mr. Chairman, our guard remains up on this question.

There is little positive I can say, Mr. Chairman, about North Korea, the third major global proliferator, whose incentive to engage in such behavior increases as its economy continues to decline. Missiles and WMD know-how are North Korean products for which there is a real market. North Korea's sales of such products over the years have dramatically heightened the WMD threat in countries of key concern, such as Iran and Pakistan. Meanwhile, countries, such as India, Pakistan, and Iran that traditionally have been seen as technology customers, have now developed capabilities that they could export to others.

Looking at the demand side, Mr. Chairman, let's focus first on nuclear programs. Last spring dramatically made clear that both India and Pakistan are well positioned to build significant nuclear arsenals. Meanwhile, Iran, too, seems to be pushing its program forward. With regard to North Korea, the Agreed Framework has frozen P'yongyang's ability to produce additional plutonium at Yongbyon, but we are deeply concerned that North Korea has a covert program. The key target for us to watch is the underground construction project at Kumchang-ni, which is large enough to house a plutonium production facility and perhaps a reprocessing plant as well.

The missile story is no more encouraging. Indeed, we expect the high level of launch activity in 1998 to continue in 1999. Last year's activity included the first launches of the North Korean Taepo Dong 1, the Pakistani Ghauri and the Iranian Shahab-3, the latter two based on North Korea's No Dong. With a range of 1,300 km, the No Dong, Shahab-3, and Ghauri significantly alter the military equations in their respective regions; each is probably capable of delivering weapons of mass destruction.

In short, theater-range missiles with increasing range pose an immediate and growing threat to US interests, military forces, and allies—and the threat is increasing. This threat is here and now.

More disturbing, Mr. Chairman, is that foreign missiles of increased range and military potential are under development. North Korea's three-stage Taepo-Dong 1, launched last August, demonstrated technology that, with the resolution of some important technical issues, would give North Korea the ability to deliver a very small payload to intercontinental ranges—including parts of the United States—although not very accurately.

P'yongyang is also working on another missile—the Taepo Dong-2. With two stages, the Taepo Dong-2, which has not yet been flight-tested, would be able to deliver significantly larger payloads to mainland Alaska and the Hawaiian Islands and smaller payloads to other parts of the United States. In other words, the lighter the payload, the greater the range. With a third stage like the one demonstrated last August on the Taepo Dong-1, this missile would be able to deliver large payloads to the rest of the US. The proliferation implications of these missiles are obviously significant.

Foreign assistance is a fundamental factor behind the growth in the missile threat. For example, foreign assistance helped Iran save years in its development of the Shahab-3 missile, which is based on the North Korean No Dong and, as I noted earlier, includes Russian assistance. Moreover, Iran will continue to seek longer range missiles and to seek foreign assistance in their development.

If Iran follows a development time line similar to that demonstrated with the Shahab-3, which included significant foreign assistance, it would take Iran many years to develop a 9,000 to 10,000 km range ICBM capable of reaching the United States. But Iran could significantly shorten the acquisition time—and warning time—by purchasing key components or entire systems from potential sellers such as North Korea.

Iraqi capabilities to develop missiles also continues to be a concern. Iraq was ahead of Iran before the Gulf war, and if sanctions were lifted, we would have to assume that Iraq would seek longer-range capabilities.

Against the backdrop of an increasing missile threat, Mr. Chairman, the proliferation of chemical and biological weapons takes on more alarming dimensions. At least sixteen states, including those with the missile programs mentioned earlier, currently have active CW programs, and perhaps a dozen are pursuing offensive BW programs. And a number of these programs are run by countries with a history of sponsoring terrorism.

The Threat of Terrorism

On terrorism, Mr. Chairman, I must be frank in saying that Americans increasingly are the favored targets. US citizens and facilities suffered more than 35 percent of the total number of international terrorist attacks in 1998. This is up from 30 percent in 1997, and 25 percent in 1996.

Looking out over the next year, Mr. Chairman, let me mention two specific concerns. First, there is not the slightest doubt that Usama Bin Ladin, his worldwide allies, and his sympathizers are planning further attacks against us. Despite progress against his networks, Bin Ladin's organization has contacts virtually worldwide, including in the United States—and he has stated unequivocally, Mr. Chairman, that all Americans are targets.

Bin Ladin's overarching aim is to get the United States out of the Persian Gulf, but he will strike wherever in the world he thinks we are vulnerable. We are anticipating bombing attempts with conventional explosives, but his operatives are also capable of kidnappings and assassinations.

We have noted recent activity similar to what occurred prior to the African embassy bombings, Mr. Chairman, and I must tell you we are concerned that one or more of Bin Ladin's attacks could occur at any time.

One of my greatest concerns is the serious prospect that Bin Ladin or another terrorist might use chemical or biological weapons. Bin Ladin's organization is just one of about a dozen terrorist groups that have expressed an interest in or have sought chemical, biological, radiological, and nuclear (CBRN) agents. Bin Ladin, for example, has called the acquisition of these weapons a "religious duty" and noted that "how we use them is up to us." Earlier I referred to state sponsorship of terrorism, so let me take this opportunity to say, with respect to Iran, that we have yet to see any significant reduction in Iran's support for terrorism. President Khatami took office in August 1997, but hard-liners, such as Supreme leader Khamenei, continue to view terrorism as a legitimate tool of Iranian policy and they still control the institutions that can implement it.

The Threat of International Narcotics and Organized Crime

Turning now to the problem of international narcotrafficking and organized crime—I must tell you that the threat remains significant, despite many successes, particularly in the fight against cocaine trafficking. The illicit narcotics trade adapts quickly to law enforcement pressures, new markets, and shifting supply patterns. Three developments particularly concern me.

First, there is good news and bad news on coca cultivation. In Peru—which historically has accounted for more than half of the Andean total cultivation has declined by more than half over the past three years. Cultivation in Bolivia, historically the second largest coca producer, has also dropped substantially. The bad news, however, is that these declines are largely offset by significant increases in coca cultivation and production in Colombia—much of which is in high-risk insurgent-controlled territory making Colombia's eradication efforts more problematic.

- To President Pastrana's credit, he is trying to engage the insurgents in talks intended partly to seek their help in eradication efforts—the first time a Colombian President has taken such a bold and risky step.

Second, drug shipments are increasing overland through Central America to Mexico, and from there across the southwest border into the United States.

Finally, Mr. Chairman, opium production—the source of all refined heroin—has ballooned in Afghanistan. This country now accounts for almost 40 percent of potential worldwide opium production and may be approaching Burma as the top heroin exporter in the world.

Now, harder to track than drugs—but every bit as insidious—is international organized crime. In Russia, crime groups have permeated the financial sector, and bad bank loans, some made at the behest of criminal groups, have weakened individual banks and the Russian banking system.

Here's my principal concern, Mr. Chairman: the potential profitability of smuggling items related to weapons of mass destruction may lead to organized criminal involvement in brokering deals, financing transactions, or facilitating the transport of WMD materials to rogue states and terrorist groups.

The Threat of Information Warfare and the Year 2000 Problem

In another arena, Mr. Chairman, 1998 made clear to me that the increasing digital domination of our lives in the Information Age is creating a vulnerability of a different kind the potential threat to our national security posed by information warfare.

Several countries have or are developing the capability to attack an adversary's computer systems. Developing a computer attack capability can be quite inexpensive and easily concealable: it requires little infrastructure, and the technology required is dual-use.

For our part, providing timely warning of an attack against US computer systems is a tough technical challenge. It will require close coordination with law enforcement and the private sector to succeed, and that is what we are working hard to achieve.

And as we close the 20th century, Mr. Chairman, there is one more computer-based threat on my mind—the inability of some foreign countries to deal adequately with the Year 2000 problem.

In our judgment, foreign countries trail the United States in addressing the Y2K problems by at least several months, and in many cases by much longer. The lowest level of Y2K preparedness is evident in Eastern Europe, Russia, Latin America, the Middle East, Africa, and several Asian countries, including China. Y2K remediation is underfunded in most countries.

These uneven efforts account for several potential threats to our interests. Global linkages in telecom-munications, financial systems, air transportation, the manufacturing supply chain, oil supplies, and trade mean that Y2K problems will not be isolated to individual countries, and no country will be immune from failures in these sectors. There is potential for civil unrest in some countries, particularly if critical service sectors are disrupted for extended periods. Energy flows could be interrupted in some countries. Europe, for example, gets more than one-third of its natural gas from Russia and could be affected if Gazprom has Y2K problems. Some military activities, including those of our allies, depend on the secure and uninterrupted flow of digital information, making overall readiness a potential casualty of Y2K.

CHALLENGE: RUSSIA AND CHINA

Daunting as these challenges are, Mr. Chairman, we cannot, in focusing on them, overlook some more traditional concerns in two nations of critical importance to the United States: Russia and China.

Russia

Let me start with Russia. Last year I reported to you my view that Russia's future direction—whether it develops as a stable democracy, reverts to the autocratic and expansionist impulses of its past, or degenerates into instability—remained an open question. My concerns about Russia's direction are greater today than they were a year ago—largely because Russia's deteriorating economy elevates the "uncertainty quotient" in a number of key areas.

Just one year ago, Russia had its problems, but it had a basic sense of direction and seemed to be moving forward, however fitfully. Now, however, Prime Minister Primakov is struggling with mammoth problems. To his credit, he has built a good relationship with the legislature and gained passage of some long overdue legislation. But the nation is heading into a political transition, facing difficult economic choices, and possibly entering a period in which it debates its future political direction. This is playing out against continuing instances of lawlessness and growing public sentiment for a stronger hand at the helm. This could be a dangerous path for a country with Russia's authoritarian history, even though Russia has now held successful elections and adopted a constitution.

The sense of drift is accentuated by the focus most political leaders already have on the December 1999 Duma elections and the June 2000 Presidential election. Very few are disposed to take bold steps or new initiatives that might risk additional public "pain" right now.

- Meanwhile, President Yeltsin's health problems limit his involvement in decisionmaking and place on Prime Minister Primakov much of the responsibility for the day-to-day management of the country.

As the government ponders how to proceed, the economic indicators grow more worrisome. Russian consumers have been hit hard by inflation—prices have shot up 90 percent since late July—imports of consumer goods have now fallen sharply, unemployment has inched up to nearly 12 percent and is spreading to the emerging middle class, and the economy will probably contract by 6 to 8 percent this year.

This changed political dynamic and the economic slide highlights the foundation of my increased concern: Politically, Russia is increasingly unpredictable, and the worsening economic situation affects all aspects of the Russian scene, as the desperate search for revenue streams is exacerbating a number of serious problems:

- For example, it has magnified the proliferation threat across the board, as growing financial pressures raise incentives to transfer sensitive technologies—especially to Iran.

- It has also highlighted the patchwork, inconsistent nature of Moscow's relations with Russia's 89 regions—particularly in the delineation of fiscal powers and responsibilities. Alarm bells rang in Moscow as dozens of regions initially responded to the economic crisis by imposing price controls and limiting the flow of foodstuffs and other goods outside their regions.

China

Turning now to China, my concerns bear some resemblance to those about Russia, but in China's case, the trajectory is clearly different. China is a great power on the rise—diplomatically, militarily, and economically. There is no doubt that China has the potential to affect our security posture in Asia, but the extent to which its ambitions and growing capabilities represent a challenge or threat to US interests is still an open question.

The Chinese have signaled in summit meetings and elsewhere that they want constructive bilateral relations. But at the same time, they remain fundamentally suspicious of US intentions toward China, and—like Russia—seek to constrain any increase in US global influence.

Meanwhile, China's military modernization program continues apace, despite slowing economic growth. The Chinese program is assisted by sustained levels of defense spending and the availability of weapons and technologies from the former Soviet bloc. Its focus is on air, naval, and strategic nuclear modernization.

- China is increasing the size and survivability of its retaliatory nuclear missile force, even though it is unlikely to make the resource commitment needed to approach the force levels of either the United States or Russia.
- China is also developing and acquiring air and naval systems intended to deter the United States from involvement in a Taiwan Strait crisis and to extend China's fighting capability beyond its coastline.

Although China does not want a conflict over Taiwan, it refuses to renounce the use of force as an option and continues to place its best new military equipment opposite the island.

China's future is also uncertain because of its pressing domestic challenges. On the economic side, China's major concern this year will be sustaining economic growth, which officially reached almost 8 percent last year. China has not been immune from the global financial crisis, and much slower growth this year would threaten labor peace and increase pressure to devalue the currency—a step that would fuel a new round of financial turmoil in Asia.

These economic uncertainties have heightened China's fear of civil strife, and the recent arrests of several pro-democracy dissidents leave no doubt that China's leaders are determined to sustain the Communist Party's monopoly on political power.

CHALLENGE: REGIONAL TROUBLEMAKERS

Mr. Chairman, I'd like now to draw your attention to a group of hostile countries that remain determined to challenge our interests at every turn.

The Threat from Iraq

Needless to say, Iraq is high on this list. For eight years, Saddam has been scarred by military defeat, diplomatic setbacks, and UN sanctions. But he remains in power, and therefore, remains a threat.

- A fresh reminder of the threat has been Baghdad's return to anti-Kuwait themes not heard since 1994. Tariq Aziz in January, for example, called the Kuwaiti border issue "a mine that may explode in the future."
- In early January Saddam called on the Arab people to overthrow governments that support US policy.

Such threats to Kuwait and moderate Arab regimes are signs of Saddam's frustration with containment. Such threats also are classic examples of Saddam's heavy-handed approach to the world—one that exasperates Arab regimes.

- While noting their sympathy for the Iraqi people, Arab regimes have reiterated that Saddam is responsible for the consequences of his defiance.
- The Arab League Ministerial meeting in January and the November Damascus Declaration—from GCC states plus Syria and Egypt—showed that Saddam's defiance and bluster are backfiring.

We fully expect that he will continue his confrontational approach this year. It stems from Saddam's frustration that Allied airstrikes have not triggered a decisive backlash against UN sanctions from Security Council members and Arab governments.

- His challenge to the no-fly zones, for example, is an effort to deepen divisions within the Security Council and to inspire greater opposition to American and British policy.

With his diplomacy and his challenges to the No-Fly zone failing to deliver the breakthrough he seeks, Saddam will try other tactics to end sanctions.

- Over the years, I have talked about the capabilities of his military and his hidden weapons of mass destruction, as well as Saddam's ability to launch terrorism.

- Many of these capabilities remain available to him as he grows more frustrated and desperate to break out of containment.
- They remind us how dangerous Saddam is and why only his fall from power will free the region from this abiding threat.
- In this context, one important result of Operation Desert Fox was to damage the missile infrastructure that would support future Iraqi WMD development.
- But more importantly from my perspective, Baghdad learned from Desert Fox that Washington's will to address the Iraqi threat has not faded and that we know how to reach the things Saddam cares about most—the instruments of his power.

How secure is Saddam's rule, Mr. Chairman? There is good news and bad news on that score. Over the last eight years UN sanctions, and other pressures have complicated Saddam's efforts to maintain firm control over the country. Economic difficulties and the Shia insurgency in southern Iraq have helped undermine morale in the regular Army, and perhaps in the Republican Guard. And as you have heard, Operation Desert Fox at least briefly had some disruptive impact on the Republican Guard and Security Services.

Balancing such pressures on Saddam's regime, however, is a still-formidable security apparatus. The overlapping security services are pervasive and ruthless, leaving few vulnerabilities that can be exploited by those opposed to his rule. The security services are not infallible and Saddam has made many enemies inside Iraq, but his regime is not, as some have claimed, a house of cards.

The Threat from North Korea

Dangerous as Saddam is, Mr. Chairman, I can hardly overstate my concern about North Korea. In nearly all respects, the situation there has become more volatile and unpredictable. The regime is still struggling with serious food shortages, last year's grain harvest having been more than 1 million tons short of minimum grain needs. Very few heavy industrial plants are in operation. Living conditions for most North Koreans are miserable. Incredibly, this misery coexists with the robust WMD program I mentioned a few minutes ago.

Fresh signs of social decay have increased our concern about stability in North Korea. Crime and indiscipline are commonplace even in the military and security services. Citizens from all walks of life, including members of elite groups, are more apt to blame Kim Chong-il for systemic problems, including poor living conditions.

All of this will encourage the North to rely still more heavily on risky brinkmanship in its dealings with the United States. P'yongyang has a history of precipitating crises that it thought it could control to increase US engagement in bilateral relations.

A key area where this will play out in the coming year is US efforts to inspect the underground construction project at Kumchang-ni, which may be intended to house a nuclear facility.

The key point, Mr. Chairman, is that North Korea remains a serious military threat, despite dire economic conditions. In addition to the WMD capabilities I mentioned earlier, P'yongyang continues to devote considerable resources to its mainline military, which can still initiate a full-scale war on the Peninsula and inflict massive damage on South Korea and the 37,000 American troops deployed there. We see no indication that Kim Chong-il has abandoned the goal of ultimately bringing the entire Peninsula under his control.

The Threat from Iran

Turning now to Iran: Mr. Chairman, last year I described Iran as a still dangerous state in which some positive changes were taking place—changes that could—and I stressed *could* lead to a less confrontational stance toward the United States.

But Iran has had a tumultuous year, and my sense is that it is more likely to face serious unrest in 1999 that at any time since the revolution 20 years ago. The situation is very fluid, and the more moderate elements represented by Iran's President Khatami are on the defensive to a greater degree than ever before in their struggle with the country's conservatives. Some of President Khatami's domestic reforms have come under intense attack by conservatives. And the current jump in political violence, including the recent murders of several dissidents, suggests that some conservative elements have decided to revert to force to impose their will.

Khatami now has an opportunity to use the investigation of these murders, in which hardline elements appear implicated, to put his opponents on the defensive. He needs to regain the momentum he demonstrated in his first few months to make concrete gains against the conservatives. He could do so by using the investigation to push for change in the MOIS and, combined with large turnout in nationwide elections later this month, could reaffirm his popular mandate to push for reform.

But his efforts to do this will play out against a background of severe economic stress in Iran, largely the result of the slump in global oil prices. This is making it harder for Khatami to deliver on his reform promises—with high unemployment also contributing to the potential for civil unrest in the country.

Several troublesome developments involving Iran could unfold this year. First, Mr. Chairman, we need to bear soberly in mind that reformists and conservatives agree on at least one thing: weapons of mass destruction are a necessary component of defense and a high priority. Thus, as I stated earlier, we need to be vigilant against the possibility of proliferation surprise.

India and Pakistan

Moving further East, Mr. Chairman, I must tell you that India and Pakistan continue to have fragile governments committed to potentially destabilizing nuclear and missile programs.

In India, the Hindu-nationalist led coalition is struggling with internal strains, a resurgence of extremism, and rising expectations that contrast sharply with a slowing economy and weak policies.

In Pakistan, the Sharif government is hampered by enormous economic problems and is contending with rising Islamic sentiment that includes an extremist fringe inspired by the Taliban example in Afghanistan.

Meanwhile, both India and Pakistan continue to resist curbing WMD programs to escape economic sanctions. Neither side has established a clear nuclear-use doctrine, which makes deterrence unstable. And the bilateral dialogue between the two rivals does not appear promising. Further nuclear tests are a distinct possibility and testing of advanced new missiles seems a certainty. Kashmir remains a dangerous flashpoint. While neither side appears to want war, and our diplomats are working hard to ease tensions, the two sides could easily stumble into conflict by misinterpreting intentions or military posture.

The Balkans

Similarly, I must report a guarded outlook for the Balkans in 1999. Kosovo remains a tinderbox, and a constitutional struggle between Serbia and Montenegro could lead to a violent confrontation. In Bosnia, the Dayton process has brought stability and ended violence, but sharpening ethnic divisions may mean harder going for Dayton this year. Throughout the region political, economic, and social progress is unsustainable without direct international involvement in virtually every aspect of policy formation and resource allocation.

Kosovo is the most acute problem. The Kosovo Liberation Army will emerge from the winter better trained, better equipped, and better led than last year. With neither Belgrade nor the Kosovar Albanians willing to compromise at this point, spring will bring harder fighting and heavier casualties, unless the International Community succeeds in imposing a political settlement. The fragility of any political solution is likely to generate pressure for the International Community to deploy ground forces to enforce implementation and deter new fighting.

Kosovo—a province of Serbia—has long been a flashpoint between the Serbian and Albanian communities in what is now the Federal Republic of Yugoslavia. For Serbs, Kosovo is the birthplace of the Serbian nation and the location of many of the countries most famous and revered religious and historical sites. The source of tension is that over time Serb migration from the economically depressed province combined with a high birth rate among the ethnic Albanian community has resulted in the Serbs becoming a minority—they now account for less than 10 percent of the population of about 2 million.

Despite these demographic pressures, tensions between the two communities were contained through the seventies and eighties. During this period, Kosovo's Albanian majority enjoyed substantial autonomy and had representation in the main Federal Yugolsav bodies. Kosovo also had its own constitution, provincial assembly, interior ministry and wide administrative authority. This all changed in 1989 when Slobodon Milosevic—looking for an emotional and patriotic issue to rally public support behind his bid for power—posed as the defender of Serb interests in Kosovo. He abrogated Kosovo's autonomy on the wildly exaggerated grounds that the shrinking Serb population was being discriminated against by the Albanian majority.

In place of autonomy, Milosevic imposed his version of apartheid—shutting down ethnic Albanian schools and local administrative bodies and forcing ethnic Albanians out of government jobs and state-run businesses. Ethnic Albanian leaders initially responded to this repression by organizing non-violent resistance and seeking to reach a compromise with Belgrade. These efforts, however, only resulted in more repression.

By 1996, a loosely organized insurgency—the Kosovo Liberation Army or KLA—had emerged—dedicated to overthrowing Belgrade's rule by force. The KLA grew quickly and was able last spring to mount low-level attacks against Serb police forces and expand its presence throughout the province, even exercising effective control over some areas in central Kosovo.

Alarmed by the growing threat posed by the KLA, Belgrade launched a major counter-insurgency operation that lasted until late October. Serb security forces succeeded in pushing the KLA out of many areas, but they were unable to inflict a mortal wound. The KLA suffered relatively light casualties and its command structure remained largely intact. The Albanian civilian population was not so fortunate, bearing the brunt of the Serbs scorched earth campaign.

The agreements Belgrade signed last October stemmed the fighting only temporarily. The KLA used the cease-fire and the presence of international verifiers to reoccupy all the territory it lost last year, and it has kept up a continuous series of small-scale attacks against Serb security forces. Belgrade, for its part, has failed to comply with many of the provisions of the October agreements, including those pertaining to troop withdrawals—maintaining considerably more police in Kosovo than permitted under the agreements reached with NATO. The large presence of so-called special police—the most brutal of the Serb forces in Kosovo—has served as a lightening rod for KLA attacks.

Mr. Chairman, we are on the verge of a dramatic deterioration of the Kosovo crisis as the limitations of winter weather pass. The cease-fire negotiated last October is near collapse. The number of attacks by both sides is increasing as are the casualties.

Both sides are now preparing for much heavier fighting in the spring. The KLA has used the cease-fire to improve its training and command and control, as well as well as to acquire more and better weapons. As a result the KLA is a more formidable force than the Serbs faced last summer. We estimate that there are several thousand KLA regulars augmented by thousands more irregulars, or home guards. Moreover, funds pouring into KLA coffers from the Albanian Diaspora have increased sharply following the massacre at Racak.

We assess that if fighting escalates in the spring—as we expect—it will be bloodier than last year's. Belgrade will seek to crush the KLA once and for all, while the insurgents will have the capability to inflict heavier casualties on Serb forces. Both sides likely will step up attacks on civilians. There is already evidence that the KLA may be retaliating for the slaying of Albanian civilians at the hands of Serb security forces by attacking Serb civilians. The recent attacks against Serb bars and restaurants in Pristina and Pec could be the beginning of a pattern of tit-for-tat retaliation that will grow more severe as fighting intensifies. Heavier fighting also will result in another humanitarian crisis, possibly greater in scale than last year's, which created 250,000 refugees and internally displaced persons along with hundreds of destroyed buildings and homes.

The Aegean, Haiti, and Africa

The outlook is better in the Aegean, Mr. Chairman, where tensions remain but the chances of an immediate armed confrontation between Greece and Turkey have receded, now that Cypriot President Clerides has agreed to divert Russian SA-10 missiles to Crete.

In Haiti, progress toward strengthening democratic rule suffered a series of setbacks last year and we could see an up-tick in unrest, violence, and crime as Haitians struggle to meet basic needs.

And Africa continues to present huge challenges as it struggles to build stronger political and economic institutions, but erupts increasingly into violence.

CONCLUDING REMARKS

So, Mr. Chairman, the world seen from my window is far from placid. It is becoming a vastly more challenging place for those of us whose job it is to warn our nation's leaders and to protect American lives.

The questions are growing in number, the problems are more complex, and the issues are increasingly tangled together in intricate patterns. Many of our targets are paying closer attention to information security, and many are adding emphasis and resources to deny and deceive our intelligence gathering capabilities. Moreover, media leaks give our adversaries a road-map to find and defeat our sources and methods.

With all of this in mind, we are working hard to improve our operational reach and analytic depth; to reinvigorate our ability to get the best human and technical intelligence possible; to ensure that our analytic corps has the sophistication to grapple with the growing intricacy of the threats;

As we do this Mr. Chairman, rest assured that we will give you the good news and the bad news with equal dedication. Our overarching aim is to ensure that our nation has the intelligence it needs to to anticipate and counter the threats I've discussed here this morning.

Now, I will be glad to take your questions.

CHAPTER 12

CORNERSTONES OF INFORMATION WARFARE

FOREWORD

As information systems permeate our military and civilian lives, we are crossing a new frontier—the Information Age. It will define the 21st century and influence all we do as an air force. Information Warfare has become central to the way nations fight wars, and will be critical to Air Force operations in the 21st century. This means, of course, that today we must invest in our people, planning, equipment, and research so our ambitions can become reality. We will involve every Air Force person in this effort, generating a wave of momentum that will carry us into the next millennium.

Information Warfare is not the exclusive domain of the Air Force, or any other service. Information technology advances will make dramatic changes in how this nation fights wars in the future. They will allow a commander's vision and view of the battlespace to be shared at the lowest level. Because of this, every practitioner of the profession of arms has a responsibility to understand the impact of information warfare on their service. From our unique perspectives as soldier, sailor, marine, or airman, we can then forge a common understanding of how to use information warfare to enhance joint warfighting capabilities.

Ronald R. Fogleman
General, USAF
Chief of Staff

Sheila E. Widnall
Secretary of the Air Force

CORNERSTONES OF INFORMATION WARFARE

The competition for information is as old as human conflict. It is virtually a defining characteristic of humanity. Nations, corporations, and individuals each seek to increase and protect their own store of information while trying to limit and penetrate the adversary's. Since around 1970, there have been extraordinary improvements in the technical means of collecting, storing, analyzing, and transmitting information. Reams have been written about the impact of this technical revolution on the conduct of war, particularly since DESERT STORM. However, most of the literature focuses primarily on technical developments, not on how these developments impact doctrine.

This paper will pose questions important to Air Force policy makers and provide answers firmly grounded on concise definitions, institutional experience, and doctrinal concepts. In the process, it will clarify why the competition for information, which predates the dawn of history, is suddenly a riveting national security topic. Closer to home, this paper will also describe how Air Force doctrine should evolve to accommodate information warfare. The ultimate goal is a sound foundation on which to base the inevitable changes in organizing, training, equipping, and employing military forces and capabilities.

WHY ARE WE TALKING ABOUT INFORMATION NOW?

Because there is a technological revolution[1] sweeping through information systems and their integration into our daily lives leading to the term 'Information Age.' information-related technologies concentrate data, vastly increase the rate at which we process and transmit data, and intimately couple the results into virtually every aspect of our lives. The Information Age is also transforming **all** military operations by providing commanders with information unprecedented in quantity and quality.[2] The commander with the advantage in observing the battlespace, analyzing events, and distributing information possesses a powerful, if not decisive, lever over the adversary.

Before continuing, we must distinguish between *information age warfare* and *information warfare*. We make this distinction because much of the literature treats information warfare and advances in information technology synonymously. Information age warfare uses information technology as a tool to impart our combat operations with unprecedented economies of time and force.[3] Ultimately, information age warfare will affect all combat operations. **In contrast, information warfare, the point of this paper, views information itself as a separate realm, potent weapon, and lucrative target.** Information, as we will show below, is technology independent. However, information age technology is turning a theoretical possibility into fact: *directly*[4] manipulating the adversary's information. This is the driving force behind this paper.

---

1 The technological basis of information systems has radically changed since roughly 1970. These changes include increases in storage density, access speed, instruction rates, and communication rates, plus decreases in size and weight. Because of these changes, information systems now permeate our lives. We consider this prima facie evidence of a revolution. However, this paper does not intend to answer the question whether these changes constitute a revolution, or merely an evolution.

2 Quality here refers to concepts such as timeliness, specificity, and accuracy.

3 Cruise missiles exploit information age technologies to put a bomb on target. This is an example of information age warfare—using information age technologies as tools to better perform combat operations.

4 Direct manipulation changes the adversary's information while completely bypass-

WHAT IS INFORMATION?

This question is elementary, but pivotal. It is impossible to discuss information warfare meaningfully without rigorously defining the central concept: information.

Information derives from phenomena. Phenomena, observable facts or events, are *everything* that happens around us. Phenomena must be perceived and interpreted to become information. Information, then, is the result of two things: perceived phenomena (data) and the instructions required to interpret that data and give it meaning.

This distinction is important, and easily encompassed by a familiar paradox: If a tree falls, but no one was around to hear it, did it make a noise? The falling tree caused pressure waves in the atmosphere, a phenomenon. Noise, the information denoting a falling tree, occurs when someone's ear detects the pressure waves, creating data, and the brain's instructions manipulate that data into the sound recognizable as a falling tree. Within that person's context, there is no falling tree until the person hears (or sees) it.

Phenomena become information through observation and analysis. Therefore, information is an abstraction of phenomena. Information is the result of our perceptions and interpretations, regardless of the means. As falling trees make clear, to define information requires only two characteristics:

Information: *data and instructions.*

Note that the definition for information is absolutely distinct from technology. However, what we can do with information, and how fast we can do it, is very dependent on technology. Technology dramatically enhances our observational means, expands and concentrates data storage, and accelerates instruction processing. We use the following term to encompass the technology-dependent elements associated with information:

Information Function: *any activity involving the acquisition, transmission, storage, or transformation of information.*

For example, the system that tells a machine to stamp eighty hubcaps is performing an information function. The sheet metal press stamping those hubcaps is not.

WHAT ARE SOME MILITARY INFORMATION FUNCTIONS?

Quality information is the counter to the fog of war. As mentioned earlier, the commander with better information holds a powerful advantage over

---

ing the adversary's perceptive, analytical, or decision processes,.

his adversary. Military operations make special demands on information functions in seeking to give the commander an information advantage.

Surveillance and reconnaissance are our powers of observation. Intelligence and weather analysis are the bases for orienting observations. We use those bases to form an Air Tasking Order, which command and control operations execute and monitor in directing the conflict. Precision navigation enhances mission performance. Together, these are the kinds of military information functions that enhance all military operations. Collectively, we use the term military information functions to describe force enhancing information functions.

Military Information Function: *any information function supporting and enhancing the employment of military forces.*

This definition serves to delineate militarily important information functions from the total universe of information functions.

WHAT IS INFORMATION WARFARE?

At the grand strategy level, nations seek to acquire, exploit, and protect information in support of their objectives. This exploitation and protection can occur in the economic, political, or military arenas. Knowledge of the adversary's information is a means to enhance our own capabilities, degrade or counteract enemy capabilities, and protect our own assets, *including our own information.* This is not new. The struggle to discover and exploit information started the first time one group of people tried to gain advantage over another.

Information warfare consists of targeting the enemy's information and information functions, while protecting our own, with the intent of degrading his will or capability to fight.[5] Drawing on the definitions of information and information functions, we define information warfare as:

Information Warfare: *any action to deny, exploit, corrupt, or destroy the enemy's information and its functions; protecting ourselves against those actions; and exploiting our own military information functions.*[6]

---

5 There is troubling offense-defense asymmetry in the scope of information warfare. The military may, consistent with the law of armed conflict, attack any militarily significant target. In the context of information warfare, this means we may target any of the adversary's information functions that have a bearing on his will or capability to fight. In stark contrast, our military may defend only military information functions. There are many information functions critical to our national security that lie outside the military's defensive purview.

6 Note, this definition depends only on the nature of the action, not how it is accomplished.

This definition is the basis for the following assertions:

Information warfare is any attack against an information function, regardless of the means. *Bombing a telephone switching facility is information warfare. So is destroying the switching facility's software.*

Information warfare is any action to protect our information functions, regardless of the means. *Hardening and defending the switching facility against air attack is information warfare. So is using an anti-virus program to protect the facility's software.*

Information warfare is a means, not an end, in precisely the same manner that air warfare is a means, not an end. *We may use information warfare as a means to conduct strategic attack and interdiction, for example, just as we may use air warfare to conduct strategic attack and interdiction.*

Militaries have always tried to gain or affect the information required for an adversary to effectively employ forces. Past strategies typically relied on measures such as feints and deception to influence decisions by affecting the decision maker's perceptions. Because these strategies influenced information through the perception process, they attacked the enemy's information indirectly. That is, for deception to be effective, the enemy had to do three things:

- observe the deception,
- analyze the deception as reality, and
- act upon the deception according to the deceiver's goals.

However, modern means of performing information functions give information added vulnerability: direct access and manipulation.[7] Modern technology now permits an adversary to change or create information without relying on observation and interpretation. Here is a short list of modem information system characteristics creating this vulnerability: concentrated storage, access speed, widespread information transmission, and the increased capacity for information systems to direct actions autonomously. Intelligent security measures can reduce, but not eliminate, this vulnerability; their absence makes it glaring.

---

7 To reiterate: Indirect Information Warfare: changing the adversary's information by creating a phenomena that the adversary must then observe and analyze. Electronic warfare, when used to degrade the adversary's ability to accurately perceive phenomena, is conducting indirect information warfare.
Direct Information Warfare: changing the adversary's information without involving the intervening perceptive and analytical functions. Replacing bar code labels with bogus ones on pallets in a deployment staging area is a form of direct information warfare.

Militaries are not inclined to trust their success to the fortunes of war. So we must direct our information warfare efforts to more than just targeting an adversary's information: we must also defend our own information, and all its operations. The Air Force depends heavily upon military information functions, making us vulnerable to information warfare. The integrity of our military information functions, as well as the information itself, bears heavily and directly on the success of our military operations.

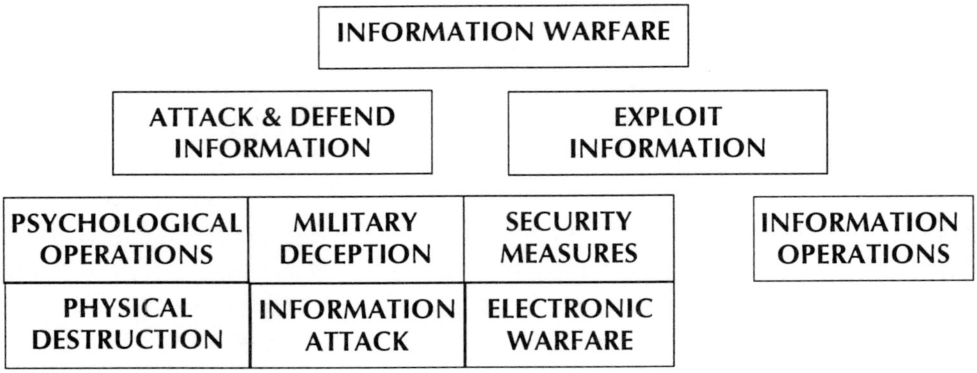

WHAT COMPRISES INFORMATION WARFARE?

Recalling the definition, information warfare consists of activities that deny, exploit, corrupt, destroy, or protect information. Traditional means of conducting information warfare include psychological operations, electronic warfare, military deception, physical attack, and various security measures.

Psychological Operations *use information to affect the enemy's reasoning.*

Electronic Warfare *denies accurate information to the enemy.*[8]

Military Deception misleads the enemy about our capabilities or intentions.[9]

Physical Destruction *can do information warfare by affecting information system elements through the conversion of stored energy to destructive power. The means of physical attack range from conventional bombs to electromagnetic pulse weapons.*

---

8 Note, this paper uses the JCS recognized label "electronic warfare" instead of "electronic combat".

9 Military deception provides an excellent example of how actions are defined by their ends, not means. Air power may conduct information warfare through military deception, as in World War II's Operation FORTITUDE NORTH, where the Allies dropped two bombs in the Pas de Calais for each bomb dropped in Normandy, a successful effort to mislead the Germans about our invasion plans.

Security Measures *seek to keep the adversary from learning about our military capabilities and intentions.*[10]

The Information Age has provided new and practical means to deny, exploit, corrupt, or destroy information,[11] as well as the vulnerabilities to make those attacks possible. Air Force doctrine does not yet acknowledge or define these assaults on information, which we call Information Attack.

Information Attack: *directly corrupting*[12] *information without visibly changing the physical entity within which it resides.*

Information attack, constrained by the definition of information, is limited to directly altering data or instructions. It is, therefore, just another means of conducting information warfare, one whose immediate effects do not include visible changes to the entity within which the information resides. That is to say, after being subjected to information attack, an information function is indistinguishable from its original state except through inspecting its data or instructions.[13]

HOW IS INFORMATION ATTACK DIFFERENT?

As previously described, there are two ways to influence the adversary's information functions: indirectly and directly.

Indirect information warfare affects information by creating phenomena, which the adversary will perceive, interpret, and act upon. Military deception, physical attack, and OPSEC traditionally achieved their ends indirectly.[14] For example, the goal of deception is to cause the adversary to make incorrect decisions; deception does this by creating an apparent reality. Generally, this entails creating phenomena for the enemy to observer Success, however, depends on several conditional events: the adversary actually observes the phenomenon, thereby turning it into

---

10 Current security measures include OPSEC, COMSE, and COMPUSEC. As information technology changes, these security measures will almost certainly evolve into entirely different forms.

11 Examples: Cutting the radar to command center landline is information denial. Insider trading is information exploitation. Altering a barcode is information corruption. Reformatting someone's hard drive is information destruction.

12 Corruption covers the spectrum from subtle alteration to thorough destruction.

13 A SOF team dropping a powerful magnet on a hard disk is an example of information attack.

14 Military deception creates an alternate reality upon which the adversary will base his decisions; psychological operations target the will to fight; physical attack creates phenomena for the adversary to perceive and convert into information; OPSEC denies the enemy knowledge of our own operations.

data; analyzes it into the desired information; and acts upon the information in the desired manner.

Direct information warfare affects information through altering its components without relying on the adversary's powers of perception or interpretation. Information attack acts directly upon the adversary's information. Since nearly all modem information functions are themselves controlled by information, information attack may be directed against most information functions.

Direct information warfare, the point of information attack, acts on the adversary's information without relying on the adversary's collection, analysis, or decision functions. It can short circuit the OODA loop[15] through creating observations and skewing orientation, or decapitate it by imposing decisions and causing actions.

A short illustration will serve to demonstrate the difference between indirect and direct information warfare applications:

Our goal, using military deception, is to make the adversary think there is a wing of combat aircraft where, in fact, there is none, and act on that information in a manner benefiting our operations.

Indirect information warfare: Using military deception, we could construct fake runways and parking areas, and generate enough other activities to present a convincing image. We rely on the adversary to observe the pseudo combat operation and interpret it as real (as opposed to detecting the fake). Only then does it become the information we want the adversary to have.

Direct information warfare: Conversely, if we use information attack to create the pseudo combat wing in the adversary's store of information, the result-deception-is precisely the same. But the means to that result, never mind the resources, time, and uncertainty, are dramatically different.

WHAT IS THE OTHER EDGE OF THE INFORMATION WARFARE SWORD?

The defensive side of information warfare security measures aimed at protecting information-prevents an adversary from conducting successful information warfare against our information functions. Current security measures such as OPSEC and COMSEC are typical means of preventing, detecting, and subverting an adversary's indirect actions on our military

15 OODA Loop: Observe, Orient, Decide, and Act. Observation converts phenomena into data. Orientation processes the data into information. Decide converts the information into orders. Act converts the orders into action. The OODA loop is continuous-as you act you observe the results, starting the process all over again.

information functions. In contrast, security measures such as COMPUSEC encompass preventing, detecting, and subverting direct information actions on our information functions. Future security measures must evolve as information technology advances. Consequently, new-measures will likely take forms entirely different from today's security measures, rooted as they are in previous security requirements. As the simple examples in this paper illustrate, we must avoid falling victim to profound, debilitating effects of direct information warfare.

WHY IS INFORMATION WARFARE IMPORTANT TO THE USAF?

Two reasons. First, because information warfare offers important means to accomplish Air Force missions. Second, because the widespread integration of information systems into Air Force operations makes our military information functions a valuable target set.

A hypothetical example using information attack shows how information warfare might accomplish a typical Air Force mission:

Interdiction prevents or delays essential resources from reaching combat units. One approach to interdiction is wrecking bridge spans using laser-guided bombs. Alternatively, we might be able to alter the adversary planners' information, falsely categorizing the bridges as destroyed, causing the planners to reroute forces and supplies. Each means performs interdiction; information attack offers the possibility of achieving our goal while consuming fewer resources or without exposing our assets to attack.

As an example emphasizing the need for robust defenses against information warfare, imagine the chaos that would ensue should an adversary manage to penetrate our time-phased force deployment database. Subtle changes in it could be sufficient to bring our power projection capabilities to a near standstill.

HOW SHOULD WE CHANGE AIR FORCE DOCTRINE TO ACCOMMODATE INFORMATION WARFARE?

Presently, Air Force doctrine recognizes air warfare and space warfare. However, the doctrine doesn't identify separate missions for air warfare or space warfare. Instead, both cut across all roles and missions. Similarly, information warfare cannot be pigeonholed as a single mission. To do so would fail to completely integrate information warfare into Air Force doctrine.

Recall that missions are operational tasks performed to achieve military objectives. Air warfare is a means, defined by the environment, to execute those missions. There are three objectives of air warfare:

- **control** the air while protecting our forces from enemy action,

- **exploit** control of the air to employ forces against the enemy, and,
- **enhance** our overall force effectiveness.

In our doctrine, the *objectives* of control, exploit, and enhance translate into the *roles* of aerospace control, force application, and force enhancement.

In many respects, one can consider information as a realm, just as land, sea, air, and space are realms[16] information has its own characteristics of motion, mass, and topography, just as air, space, sea, and land have their own distinct characteristics.[17] There are strong conceptual parallels between conceiving of air and information as realms. Before the Wright brothers, air, while it obviously existed, was not a realm suitable for practical, widespread military operations. Similarly, information existed before the Information Age. But the Information Age changed the information realm's characteristics so that widespread military operations within it became practical.

Information warfare, like air warfare, is the means defined by the environment to execute military missions. There are three objectives of information warfare:

- **control** the information realm so we can exploit it while protecting our own military information functions from enemy action,
- **exploit** control of information to employ information warfare against the enemy, and,
- **enhance** overall force effectiveness by fully developing military information functions.

The **first** objective of information warfare, to **control** the realm so we can exploit it while protecting our own military information functions from enemy action, contributes significantly to controlling the combat environment. Presently, Air Force doctrine recognizes two missions to control the combat environment: counterair and counterspace. Counterair comprises missions whose objectives are control of the air; counterspace comprises those missions whose objectives are control of space. Clarity and consistency require we term those activities dedicated to controlling information as *counterinformation*.

Counterinformation: *actions dedicated to controlling the information realm.*

---

16 realm, n: the region, sphere, or domain within which anything occurs, prevails. or dominates. [The Random House College Dictionary]

17 Land, sea, air, and space are realms within which we may conduct military operations. Each realm imposes its characteristics on operations within it. Information and its functions also may host military operations, but those operations must conform to the characteristics of the information system and its functions.

Further, counterinformation, like counterair and counterspace, has both offensive and defensive aspects. Offensive counterinformation enables us to use the information realm and impedes the adversary's use of the realm. Typical means include physical attack, military deception, psychological operations, electronic warfare, and information attack. Defensive counterinformation includes both active and passive actions to protect ourselves from the adversary's information warfare actions. Defensive counterinformation is accomplished, for instance, through physical defense, physical security, hardening, OPSEC, COMSEC, COMPUSEC, and counterintelligence.

Successful aerospace control enables us to use the air and space realms without suffering substantial losses, and inflict substantial losses on the enemy's use of those realms. Counterinformation, working with counterair and counterspace, seeks to create such an environment.

The **second** objective of information warfare is to **exploit** our control of information. In air warfare's force application role, the missions of strategic attack, interdiction, and close air support exploit air control. Similarly, information warfare might also be used to achieve the same ends. We have already cited an example of how information warfare can perform interdiction. It can also perform strategic attack:

Suppose we want to limit the enemy's long-term mobility by restricting his POL resupply. We first identify his refineries as the most suitable target to achieve this goal. Through research we further identify the specific refineries comprising most of his production capacity. For each refinery, we find there is one critical cracking tower. We mount a strike and, with admirable economy of force, put the refineries out of operation by destroying just those towers, while leaving everything else untouched. This is a classic example of strategic attack.

Same situation. Like all modern refineries, these have extensive automated control systems. These extensive information functions offer a potential target for information warfare. Early in the conflict we performed an offensive counterinformation mission by penetrating and characterizing the refinery's automated control system. In the process, we uncovered several vulnerable information dependencies, giving us the means of affecting the refineries' operations at a time of our choosing. Later in the conflict, combined with interdiction and ground maneuvers, we choose to exploit one of the vulnerabilities. We have just disabled their refineries. This, too, is a classic example of strategic attack.

Information technology is already tightly woven with our military operations, providing heretofore unimaginable amounts of information. Exploiting this information has provided us striking capabilities; relying on it inevitably creates potentially crippling vulnerabilities. This, coupled with advances in the ability—to both locate and destroy command and control (C2) nodes makes C2, more than ever, a lucrative target set. History has shown successful militaries can achieve striking success through paralyzing the enemy's ability to exercise command and control. Airmen have always considered this an important objective and expended much effort against C2.[18] For these reasons, the efforts to disrupt and destroy the adversary's command and control elements have prompted us to identify a separate mission under force application.

C2 Attack: *any action against any element of the enemy's command and control system.*

The **third** objective of information warfare is to develop information functions to **enhance** total force effectiveness. Previously we described military information functions as supporting the employment of military forces. Our current doctrine does not include such a mission. To fill that void, we will include information operations under force enhancement. Some examples of information operations are: surveillance, reconnaissance, command and control, communications, combat identification, intelligence, precision navigation, and weather. The distinguishing characteristic of the information operations mission is that it deals primarily with information as both its resource and product.

ROLES AND MISSONS OF AEROSPACE POWER			
AEROSPACE CONTROL	**FORCE APPLICATION**	**FORCE ENHANCEMENT**	**FORCE SUPPORT**
COUNTERAIR COUNTERSPACE COUNTERINFORMATION	STRATEGIC ATTACK INTERDICTION CLOSE AIR SUPPORT C2 ATTACK	AIRLIFT AIR REFUELING SPACELIFT SPECIAL OPERATIONS INFORMATION OPERATIONS	BASE OPS & DEF LOGISTICS COMBAT SUPPORT ON-ORBIT SUPPORT

Information Operations: *any action involving the acquisition, transmission, storage, or transformation of information that enhances the employment of military forces.*

Since we require relevant, accurate, and timely information for *everything* we do, information operations support the conduct of missions across all four roles', from aerospace control to force support. Information operations

---

18 Consider, for example, DESERT STORM, or Israel's domination of the air over the Bekaa Valley in 1982. Reaching further into the historical record, in the initial stages of General Allenby's 1918 Palestine Campaign, Allied aircraft attacked key Turkish telephone exchanges, destroying communications among Turkish forces. Consequently, the Fourth Army remained in bivouac for four days, unaware of the disaster befalling other Turkish forces to their west.

provide commanders the ability to observe the battlespace, analyze events, and direct forces. information operations provide logisticians the ability to know what is in inventory, and where it is needed. Information operations provide the flight lead the ability to know where the target is, its defenses, and select the most appropriate weapon.

In sum, information warfare cuts across all Air Force roles and missions. It is another means to conduct our traditional missions. However, there are three additional operational tasks that information warfare enables us to execute which are not suitably addressed by our current doctrine: counter-information, C2 Attack, and information operations. Similarly, we elected to delete two missions no longer relevant under regrouped missions: electronic combat, previously under force enhancement, is now subsumed by information warfare; surveillance and reconnaissance are now considered instances of information operations. However, this list is by no means exhaustive. As this paper's title conveys, the ideas contained herein provide the cornerstones, not the entire building. Invariably, as the Air Force fully accommodates the information technology revolution, additional operational tasks may arise which will in turn warrant adding or removing missions. To the extent these cornerstones continue to provide a valid litmus test for information warfare, all new missions need to meet and pass it.

WHAT IS THE RELATIONSHIP BETWEEN INFORMATION WARFARE AND COMMAND & CONTROL WARFARE?

The focus of information warfare is *any* information function, whether it is C2, a refinery's control system, or a telephone switching station. C2 represents only part of the universe of military information functions. The Joint Staff defines command and control:

Command and Control: *the exercise of authority and direction by a properly designated commander over assigned forces in the accomplishment of the mission. [joint Pub 1-021*

Command and control warfare (C2W) only addresses activities directed against the adversary's ability to direct the disposition and employment of forces, or those which protect the friendly commander's ability to do so. As we have illustrated, information warfare not only attacks the C2 process, but it also attacks the enemy's combat power itself. Conversely, by definition, C2W is not associated with reducing or nullifying the ability or desire of combat units to execute their orders. Tactical psychological operations and electronic countermeasures self-protection hinder the ability of units to execute orders. But they in no way affect commanders' ability to issue orders to those units, nor their ability to receive those orders.

Although extraordinarily important, the JCS's policy of Command and Control Warfare is only a particular application of information warfare. For the military to concentrate only on C2W would be ignoring other legitimate target sets. Therefore, information warfare, and its attendant organizing, training, and equipping issues, is essential to fully effective C2W.

IS INFORMATION WARFARE IMPORTANT ONLY TO THE AIR FORCE?

We have established that information warfare is important to the Air Force for two reasons. First, since our military information functions present a valuable target set, we must make commensurate defensive efforts. Second, as the examples in this paper show, information warfare is a potential means to achieve typical Air Force ends: strategic attack, interdiction, etc. More fundamentally, the Air Force *already does* information warfare through such systems as the EF-111 and Compass Call.

But in a broader sense, information warfare might be a means to conduct any mission the services already conduct—and the services are best positioned to choose the best means for their ends. Each service has its own unique operational demands. After all, the Army is best qualified to decide which means are best suited for pursuing the goals the joint Force Commander apportions to the Army.

As a result of its service-unique expertise, its own OODA loop requirements, logistics, etc., each service has information warfare concerns. In developing the doctrinal constructs in this paper, we used airpower terminology and examples. That is our background, those are the terms and the environment with which we are familiar. But the argument we present **is not** dependent on terminology. Replacing Air Force terms with Army or Navy terms would leave the conclusions unchanged.

CONCLUSION

The information revolution, startlingly fast as it is, shows no signs of slowing. As the Air Force becomes more technologically sophisticated, it becomes more technologically dependent. We need to use that technological sophistication to avail ourselves of all the opportunities that information, as a target, presents. We also need to be aware that our technical dependencies represent potentially crippling vulnerabilities. Sophisticated, robust, multi-layered defenses for our military information functions may well be what separates us from joining the sorry league of military failures.

Information, combined with modern information functions, has distinct characteristics that warrant it being considered a realm, just as land, sea, air, .and space are realms. Information warfare does not fill a discrete place in Air Force doctrine. Just like air warfare, information warfare can be part of many AFM 1-1 missions. just as when space warfare was integrated into Air Force doctrine, viewing information as a realm now leads us to add several missions:

Counterinformation: *controlling the information realm.*

C2 Attack: *any action against the enemy's command and control system.*

Information Operations: *any action involving the acquisition, transmission, storage, or transformation of information that enhances the employment of military forces.*

Since World War 1, airmen have had to control the air environment effectively to employ airpower. What is more, air and space superiority are virtually a *sine qua non* for employing ground and naval forces. Information is the next realm we must control to operate effectively and with the greatest economy of force.

At the outset we stated the competition for information is as old as man's first conflict. It involves increasing and protecting our own store of information while limiting and penetrating the adversary's. The recent explosion in information technologies is prompting the current discussion in and outside government on the topic of information warfare—targeting the enemy's information functions, while protecting ours, with the intent of degrading his will or capability to fight.

For airmen, controlling the combat environment is job One. With the advances in information technology, airmen must pursue information superiority just as they do air and space superiority. Only with these realms under our control can we effectively employ **all** our combat assets. Military information functions are essential to our combat operations—they are a tool for achieving the Joint Force Commander's campaign objectives. Targeting the enemy's information functions keeps him from achieving his.

In this paper we have laid out information warfare's doctrinal foundation. Our goal is to provide a sound and widely accepted basis from which we can adapt Air Force doctrine to the Information Age. The ultimate aim? Incorporating information warfare into the way the Air Force organizes, trains, equips, and employs.

DEFINITIONS

C2 Attack: *Any action against any element of the enemy's command and control system.*

Command and Control: *The exercise of authority and direction by a properly designated commander over assigned forces in the accomplishment of the mission.*

Counterinformation: *Actions dedicated to controlling the information realm.*

Defensive counterinformation: *Actions protecting our military information functions from the adversary.*

Direct Information Warfare: *Changing the adversary's information without involving the intervening perceptive and analytical functions.*

Indirect Information Warfare: *Changing the adversary's information by creating phenomena that the adversary must then observe and analyze.*

Information: *Data and instructions.*

Information Attack: *Directly corrupting information without visibly changing the physical entity within which it resides.*

Information Function: *Any activity involving the acquisition, transmission, storage, or transformation of information.*

Information Operations: *Any action involving the acquisition, transmission, storage, or transformation of information that enhances the employment of military forces.*

Information Warfare: *Any action to deny, exploit, corrupt, or destroy the enemy's information and its functions; protecting ourselves against those actions; and exploiting our own military information functions.*

Military Information Function: *Any information function supporting and enhancing the employment of military forces.*

Offensive counterinformation: *Actions against the adversary's information functions.*

CHAPTER 13

COPERNICUS . . . FORWARD
C4I FOR THE 21ST CENTURY

THE NAVY PUBLIC AFFAIRS LIBRARY (NAVPALIB)
A SERVICE OF THE NAVY OFFICE OF INFORMATION,
WASHINGTON DC
SEND FEEDBACK/QUESTIONS TO NAVPALIB@OPNAV-EMH.NAVY.MIL

June 1995

FOREWORD

In the 1980s companies like Apple and IBM revolutionized industry by introducing the first desktop personal computers and the Information Age was born. The Navy recognized the potential of using information as a warfighting tool and in 1990, published Copernicus. Copernicus is the Navy's initiative to make command, control, communications, computers and intelligence (C4I) systems responsive to the warfighter; to field these systems quickly; to capitalize on advances in technology; and to shape our doctrine to reflect these changes.

We must recognize that success on many past battlefields has come from innovative ways of combining available new technologies, and not solely from technological advances, per se. History is full of examples where outnumbered forces were victorious because they controlled critical information. Many times both sides had the same tools or technology available, but one was able to achieve an advantage through innovation.

We are in the midst of a time of significant change that is no less revolutionary than the advent of steam propulsion, carrier aviation or nuclear submarines. The Revolution in Military Affairs has moved information and the requirement for information dominance in the joint battlespace to center stage in our thinking about modern warfare. One important element of this revolution is Information Warfare (IW)—a powerful capability that will have profound implications in the ways Naval forces influence, deter and, if necessary, fight wars. The Naval services will, correspondingly, have a prominent role in IW. Forward presence requires the Naval services to aggressively approach IW from the perspective of both an engaged and enabling force. Unlike the other Services, the Navy must imbed IW capabilities in the fleet and be capable of conducting IW from the time we leave CONUS to the end of an extended deployment. It has been five years since the Navy published Copernicus. As the role of C4I in the Information Age progressed, Copernicus evolved. We now understand the benefits of having a systematic way to use information to influence future military operations. We have come to grips with the fact that we have crossed the threshold into the Information Age. In a time when one of the most important enablers for the Naval expeditionary forces is information, tactics flow in large part from characteristics of our information processing capability. We can only gain an advantage over our opponents if we are the first to implement offensive and defensive information tactics into our warfighting arsenal.

Because our forward posture allows the Navy to be in position when crises develop, we can use IW to slow and influence the enemy's decision making cycle, to prepare the battlespace before the start of open hostilities, and to dictate the battle on our terms. Information-based warfare, employing advanced IW with its rapid, reliable and secure communications, permits integration of battlefield information and increases effects from offensive firepower and maneuvers of our dispersed units.

Information-based warfare allows our forces to exploit new weapons technology to increase the speed of battle. Impeding the enemy's ability to communicate, attacking command and control (C2) nodes, and shutting down enemy sensors will give us the upper hand on the battlefield. Likewise, information-based warfare will give the U.S. the operational flexibility to allocate forces and fires in real time, and to defeat enemy forces at the time of our choosing. In the battlefield of the future, decisive victory will depend on having a comprehensive, global C2 system. This comprehensive analytical approach to IW which combines strategy, tactics, and doctrine fully prepares the Navy/Marine Corps team to meet the challenges of the 21st century. Implementing IW will be one our biggest challenges in the near future.

To realize this future vision, all C4I systems must be built under a JCS unified strategy. Copernicus provides this focus for the Navy and Marine Corps. Our approach demands implementing state-of-the-art technology with highly trained operators. Copernicus although not in its final form, is fielded and operational. It is a robust and dynamic architecture based on the Navy and Marine Corps' vast experience over the past 30 years digitizing the battlefield and providing global C4I in support of the National Military Strategy. We have to be able to adapt quickly to changing technology to fight and win wars in the Information Age. It is clear that information has become a major factor in warfare and will grow in importance in the next century. I challenge you all to join me as we redefine how wars are fought and won!

Admiral J. M. Boorda, USN
Chief of Naval Operations

COPERNICUS

1990: The Beginning

During the early 1980s, the introduction of the desktop personal computer and advances in telecommunications sparked the Information Age and forever changed the way nations conduct business and warfare. Late in the decade, the collapse of the former Soviet Union marked the end of the Cold War resulting in a shift in our national priorities. This shift, however, was not a move away from military leadership or capability. Instead, it was a move toward capturing the forces of change, especially information dominance and technological advancements, to move into the 21st century. Struck by the strategic implications of information dominance in boardrooms and on the battlefield, the Navy, in a white paper, defined Copernicus. In 1992, the Navy and Marine Corps made another bold move by publishing "... From the Sea." Copernicus and "... From the Sea" reflect a shift from maritime, open ocean warfighting to joint operations in the littoral. ("... From the Sea", Sept. 1992)

Copernicus, designed as a user-centered command, control, communications, computers and intelligence (C4I) information management architecture, provided a blueprint for capturing technological change. Copernicus answered critical naval C4I problems and articulated the true essence of modern command and control (C2). It laid the foundation, through its pillars, for joint and allied operations.

Copernicus . . . Forward
The Evolution

When Copernicus debuted it was revolutionary. Its planners recognized the technological limitations challenging Copernicus and selected a building block approach to accommodate innovation. By focusing on fielding systems that provide access to essential data, Copernicus allows the decision making process to migrate from upper echelons down to the tactical commander, or the shooter, realizing a goal of Copernicus—a true sensor-to-shooter environment.

Five years later, one of Copernicus' enduring characteristics is its evolutionary nature. Fielded advances in data processing and storage media technologies permit the operational commander at sea and in the field to process, store and manipulate greater amounts of information more efficiently, improving both the view of the battlespace and the decision making process.

Today's Naval forces perform an increasing number of non-combat and contingency operations as part of United Nations or coalition forces. These operations highlight the need to respond quickly to diverse threats that are not easy to predict or identify. Additionally, these operations emphasize the continuing need for Copernicus as the best means of moving from stovepipe systems to joint operations capability and systems interoperability. In this document, we reaffirm the Copernicus foundation, summarize its original vision, discuss Copernicus today and describe the vision for Copernicus' evolution into the 21st century.

COPERNICUS' FOUNDATION

There are five essential elements of Copernicus that provide architectural oversight to leverage the C4I infrastructure effectively and enhance the C4I operational perspective. These elements:

- Seamlessly blend, through common applications in one workstation, critical tactical, operational and administrative data to the warfighter, thus allowing tactical objectives to drive operations.
- Assimilate required information rapidly through standardized data formats, permitting operational commanders and users to "pull" desired information to accomplish tasks. A two-way intelligent "push" capability supplements user-pull when required and prevents information overload.
- Provide information using integrated data formats in a multimedia environment where form fits function (i.e., voice, video, imagery, and tactical data at high speeds).
- Provide a common operating environment (COE) that standardizes workstations for the operator. Workstation and user interface standardization permits greater operator proficiency while reducing training requirements.
- Use common building blocks for modular and standardized hardware design, which permit upgrades and additions to the architecture in an expeditious, cost-effective manner. Copernicus supports the warfighter at all levels:
    - The watchstander, by employing high-tech computer workstations and common interfaces.
    - The shore commanders, by developing multimedia connectivity and establishing rapidly configurable shore networks that link commanders to all echelons, across all Services, to all allies (whether temporary or enduring) across the full spectrum of warfare.
    - The Composite Warfare Commander (CWC), by employing a series of tactical information networks that change in number and nature to suit the CWCs doctrinal decisions and allow commanders to customize their C2 needs.

- The Commander Joint Task Force (CJTF), by employing networks that must be flexible to permit commanders to customize their C2, especially during joint and allied operations.

The Pillars of Copernicus

Copernicus, an interactive framework of pillars, links the C2 processes of the warfighter at all echelons of command.

The pillars of Copernicus include:

- Global Information Exchange System (GLOBIXS) support the joint and allied tactical commanders by providing access to all required information from any location through a series of wide area Defense Communications System (DCS) networks.
- The CINC Command Complex (CCC) serves as the primary gateway for communications and information flow from GLOBIXS to forward deployed warfighters via Tactical Data Information Exchange System (TADIXS). The CCC performs C2, correlation and fusion functions. A CINC decision making capability, with a focus on rules of engagement and operational intent is included. Battlespace decisions are made by the tactical commanders and shooters.
- Tactical Data Information Exchange System are the tactical networks connecting the CCCs with the Tactical Command Centers (TCCs). These tactical networks fall into four general categories: Command, Direct Targeting, Force Operations and Support. TADIXS provide enhanced digital communications links to the shooters' combat systems from the Copernicus infrastructure, enabling user-pull functionality and enough computer power and bandwidth to receive and process tactical information.
- The Tactical Command Center disseminates information to the warfighter. The TCC can be any forward deployed command center, ashore or afloat, mobile or fixed, and includes tactical centers for individual units. The TCC is the gateway for information flow between TADIXS and the shooter and weapons using Tactical Data Information Links (TADILs).

Copernicus Pillar Evolution

As Copernicus evolved, a new pillar emerged—the Battlecube Information Exchange System (BCIXS). The original pillars flowed and filtered information to and from the TCC for use in the battlespace. The Copernicus battlespace is defined as the entire military and political infrastructure that spans the range of the pillars to the TCC. The BCIXS extends the architecture to include the battlecube, the area in which shooters and weapons reside. The battlecube is a conceptual, multi-dimensional area that includes subsurface, surface, air and space as the environment for conducting warfare.

- BCIXS represents the battlecube in which tactical forces operate. BCIXS boundaries are fluid and defined by the dynamics of the battle. Shooters operating in the battlecube form the operational nodes in the BCIXS. Shooters are equipped with C4I tools that allow them to receive and process information from the Copernicus architecture.

Essential Functions of C4I

Copernicus provides the following four essential C4I functions:

- Common Tactical Picture (CTP)
- Connectivity
- Sensor-to-Shooter
- Information Warfare (IW).

Common Tactical Picture is all information spanning the spectrum from the sensor to the shooter that allows tactical commanders to understand the battlespace. CTP consists of surveillance, intelligence, identification, environmental and positioning inputs and tactical decision aids. Key factors in the CTP include timeliness, coverage, sensor revisit rates, accuracy and completeness. All users then share the same scaleable picture and can extract the pieces relevant to their specific needs and tactical situation.

Reducing fratricide during hostilities is an objective of national policy. This depends on a near-perfect tactical picture with common grid (locational) references on all platforms. The Joint Requirements Oversight Council (JROC) established the General Officer Steering Committee Combat Identification (G'sC-CI) and the Joint Combat Identification Office (JCIDO) to address these issues.

Connectivity: Connectivity links nodes throughout Copernicus to implement the sensor-to-shooter construct. Rapid and reliable connectivity is the cornerstone of all C4I provided by GLOBIXS, TADIXS and BCIXS. Connectivity is critical to the CTP because it provides the managed bandwidth for timely transmission of imagery, video, voice and data. Connectivity is critical to DII users in peace, crisis, conflict, humanitarian support and war. It is the widely-distributed, user-driven infrastructure composed of the information assets owned by the Services into which the warfighter can gain access from any location, for all required information.

Defense Information System Network (DISN) is the information superhighway of the DII and is the primary connectivity for GLOBIXS. DISN consolidates voice, video, data, imagery and record traffic stovepipes into the global, joint C4I grid. Implementation of the Integrated Tactical Strategic Data Networking (ITSDN) structure within the DISN will enable horizontal and vertical cross-connection of the global grid. DISN will also be the transport medium for TADIXS and BCIXS shore networks. The world-wide shore-based communications infrastructure will continue to play a vital role in supporting the Copernican and DII architectures.

Continuing expansion of available bandwidth to the warfighter will be a hallmark of Copernicus in the 21st century. Exploitation of fiber optic wavelength multiplexing techniques, direct satellite broadcast and wide-band transmission systems will sustain the Copernican evolution and result in increased precision, lethality and survivability of the warfighter. In addition, these technological advances will allow other quality of life enhancements for personnel including tele-medicine, tele-training, tele-education or something as simple as a phone call home from remote sites.

Sensor-to-Shooter focuses on the process of putting a weapon on target. This includes surveillance and reconnaissance, acquisition and localization, combat identification, targeting, engagement and guidance, and battle damage assessment. Historically, systems were developed to engage a specific threat with little regard for the interrelationship with other systems or supporting infrastructure. Stovepipe systems made it difficult for platforms to share information in a timely manner causing inefficiencies, especially in joint and allied operations. The sensor-to-shooter construct integrates all systems in the weapon procurement and employment process.

Information Warfare/Command and Control Warfare (C2W) is any action to exploit, manipulate or destroy an adversary's information and/or information systems while leveraging and defending friendly information and information systems to achieve information dominance. IW can be employed before and during hostilities and is fought in the information battlespace.

IW:

- Permeates strategic, operational and tactical levels,
- Encompasses political, economic, physical and military infrastructures,
- Expands the spectrum of warfare from competition to conflict,
- Redefines traditional military and national security concepts, and
- Spans the spectrum from peace through warfighting. C2W, the military implementation of IW, is the integrated use of operations security, military deception, psychological operations, electronic warfare and physical destruction to deny information to, influence, degrade or destroy an adversary's C2 capabilities, while protecting friendly C2 capabilities against such actions.

IMPACT OF INFORMATION TECHNOLOGY

Trends

Information will continue to be a critical driver of warfare, changing the fundamental way warfare is conducted. The same technological advancements that have provided business users greater computing power in smaller packages at lower cost have improved decision making within the battlespace. Technology has profound effects in the battlespace:

- Decreased time from identification to engagement,
- Improved precision and rates of fire cause dispersion of all forces,
- Enhanced digitization and integration of battlecube information increases effects from fire and maneuver of dispersed units,
- Improved situational awareness, and
- Better campaign assessment tools to evaluate courses of action.

Warfare is changing because of enhanced battlecube effectiveness. Massing of troops is giving way to massing of firepower.

Warfighting Challenges

Even though technology enhances battlecube effectiveness, there will always be challenges to overcome. For example, during battle there is a tendency of each Service to resort to its individual operating rhythm. C2 systems must integrate and synchronize the operations of forces to optimize individual Service's strengths and synergize capabilities. Other cultural differences which distinguish our armed Services such as organization and doctrine must continue to be modified to accommodate technological advancements and to improve battlecube effectiveness. Intelligence gathering in the battlecube must be wholly integrated to realize the system's full potential. Finally, there must be adequate communications capability to ensure effective C2.

Information Technology Challenges

Besides the warfighting challenges that exist, there are information technology challenges too. For example:

- The acquisition cycle is slower than the pace of technology expansion. The Copernicus building block approach has allowed the Navy/Marine Corps team to insert advanced technology during system upgrades; however, the acquisition cycle still cannot keep pace with the speed of advancing technology. Standardizing interfaces and replacing military standards with commercial standards are positive steps toward obtaining needed bandwidth and greater price performance of computing capability. Other steps can and are being taken to capitalize on technology in the areas of software development and the application of modeling and simulation (M&S).

- The Services must correctly define requirements early in the acquisition process. Affordable software requires reduction of costly error correction in later stages of software development. Using software engineering tools and techniques, systems can be modeled and tested for compatibility and interoperability before the engineering process begins. M&S and code development, using formal methods, can be an effective means of reducing errors, accelerating the acquisition process and reducing costs.
- As information requirements increase, not only must bandwidth in the electromagnetic spectrum be managed, but data transmission techniques must be developed to minimize or optimize the amount of data to be transmitted. This must be accomplished without degrading the quality of information in that data. Techniques which can be employed include demand assigned multiple access (DAMA) and time division multiple access.
- C4I systems can now inundate the warfighter with data but must transition to providing information and knowledge of the battlecube.
- As our warfighting capability migrates from weapons and platforms to the information systems that drive warfare, our susceptibility to IW is increased.
- To maximize its timeliness and effectiveness, information dissemination in the battlecube requires a more direct organizational structure. The focus must shift from sensor-to-command to sensor-to-shooter. With mission objectives and rules of engagement established, warfighters and shooters will perform their missions and commanders will command by negation.

Flowing Information Forward: Building Sensor-to-Shooter

The original Copernicus pillars have also evolved to support the shift to "Forward . . . From The Sea" and "Operational Maneuver From The Sea." In "Forward . . . From The Sea," there was a recognition that the most important role for Naval forces, short of war, is to be engaged in forward areas, preventing conflicts and controlling crises. ("Forward.. from the Sea", 1994) The Copernicus evolution reflects the need for the C4I infrastructure to support the architecture, down to the shooter and the weapon. Deployed forward, Naval C4I gives the joint commander C2 on arrival.

Copernicus supports worldwide C4I coverage to the shooter. Fixed and mobile elements now provide the shooter the same information previously available only to decision makers in command centers.

Conceptually, platforms are linked by moving information around the information spectrum. The information spectrum consists of three integrated grids.

- Surveillance Grid: A capabilities grid blanketing the battlespace instead of a series of single sensors. This grid consists of national, theater and platform sensors that the warfighter can access directly or through GLOBIXS and TADIXS.
- Communications Grid: An overlaying wide area network of pathways that use multiplexing and digital technology to move data and information into and around the battlespace. Copernican connectivity facilitates the movement of information among operators and analysts.
- Tactical Grid: A tactical network of communications links that ties together all units of a force regardless of the platform or component. This grid connects the Combat Direction Systems (CDSs) among units' TCCs to provide fire-control grade information across the battlecube to the shooters. The BCIXS can "plug" and "play" to access C4I information directly by using TADILs tied to higher echelon TCCs and the tactical grid itself.

COPERNICUS EVOLVING

Strategy to Match the Change—"Forward . . . From The Sea"

The Navy and Marine Corps team's focus in "Forward. . . From The Sea" moves operations into the littoral and is consistent with the Navy and Marine Corps' traditional forward presence role. Operating in the "blue water" environment, a battle group commander could identify threats and have time to react in a battlespace diameter that spanned 3000 nm. Threat cueing and advanced warning could occur long before engaging hostile forces. The time advantage is lost for forces operating in littoral areas which might span a radius of less than 100 nm. Transitioning to littoral operations required a fundamental shift in how we think about C4I from operations at sea, to mobile versus fixed units. Proliferation of cruise missiles, mines and even more common devices such as cellular phones provide potentially hostile forces with capabilities that demand an inordinate amount of attention and could delay friendly operations.

C4I/CDS Integration

Reduced reaction times, combined with increasingly capable threat weapons, makes full C4I/CDS integration a critical objective of Copernicus. Fewer nodes and integration of C4I/CDS accelerates the decision making process and assists the joint warfighter in achieving information dominance over the enemy. Copernicus is accomplishing this integration by prescribing the interfaces between C4I systems and the CDS, empowering platforms to react immediately to threats. These interfaces depend on common standards and protocols so that systems in the architecture can transfer data.

The first major step in fielding Copernicus was implementing the Joint Maritime Command Information Strategy (JMCIS). The JMCIS architecture links C2 systems into functional categories and creates an environment for Services to field interoperable systems with common user interfaces. Already, JMCIS migrated several stovepipe systems into one workstation to produce a CTP. More Navy and Marine Corps C4I systems will continue to migrate into the JMCIS architecture as Copernicus evolves. JMCIS forms a kernel of the Global Command and Control System (GCCS). GCCS supports an open system environment for automated information processing at all warfighting levels of the Department of Defense (DOD). The GCCS, in a departure from traditional developmental programs, promotes a rapid migration strategy that cost-effectively and continuously builds on changing technology and user information needs. A major DOD integration initiative is the selection of those migration systems that will ultimately lead to the creation of standard DOD systems.

COPERNICUS AND THE WARFIGHTER

Organization and Doctrine

For the warfighter to benefit from improved information availability and processing power, the Services must streamline their command structures. Joint doctrine must be developed to ensure technology and investment are captured for the warfighter. Evolving to a sensor-to-shooter construct, information management systems must correlate and fuse data and automatically update the CTP. Rules of engagement and operational intent combined with real-time sensor information will guide warfighters in executing their missions. Rapid information exchange enables decisions at lower command levels where timeliness is paramount. Direct access to information allows shooters to engage targets of opportunity more rapidly.

Decision Process

The C2 decision making process has four phases: Observe, Orient, Decide, and Act—the OODA loop. The OODA loop drives decision implementation within the battlespace. Sensors observe reality. Processors and displays supply decision makers with the means to visualize and orient themselves to the scenario. Perceptions lead to the commanders' intentions and allow decision makers to decide on a course of action. Following the decision comes the execution, or the act. The benefits of technology allow simultaneous C2 decision making processes, empowering the warfighter with faster, better and more direct access to the decision making process.

Command and Control

Decision makers can be given all available situational information; still, understanding the total scenario involves knowing why hostile forces are acting. When opposing forces engage, the battle progresses at a pace set by each combatant's actions. C2 becomes difficult at best. Under such conditions, commanders must rely on training, doctrine and knowledge of the enemy. Knowledge goes beyond just situational awareness and includes the enemy's motives and doctrine. It is important that C2 systems have the capability to transcend situational awareness and free the commander for higher understanding tasks. Actions, visualization, collection and correlation drive the C2 engine toward understanding. Ascending the cognitive hierarchy will allow decision makers to create strategy, plan missions and rehearse based on an understanding of how the enemy is thinking.

Real-Time Mission Planning

When commanders conduct operations with a streamlined command structure and improved C2, there are mission planning implications. Mission planning maximizes the enhanced capabilities provided by advanced technologies. Increased intelligence in the mission planning process provides the mechanism for the sensor-to-shooter construct to build a mission planning capability into the launch platform, and even into weapons. Real-time mission planning allows weapon in-flight reprogramming, updating and terminal homing based on sensor inputs.

Real-time mission planning is enhanced by M&S, which provides tactical commanders planning option alternatives based on scenario variables and produces rapid answers to "what-if" questions. Eliciting and storing details of mission plans is especially useful in the battlecube when the shooter is relying on the system to provide executable plans for targets of opportunity.

Information Spectrum

An objective of Copernicus is to provide worldwide, seamless communications by integrating the surveillance, communications and tactical grid systems within the Information Spectrum. These capabilities will provide worldwide commanders the operational coverage needed to perform their missions. Interoperability will be fully realized when the surveillance, communications and tactical grids seamlessly transfer information on a user-pull basis across boundaries.

Currently, each Service uses components of the Information Spectrum to obtain data; however, due to insufficient or non-interoperable communications links, the data is still not transferred seamlessly. The Navy is in the process of eliminating stovepipe systems and implementing Copernicus using COTS/GOTS technology.

Interoperability lowers the relative cost of information by maximizing a system's ability to reach more users. The grid concept maximizes friendly force use of the Information Spectrum in the battlecube. If mission planners can conceive a grid for Joint forces, they can also conceive analogous hostile grids and devise methods to counter them.

IW/C2W to Preserve National Security

The revolution in military affairs moved the requirement for information dominance to center stage in modern warfare. Our growing reliance on information systems and the global information network results in emergent national security requirements. The Naval services must meet the challenge of these new requirements by supremacy in IW.

IW is rapidly becoming a primary tool to discourage and deter potential adversaries. By leveraging information technologies, we can shape opponents' information to confuse or convince them that we will win if conflict is initiated. IW seeks to avoid hostilities or gain an information advantage before shots are fired, missiles are launched or Marines are landed.

Should deterrence fail, IW can be implemented at the operational and tactical levels through C2W to disrupt the adversaries' decision making process. C2W attacks the various phases of the opponents' decision cycle which slows their decision making, generates confusion, magnifies uncertainty and results in their inability to take effective action. Conversely, we must protect our decision making processes to maintain the tactical advantage. C2W provides an efficient, potentially non-lethal capability to neutralize an adversary's warfighting capability through a measured response. Naval forces deployed in a 'presence' role will be optimally positioned to conduct C2W.

The Joint/Allied Perspective

Future wars will be fought with joint, allied and coalition forces. Alliances will have the potential to produce a powerful synergy of forces and capabilities. However, individual strengths can easily be offset if the interfaces among forces are not transparent to the commanders and warfighters. Technology infusion, standards and protocols, concepts of operations, a flexible architecture and a common command structure are required to bring diverse forces together. "The joint force is the source for service capabilities not the result of individual service capabilities coming together." ("C4I For the Warrior", 12 June 1993)

"C4I For The Warrior" is the conceptual roadmap for achieving global joint C4I interoperability that will allow any warfighter to perform any mission, any time, any place and is responsive, reliable, secure and affordable. In "C4I For The Warrior," information exchange must incorporate interoperable technologies to fuse and automatically update information for joint users to "pull" when required. The command hierarchy must be flattened for the warfighter to benefit from the revolution in technology and information availability. Information flow to the warfighter must be as direct as possible, with command being exercised by negation to accelerate the decision and action processes. Each Service has its own strategy for achieving global, joint C4I interoperability:

- Navy/Marine Corps—Copernicus
- Army—Enterprise
- Air Force—Horizon

"Enterprise" takes a "holistic, process-oriented view of C4I systems development, weapon and weapon support systems development requirements definition, systems acquisition, systems integration, systems improvement, systems employment, and sustainment across the tactical, sustaining base, and strategic operations." ("Enterprise")

"Horizon" provides the warfighter with responsive, advanced C4I services. It is a charge for leading the Air Force into an era of technological innovation and better satisfying the warrior's requirements. "Horizon" charts the course to orient Air Force thinking toward providing warfighters with C4I support in an expeditionary environment and to seek advantages in the coming age of information warfare. (Lt.Gen. Carl O'Berry, "Horizon")

The Navy/Marine Corps team enthusiastically embraces these architectures. Together with each Service, we will fulfill the Joint Chief of Staff's "C4I For The Warrior" vision. Each Service can maximize their operational specialties by working together to achieve battlespace dominance. "C4I For The Warrior" requires interoperable systems that can move information among mobile units. Mobile units that characterize the battlecube will share information across the information grids, using space as the common information transfer medium. The Services will fight synergistically, seamlessly transmitting critical information to become fully integrated in the battlecube.

COPERNICUS. . . FORWARD

Copernicus in the 21st Century: C4I for the Warrior

In the 21st century, Naval forces will achieve C4I for the warrior through the implementation of Copernicus. While the planning horizon extends into the 21st century, Copernicus emphasizes action and near-term results that can immediately benefit the warfighter. By designing for continuous change, Copernicus creates an evolving systems environment that focuses on the process of how we get there from here rather than defining the ultimate destination.

Future Joint Warfighting

Future warfare will take on a new dimension with IW being employed before and during hostilities. Forward deployed forces will conduct continuous surveillance and intelligence collection, providing critical battlespace information. IW preeminence will start early in a crisis. It will shape the conflict and reduce the adversary's warfighting capabilities before hostilities begin. This gives our forces the clear advantage. Technology proliferation makes a clear requirements statement for a robust, interoperable, highly responsive C4I system to provide timely and accurate data and machine-assisted planning options to joint force decision makers.

Joint Battlespace

The Services must work together without boundaries so that offensive actions can be executed with maximum effectiveness and economy of resources. Smaller, diverse forces will use C4I systems to integrate their operating rhythms, creating a distinct advantage over adversaries. Communications will be seamless over multimedia interfaces. Workstations will correlate and fuse data, and users will pull information necessary to the mission. Information will be received in a common format, through standard graphical user interfaces. Copernicus empowers users to act on real-time information, which enables swift, decisive moves to dominant the battlespace.

In Summary Copernicus . . . Forward

The Navy/Marine Corps team continues to ensure the architecture remains a viable and evolving construct that fully supports the warfighter. Copernicus continues to adapt to new technologies and requirements. As a result of fielded systems that support the Copernicus architecture, the Naval forces' goal of true joint and allied interoperability is becoming a reality as is the establishment of an IW strategy and capability. The Joint force of the future will require C2 on arrival. This capability is being addressed by each Service as we all strive toward coherent joint operations. Army "Enterprise," Air Force "Horizon" and Copernicus represent tremendous leaps forward in achieving this goal. The integration of these architectures are the mainstay of "C4I For The Warrior." Examples of systems that work together to support the C4I needed for the battlespace of tomorrow include: JMCIS-based C2 systems like the Naval Tactical Command System-Afloat (NTCS-A) and Marine Air-to-Ground Task Force (MAGTF) C4I and their compliance with the GCCS Common Operating Environment, UHF Follow-on, JTIDS-CEC, and ongoing TBMD efforts.

The anticipated speed of future battles dictates fundamental changes in the way Joint forces organize, plan and execute warfighting. New concepts of operation and doctrine will force C4I systems and architectures from linear, centralized constructs to simultaneous, adaptable systems allowing almost continuous planning, execution and replanning in near real time and real time. The Navy/Marine Corps team's implementation of the Copernicus vision is now paying dividends. Our ability to adapt new technology and continually improve systems to support the battlespace of the future is a proven element of that vision and our acquisition strategy.

The enduring success of Copernicus is assured by its inherent ability to capture change. Copernicus continues to reduce redundancies and accommodate evolving requirements, technological innovations, systems improvements, standards-based interoperability at the lowest possible level, and rapid refinements in policy and doctrine such as "Forward . . . From the Sea" and "C4I For The Warrior." As the Services continue to improve and refine their joint-oriented C4I efforts, Copernicus will provide the warfighter a leading-edge architecture.

For additional information, contact
Chief of Naval Operations (OPNAV)
Space—Command and Control—Information Warfare

Strategic Planning Office (N6C)
2000 Navy Pentagon Washington, D.C. 20350-2000
Phone: 703-614-4770
Fax: 703-653-7524

CHAPTER 14

DEPARTMENT OF THE NAVY INFORMATION SYSTEMS SECURITY (INFOSEC) PROGRAM

DEPARTMENT OF THE NAVY
OFFICE OF THE SECRETARY
WASHINGTON, D.C. 20350-1000

SECNAVINST 5239.3
NISMC
14 JULY 1995

From: Secretary of the Navy

Subj: DEPARTMENT OF THE NAVY INFORMATION SYSTEMS SECURITY (INFOSEC) PROGRAM

Ref:

(a) DODD TS3600.1 of 21 Dec 92, Information Warfare (NOTAL)

(b) P.L. 100-235 of 8 Jan 88, Computer Security Act of 1987

(c) OMB Circular A-130 of 15 Jul 94, Management of Federal Information Resources (NOTAL)

(d) NSTISSID No. 500 of 25 Feb 93, Telecommunications and Automated Systems Security Education, Training and Awareness

(e) NSTISSD No. 501 of 16 Nov 92, National Training Program for Information System Security (INFOSEC) Professionals

(f) NSTISSD No. 502 of 5 Feb 93, National Security Telecommunications and Automated Information Security

(g) NSTISSP No. 6 of 8 Apr 94, National Policy on Certification and Accreditation of National Security Telecommunications and Information Systems, (NOTAL)

(h) DODD 5200.28 of 21 Mar 88, Security Requirements for Automated Information Systems (AISs) (NOTAL)

(i) DODD C-5200.5 of 21 Apr 90, Communication Security (COMSEC) (NOTAL)

(j) DODD C-5200.19 of 23 Feb 90, Control of Compromising Emanations (NOTAL)

(k) DODD 5000.2 of 23 Feb 91, Defense Acquisition Policies and Procedure (NOTAL)

(l) CJCSI 6510.01 of 1 Sep 93, Chairman of the Joint Chiefs of Staff Instruction, Joint and Combined Communications Security (NOTAL)

(m) NSTISSI 4009 of 5 Jun 92, National Information Systems Security (INFOSEC) Glossary (NOTAL)

(n) SECNAVINST 5000.2A of 9 Dec 92, Implementation of Defense Acquisition Management Policies, Procedures, Documentation, and Reports (NOTAL)

(o) SECNAVINST 5231.1C of 10 Jul 93, Life Cycle Management Policy and Approval Requirements for Information System Projects

(p) SECNAVINST 5200.32A of 3 May 93, Acquisition Management Policies and Procedures for Computer Resources

Encl:

(1) List of Acronyms

(2) Glossary of Terms

---

1. *Purpose*. Emerging as an overarching strategy, the discipline of Information Warfare (IW), as promulgated by reference (a), encompasses not only actions that may be taken to potentially affect an adversary's information or information systems but also addresses those defensive aspects necessary to ensure that U.S. information or information systems are protected against attack. Under this defensive portion of IW, Information Systems Security (INFOSEC) is a subset of information assurance which addresses actions taken to protect U.S. information and information systems. The Department of the Navy (DON) INFOSEC policy and procedures, when properly applied, will be a major component of the efforts necessary to ensure the defense of DON information and information systems. The purpose of this instruction is:

a. To establish DON policy for the INFOSEC Program within the IW discipline and to define the organizational responsibilities for implementation of the security disciplines of Communications Security (COMSEC), Computer Security (COMPUSEC), and Emanations Security (TEMPEST) per references (a) through (l).

b. To provide the basic policy and guidelines necessary for consistent and effective application of resources in ensuring the security of national security systems as well as the security and privacy of DON systems/information under the Computer Security Act of 1987 (reference (b)).

2. *Cancellation*. SECNAVINST 5239.2.

3. *Definitions*. Reference (m) contains the official glossary of INFOSEC terms. Additional terms and acronyms relevant to this instruction are contained in enclosures (1) and (2).

4. *Objectives*

a. To establish a DON INFOSEC program that addresses the defensive aspects of IW as promulgated by reference (a).

b. To ensure that information processed, stored or transmitted by DON information systems is adequately protected with respect to confidentiality, integrity, availability and privacy.

c. To integrate the technical and management processes of the various security disciplines (COMSEC, COMPUSEC, and TEMPEST) into a cohesive INFOSEC program.

d. To establish and implement programs that mandate the certification and accreditation of information systems under DON control.

e. To require a life cycle management approach to implementing INFOSEC requirements.

f. To establish standardized INFOSEC training within the DON.

5. *Scope*

a. This instruction applies to:

(1) All DON activities.

(2) All DON-sponsored contractors who own, procure, use, operate, or maintain information systems at government or contractor facilities.

(3) All information systems and other system resources designed, developed, procured, or managed by DON activities; and by their contractors.

(4) Information systems operated, but not owned, by DON (e.g., Joint Staff, Department of Defense (DoD)).

b. This instruction applies to the protection of all elements of the information systems. Of all the elements, COMSEC, COMPUSEC, and TEMPEST are managed per this instruction. The other Information Security activities, such as protection of information against accidents, disasters, human error, physical, personnel, and operational security are managed under separate instructions.

6. *Background*. The DON has recognized the urgent need to integrate all available security capabilities into an unified system-oriented engineering approach to provide responsive, cost effective security measures for our information systems. The DON also recognizes that a thorough and consistent approach to INFOSEC for the protection of our information systems is key to the accomplishment of our defense mission and to the protection of lives, property and technology. Mission and organizational realignments have occurred at multiple levels to ensure an unified approach to DON INFOSEC requirements:

a. *INFOSEC Program Implementation*: The Chief of Naval Operation (CNO) and the Commandant of the Marine Corps (CMC) have established centralized program development and implementation of the DON INFOSEC Program at their levels.

b. *INFOSEC Program Execution*: Commander, Space and Naval Warfare Systems Command (COMSPAWARSYSCOM) has created a Program Management organization with DON-wide responsibilities for establishing and executing of an unified system-oriented engineering approach to INFOSEC.

7. *Policy*

a. *Precedence*. Policy and requirements set forth by higher authority take precedence over the policy established in this instruction.

b. *Fundamental INFOSEC Policy*

(1) Data processed, stored and transmitted by information systems shall be adequately protected with respect to requirements for confidentiality, integrity, availability and privacy.

(2) All DON information systems shall be protected by the continuous employment of appropriate safeguards.

(3) Classified information processed or stored by DON information systems shall be safeguarded as required by that level of classification.

c. *Certification and Accreditation*. The appropriate Designated Approving Authority (DAA) shall accredit every DON information system before operation. The accreditation statement shall identify the required confidentiality, integrity, and availability services and constraints under which the system can operate including data sensitivity, user authorization, physical and system configuration. For those information systems supporting cryptologic functions, Sensitive Compartmented Information (SCI)/Intelligence data, or Single Integrated Operations Plan (SIOP) data, accreditation requests shall be forwarded to the appropriate authority.

(1) Certification of DON information systems shall be performed and documented by competent personnel in accordance with specified criteria, standards and guidelines.

(2) Accreditation of DON information systems shall be performed by competent management personnel in a position to balance operational mission requirements and the residual risk of system operation. All accreditation decisions shall be documented and contain a statement of residual risk.

(3) Accreditation of DON information systems shall be performed when information systems are interconnected to other previously accredited information systems and networks. The DAA shall ensure that operation of the resultant system does not incur any additional unacceptable risk.

d. *Life Cycle Management.* This instruction shall be reviewed for applicability to all DON information systems being acquired in accordance with references (b), (n), (o), and (p). The INFOSEC policy and requirements are applicable throughout the life cycles of all DON systems.

(1) A System Security Plan (SSP) should be developed and maintained for all computer systems (reference (b)). This plan includes the protection strategy planned, including the certification and accreditation processes.

(2) At each milestone decision point, INFOSEC requirements shall be discussed in sufficient detail and tailored to the milestone under review and the complexity of the project. The discussion shall specifically address the issues of confidentiality, integrity and availability.

e. *Training.* The ability to provide comprehensive assurance that DON information is adequately protected is directly related to the qualifications of the individuals operating DON information systems. In accordance with references (b), (d) and (e) all individuals operating DON information systems will be afforded appropriate training and awareness information commensurate with their duties, responsibilities and the level of information protection required.

8. *Responsibilities*

a. *The Assistant Secretary of the Navy (ASN) for Research. Development, and Acquisition (RD&A)* shall:

(1) Issue the appropriate DON policies and guidance providing implementation details and procedures for the INFOSEC program.

(2) Oversee the DON acquisition process as it relates to the INFOSEC program.

b. *The Deputy Assistant Secretary of the Navy (DASN) for Command and Control. Communications, Computers, and Intelligence/Electronic Warfare/Space (C4I/EW/Space)* shall review resource requirements necessary to implement and execute the DON INFOSEC program.

c. *The Commander, Naval Information Systems Management Center (NISMC)* shall:

(1) Serve as the coordinator for DON INFOSEC Policy.

(2) Ensure that INFOSEC activities, with respect to the scope of programs covered by reference (p), are integrated into the overall DON major system acquisition process.

(3) Ensure that INFOSEC is integrated into the DON Life Cycle Management process.

(4) Monitor the implementation of this instruction.

d. *The Chief of Naval Operations (CNO)* shall:

(1) Serve as the Program/Resource Sponsor for the DON INFOSEC Program.

(2) Manage the Navy INFOSEC Program including DON program development, implementation, planning, programming, and budgeting.

(3) Establish and validate Navy INFOSEC requirements; coordinate INFOSEC requirements of joint military department concern with the Joint Staff in accordance with reference (k).

(4) Establish senior INFOSEC advisory boards with CMC representation to advise on DON INFOSEC policy and guidance as required.

(5) Coordinate the development and implementation of appropriate guidance documents.

(6) Provide Navy representative to the National Security Telecommunications and Information System Security Committee (NSTISSC), Subcommittee for Telecommunications Security (STS), and Subcommittee for Information System Security (SISS).

(7) Serve as the DAA for Navy-wide and joint service information systems (where Navy is the assigned lead) and ensure that DAAs are identified for other Navy information systems.

(8) Advise NISMC of INFOSEC issues that may have a DON or DoD-wide impact.

(9) Develop a DON INFOSEC training program.

(10) Develop and maintain the DON INFOSEC Master Plan in coordination with CMC and Systems Commands.

(11) Coordinate DON INFOSEC requirements for the DON Service Cryptologic Element (SCE) security program with the National Security Agency (NSA).

(12) Coordinate DON INFOSEC requirements for the DON Sensitive Compartmented Information (SCI)/Intelligence program, and the DON portion of the DoD Intelligence Information System (DODIIS) with the Defense Intelligence Agency (DIA).

(13) Provide recommendations to NISMC for the revision of other DoD and DON documents to standardize DON INFOSEC across all activities.

e. *The Commandant of the Marine Corps (CMC)* shall:

(1) Advise NISMC of Marine Corps INFOSEC issues that may have a DON or DoD-wide impact.

(2) Ensure that DAAs are identified and security services provided for Marine Corps information systems.

(3) Endorse and forward Marine Corps validated INFOSEC software and equipment procurement requirements to CNO for equal consideration during the development of the DON Program Objective Memorandum (POM).

(4) Submit validated Marine Corps INFOSEC requirements to CNO for inclusion in the DON INFOSEC Master Plan.

(5) Establish senior INFOSEC advisory boards with CNO to advise on DON INFOSEC policy and guidance as required.

(6) Provide recommendations to NISMC for the revision of other DoD and DON documents to institutionalize INFOSEC.

(7) Provide CMC representation to the NSTISSC, STS, and SISS.

f. *The Chief of Naval Research (CNR)* shall:

(1) Manage the DON INFOSEC Program within the Office of Naval Research (ONR).

(2) Advise NISMC of ONR INFOSEC issues that may have a DON or DoD-wide impact.

(3) Provide appropriate representation to INFOSEC advisory boards.

(4) Ensure that a DAA and security support are provided for each ONR information system.

(5) Consolidate and forward the ONR COMSEC equipment requirements to CNO for validation and consideration during development of the POM.

(6) Provide the technical base for the DON INFOSEC Research and Development (R&D) Program within ONR.

(7) Designate a DON Center for Computer High Assurance Systems (CCHAS). As part of the DON INFOSEC Program, the CCHAS shall:

(a) Maintain liaison with NSA regarding INFOSEC R&D.

(b) Conduct R&D in evaluation methodologies for application systems.

(c) Provide INFOSEC research support to other DON activities.

g. *The Commanders, Naval Systems Commands (SYSCOMs)* shall:

(1) Submit INFOSEC POM recommendations to CNO for validation and consolidation in the DON INFOSEC Master Plan. Marine Corps recommendations shall be submitted via CMC.

(2) Coordinate INFOSEC integration into information systems with COMSPAWARSYSCOM.

(3) Ensure that each information system acquisition or deployment under the command's purview adheres to the DON life cycle management policy.

h. *Commander, Space and Naval Warfare Systems Command (COMSPAWARSYSCOM).* In addition to the responsibilities set forth for SYSCOMs, COMSPAWARSYSCOM shall:

(1) As the technical lead for DON INFOSEC, provide systems engineering and integration support to the systems commands for all DON information systems with INFOSEC requirements.

(2) Budget for DON INFOSEC programs as defined in the INFOSEC Master Plan.

(3) Integrate INFOSEC engineering and integration into the Warfare Systems Engineering Process.

(4) Develop and manage the DON INFOSEC R&D Program.

(5) Develop and acquire DON standard and specified INFOSEC products in accordance with the DON INFOSEC Master Plan, ensuring that a Certification Authority and an In-Service Engineering Activity (ISEA) and/or Software Support Activity (SSA) are assigned.

(6) Budget for operations and maintenance funding for fielded DON centrally procured INFOSEC products and systems throughout their life-cycle.

(7) Establish Memorandum of Agreement (MOA) with NSA, as necessary, to facilitate the embedding and/or development of INFOSEC equipment/models.

(8) Provide direct liaison with NISMC on INFOSEC acquisition issues.

i. *The Director, Naval Criminal Investigative Service (NCIS)* shall:

(1) Investigate fraud, waste, abuse and other criminal violations involving DON information systems.

(2) Maintain a staff skilled in the investigation of computer crime. This staff may be augmented, when necessary, by personnel provided by other DON activities.

(3) Collect threat information and disseminate as appropriate.

9. *Action*. All action addressees shall implement this guidance within their organizations. All developing and operating activities shall budget for, fund and execute the actions necessary to comply with this instruction and the implementing documents that support it.

W.C. Bowes
Principal Deputy Assistant
Secretary of the Navy
(Research, Development and Acquisition)

Distribution:
SNDL

A1 (Immediate Office of the Secretary)(AAUSN, ASN(RDA) and DASN (C41/EW/Space) only)

A2A (Department of the Navy Staff Offices) (CNR and NAVCOMPT only)

A3 (Chief of Naval Operations) (N6 and N09 only) (25)

A6 (CMC) (75)

D30 (NAVINFOSYSMGTCEN) (50)

FKA1 (Syscoms Commands)

MARCORPS PCN 71000000000 and 71000000100

Copy to:
SNDL

22A (Fleet Commanders)

24 (Type Commanders) (less 24E)

26H (Fleet Training Group and Detachment)

A2A (Department of the Navy Staff Offices) (less CNR and NAVCOMPT)

B1B (Offices of the Secretary of Defense)

B3 (College and University)

B5 (U.S. Coast Guard)

C25A (OPNAV Support Activity Detachment (Ft. Ritchie, only))

C4L (Navy Laboratories)

C4EE (Center for Naval Analyses)

FKP14 (FCDSSA)

FKP16 (NAVSSES)

FKP18 (NAVSEAASA)

FKP20 (AEGIS TRACEN)

FL4 (Regional Data Automation Center, Data Automation Facility)

FT74 (NROTCU)

W (Department of the Navy Echelon 2 Activities) (less Syscoms) (5)

SECNAV/OPNAV Directives Control Office
Washington Navy Yard Building 200
901 M Street SE
Washington, DC 20374-5074 (20 copies)

Order from:
Naval Inventory Control Point
Cog "I" Material
700 Robbins Avenue
Philadelphia, PA 19111-5098

Stocked: 50 copies

LIST OF ACRONYMS

AIS	Automated Information Systems
ASN	Assistant Secretary of the Navy
CCHAS	Center for Computer High Assurance Systems
CMC	Commandant of the Marine Corps
COMPUSEC	Computer Security
COMSEC	Communications Security
CNO	Chief of Naval Operations
CNR	Chief of Naval Research
DAA	Designated Approving Authority
DASN (C41/ EW/Space)	Deputy Assistant Secretary of the Navy (Command, Control, Communications, Computers and Intelligence/ Electronic Warfare/Space)
DIA	Defense Intelligence Agency
DoD	Department of Defense
DODIIS	Department of Defense Intelligence Information System
DON	Department of the Navy
INFOSEC	Information Systems Security
IS	Information System
ISEA	In-Service Engineering Activity
IW	Information Warfare
LCM	Life Cycle Management
MARCOR-SYSCOM	Marine Corps Systems Command
MOA	Memorandum of Agreement
NCIS	Naval Criminal Investigative Service
NISMC	Naval Information Systems Management Center
NSA	National Security Agency

NSTISSC	National Security Telecommunications and Information System Security Committee
ONR	Office of Naval Research
POM	Program Objective Memorandum
RD&A	Research, Development and Acquisition
R&D	Research and Development
SCI	Sensitive Compartmented Information
SECNAV	Secretary of the Navy
SIOP	Single Integrated Operations Plan
SISS	Subcommittee on Information Systems Security
SPAWAR	Space and Naval Warfare Systems Command
SSA	Software Support Activity
SSP	System Security Plan
STS	Subcommittee on Telecommunications Security
SYSCOM	Systems Command

GLOSSARY OF TERMS

Many of the terms used in this instruction are drawn from different specialization areas, and their special meanings in the context of the DON INFOSEC Program may differ from the common English usage. This enclosure explains how such terms are commonly used in the INFOSEC Community.

1. *Accreditation.* Formal declaration by a Designated Approving Authority (DAA) that an AIS is approved to operate in a particular security mode using a prescribed set of safeguards.

2. *Accreditation Authority.* Synonymous with designated approving authority (DAA)

3. *Assessment.* The laboratory and field examination of products, components, or application systems against NSA or service defined criteria for an anticipated operating environment. Assessment involves a thorough examination of security mechanisms but not necessarily all assurances. See Evaluation.

4. *Availability.* The property that ensures the information system's data, services, and resources are available to authorized users reliably, consistently, and in a timely manner.

5. *Certification.* Comprehensive evaluation of the technical and nontechnical security features of an AIS and other safeguards, made in support of the accreditation process, to establish the extent to which a particular design and implementation meets a set of specified security requirements.

6. *Certification Authority.* The official (e.g., Program Manager, etc) responsible for an information system's certification process and the signature authority for the certification statement.

7. *Communications Security (COMSEC).* Measures and controls taken to deny unauthorized persons information derived from telecommunications and ensure the authenticity of such telecommunications. NOTE: Communications security includes crypto security, transmission security, emission security, and physical security of COMSEC material.

8. *Computer Security (COMPUSEC).* Measures and controls that ensure confidentiality, integrity, and availability of the information processed and stored by a computer.

9. *Critical.* Relating to information system data, services, or other resources whose availability is necessary for successful execution of the DON mission supported by the information system.

10. *Data.* Text, video, audio, imagery, and telemetry used as information or in control functions in an information system. For this instruction, data may reside in any form or media, including human and machine readable, such as magnetic disk and tape, optical disk, paper, punch cards, paper tape, video displays, audio forms, electrical signals, radio-frequency signals, and optical signals. See *Information.*

11. *Designated Approving Authority (DAA).* Official with the authority to formally assume the responsibility for operating an AIS or network at an acceptable level of risk.

12. *Evaluation.* The examination of products or components against defined criteria. Evaluation has three forms: trusted product evaluation, in which products or components are evaluated against a specified class of the Trusted Computer System Evaluation Criteria or its interpretations; cryptographic product evaluation, in which products or components are evaluated to provide appropriate cryptographic mechanisms; and TEMPEST product evaluation, in which products are evaluated to meet national standards for emissions security.

13. *Firmware.* Equipments or devices within which computer programming instructions necessary to the performance of the device's discrete functions are electrically embedded in such a manner that they cannot be electrically altered during normal device operations.

14. *Hardware.* The electric, electronic, and mechanical equipment used for processing data.

15. *Information.* The meaning assigned to data based on the data's representation. See *Data.*

16. *Information System (IS).* Any telecommunications and/or computer related equipment or interconnected system or subsystems of equipment that is used in the acquisition, storage, manipulation, management, movement, control, display, switching, interchange, transmission, or reception of voice and/or data, and includes software, firmware, and hardware.

17. *Information Systems Security (INFOSEC).* The protection of information systems against unauthorized access to or modification of information, whether in storage, processing or transit, and against the denial of service to authorized users or the provision of service to unauthorized users, including those measures necessary to detect, document, and counter such threats.

18. *INFOSEC Master Plan.* The INFOSEC Master Plan describes the current DON INFOSEC baseline and threats; projects INFOSEC requirements and threats into the future; analyzes potential sources for meeting those requirements; and prioritizes SPAWAR INFOSEC investments to lower DON INFOSEC risks.

19. *Integrity.* The security property that ensures that data is not modified by unauthorized users, services, or components or that ensures a critical process cannot be affected by unauthorized interaction.

20. *National Security Systems.* Telecommunications and automated information systems operated by the U.S. Government, its contractors, or its agents, that contain classified information or, as set forth in 10 U.S.C. Section 2315, that involves intelligence activities, involves cryptologic activities related to national security, involves command and control of military forces, involves equipment that is an integral part of a weapon or weapon system, or involves equipment that is critical to the direct fulfillment of military or intelligence missions.

21. *Network.* A network is composed of a communications medium and all components attached to that medium whose responsibility is the transference of information. Such components may include Automated Information Systems (AISs), packet switches, telecommunications controllers, key distribution centers, and technical control devices.

22. *Personnel Security.* The procedures established to ensure that personnel with access to sensitive or classified data, services, and resources have the required investigation, clearances, and access authorizations. In addition, personnel security includes security training and awareness specific to information systems users.

23. *Physical Security.* The use of barriers, guards, locks, badges and related measures to control physical access to an information system, its resources, and facilities.

24. *Risk.* A combination of the likelihood that a threat shall occur, the likelihood that a threat occurrence shall result in an adverse impact, and the severity of the resulting adverse impact.

25. *Security Safeguards.* The protective measures and controls that are prescribed to meet the security requirements specified for an AIS. These safeguards may include, but are not necessarily limited to, hardware and software security features; operation procedure; accountability procedures; access and distribution controls; management constraints; personnel security; and physical structures, areas, and devices.

26. *Service.* An automated capability provided by an information system. For the purposes of this instruction, services can be categorized as data processing services and data transfer services.

27. *Software (or Computer Software).* A combination of associated computer instructions and computer data definitions required to enable the computer hardware to perform computational or control functions.

28. *TEMPEST.* Short name referring to investigation, study, and control of compromising emanations from information systems and telecommunications equipment.

CHAPTER 15

DEPARTMENT OF THE AIR FORCE FACT SHEET

USAF FACT SHEET 95-20 Information Warfare

Information Warfare and its Importance

There are many views on what constitutes information warfare, but the U.S. Air Force defines it as "any action to deny, exploit, corrupt or destroy the enemy's information and its functions while protecting Air Force assets against those actions and exploiting its own military information operations." Therefore, information warfare is any action which attacks, protects or uses military information functions or operations.

During the past two decades, the technological basis of information systems has changed radically. Communication rates, storage density, access speed and instruction rates have all increased, while the size and weight of information systems have decreased. Information systems now permeate modern life. Likewise, these improvements are transforming military operations by providing commanders with information unprecedented in quantity and quality. The commander having an advantage in observing the battlespace, analyzing events and distributing that information possesses a powerful, if not decisive, lever over the adversary.

Modern information systems not only provide the commander with many new capabilities, they also present new vulnerabilities. The widespread integration of information systems into Air Force operations makes information operations a valuable target to an adversary.

Information warfare is not new, but the rapid technological advances in information systems has vastly increased the value of information in warfare, thereby intensifying Air Force interest in it. Senior Air Force leadership has focused significant intellectual resources on this topic.

Information Warfare and Current Air Force Doctrine

Presently, Air Force doctrine recognizes air warfare and space warfare. They are defined by the environment in which they are conducted. There are not separate missions for either; rather, both cut across all roles and missions. The Air Force views information warfare in the same fashion. Information is a new realm in which military operations can be conducted. Consequently, information warfare cannot be integrated into Air Force doctrine as a separate mission. However, if information is a separate realm, like air and space, information warfare's place in doctrine can be modeled on air warfare.

There are three objectives of air warfare: control the air while protecting one's own forces from enemy action through the air; exploit control of the air to employ forces through the air against the enemy; and use the air to enhance one's own overall force effectiveness. Similarly, there are three objectives of information warfare: control the information realm while protecting one's own military information operations from enemy action; exploit control of information to employ information warfare against the adversary; and enhance one's own overall force effectiveness by fully developing military information operations.

New Missions

This understanding leads to three new missions: counterinformation; command and control attack; and information operations.

While counterair and counter-space missions comprise activities whose objectives are control of the air and space, counterinformation comprises activities dedicated to controlling information, maintaining access to information and securing the integrity of one's own information systems, while denying or disrupting the enemy's access to information. Counterinformation consists of both offensive and defensive aspects. Offensive counterinformation uses the information realm and impedes the adversary's use of the realm. Typical means include physical attack, military deception, psychological operations, information attack and electronic warfare. Defensive counterinformation includes both active and passive actions to protect national resources from the adversary's information attacks, typically including physical defense, physical security, operations security, communications security, computer security and counterintelligence. Both offensive and defensive counterinformation allow the commander to control the information environment.

Such control enables force application by creating an environment within which friendly forces may attack targets while denying the adversary the ability to conduct such operations against them. Force application currently includes the typical strategic attack, interdiction and close air support missions. Information warfare offers another mission: Command and control attack which is defined as any action taken to adversely affect the adversary's command and control. Its prime objective is to cut off the enemy's leadership from its forces.

The third mission of information warfare is to develop and leverage information functions to enhance total force effectiveness. This is accomplished through information operations, which includes surveillance, reconnaissance, combat identification, weather, precision navigation, intelligence, communications and command and control. This new mission will enhance the employment of military forces through the acquisition, processing, storage or transmission of information.

Information Warfare and Joint Operations

The Air Force has a vested interest in information warfare since many of its significant targets are traditional air power targets and because information is vital to support and conduct air power functions. The Air Force must make information warfare part of all its operations. However, the Air Force view of information warfare recognizes that it is not unique to the Air Force. From the joint force commander's perspective, just as with air, space, land and sea, dominating the information arena is vital for accomplishing campaign objectives. Each service has its own unique operational demands, as well as its own critical information functions to defend, exploit and share.

Proposed Air Force Doctrine

Proposed Air Force Doctrine			
Aerospace Control	Force Application	Force Enhancement	Force Support
Counterair	Strategic Attack	Airlift	Base Operability and Defense
Counterspace	Interdiction	Air Refuelling	Logistics
Counterinformation	Close Air Support	Spacelift	Combat Support
	Command and Control Attack	Special Operations	On Orbit Support

		Information Operations	
Surveillance	Command and Control	Communications	Combat Identification
Reconnaissance	Intelligence	Weather	Precision Navigation

Point of Contact

Headquarters U.S. Air Force, Doctrine Division Office; 1480 Air Force Pentagon; Washington, D.C. 20330-1480; DSN 225-9066 or (703) 695-9066.

November 1995

CHAPTER 16

8 January 1997

REPORT OF THE DEFENSE SCIENCE BOARD TASK FORCE ON INFORMATION WARFARE—DEFENSE (IW-D)

November 1996
OFFICE OF THE UNDER SECRETARY OF DEFENSE
FOR ACQUISITION & TECHNOLOGY
WASHINGTON, D.C. 20301-3140

This report is a product of the Defense Science Board (DSB). The DSB is a Federal Advisory Committee established to provide independent advice to the Secretary of Defense. Statements, opinions, conclusions, and recommendations in this report do not necessarily represent the official position of the Department of Defense.

This report is UNCLASSIFIED.

DEFENSE SCIENCE BOARD
OFFICE OF THE SECRETARY OF DEFENSE
3140 DEFENSE PENTAGON
WASHINGTON, DC 20301-3140

25 November 1996

MEMORANDUM FOR UNDER SECRETARY OF DEFENSE (ACQUISITION & TECHNOLOGY)

SUBJECT: Report of the DSB Task Force on Information Warfare (Defense)

I am pleased to forward the final report of the DSB Task Force on Information Warfare (Defense), which was chaired by Mr. Duane P. Andrews. You asked the Task Force to focus on protection of information interests of national importance through establishment and maintenance of a credible information warfare (IW) defensive capability in several areas, including deterrence and to make recommendations regarding the creation and maintenance of specific aspects of a national information warfare defense capability.

The Task Force recommends a series of over 50 actions designed to better prepare the Department for this new form of warfare beginning with identification of an accountable focal point within the Department for all IW activities and ending with the allocation or reallocation of approximately $3 billion over the next 5 years to implement these recommended actions.

[Signature]

Craig I. Fields

Chairman

DEFENSE SCIENCE BOARD

OFFICE OF THE SECRETARY OF DEFENSE
3140 DEFENSE PENTAGON
WASHINGTON, DC 20301-3140

21 November 1996
Dr. Craig Fields
Chairman
Defense Science Board
3140 Defense Pentagon
Washington, DC 20301-3140

Dear Dr. Fields

Attached is the report of the DSB Task Force on Information Warfare (Defense).

We conclude that there is a need for extraordinary action to deal with the present and emerging challenges of defending against possible information warfare attacks on facilities, information, information systems, and networks of the United States which would seriously affect the ability of the Department of Defense to carry out its assigned missions and functions. We have observed an increasing dependency on the Defense Information Infrastructure and increasing doctrinal assumptions regarding the continued availability of that infrastructure. This dependency and these assumptions are ingredients in a recipe for a national security disaster.

I should also point out that this is the third consecutive year a DSB Summer Study or Task Force has made similar recommendations to better prepare the Department for the challenges of information warfare.

Accordingly, we recommend a series of over 50 actions designed to better prepare the Department for this new form of warfare beginning with identification of an accountable focal point within the Department for all IW activities and ending with the allocation or reallocation of approximately $3 billion over the next 5 years to implement these recommended actions.

We will be, of course, happy to provide any further assistance you may desire.

Sincerely,

[Signature]

Duane P. Andrews

Attachment

PREFACE

The Defense Science Board Task Force on Information Warfare (Defense) was established at the direction of the Under Secretary of Defense for Acquisition and Technology. By USD(A&T) Memorandum for the Chairman, Defense Science Board, dated October 4, 1995, the Task Force was directed to "focus on protection of information interests of national importance through the establishment and maintenance of a credible information warfare defensive capability in several areas, including deterrence." Specifically, the Task Force was asked to:

- Identify the information users of national interest who can be attacked through the shared elements of the national information infrastructure.
- Determine the scope of national information interests to be defended by information warfare defense and deterrence capabilities.
- Characterize the procedures, processes, and mechanisms required to defend against various classes of threats to the national information infrastructure and the information users of national interest.
- Identify the indications and warning, tactical warning, and attack assessment procedures, processes, and mechanisms needed to anticipate, detect, and characterize attacks on the national information infrastructure and/or attacks on the information users of national interest.
- Identify the reasonable roles of government and the private sector, alone and in concert, in creating, managing, and operating a national information warfare-defense capability.
- Provide specific guidelines for implementation of the Task Force's recommendations.

For the purpose of this report, the terms national and national-level are assumed to include Federal, state and local governments, academia, associations, public interest organizations, and the private sector.

This report presents the conclusions and recommendations of the Task Force based on study efforts of the Task Force and Panels created by the

Task Force to address specific areas of interest. The report is organized as follows:

- Executive Summary.
- Section 1, Introduction, provides background information.
- Section 2, Environment, describes factors pertinent to the study effort.
- Section 3, Observations, provides the major findings of the Task Force.
- Section 4, What Should We Defend?, identifies the information users of national interest and scope of interests to be defended.
- Section 5, How Should We Defend?, suggests processes and procedures necessary to defend the users against the threats. It includes a discussion of required indications and warning, tactical warning, attack assessment, and continuity of operations organizations and procedures.
- Section 6, Recommendations, presents recommendations, and provides specific guidelines for implementing the recommendations. It includes a discussion of the reasonable roles of government and the private sector and concludes with resources, in addition to current INFOSEC budgets, required to implement the recommendations.
- Section 7, Summary, briefly summarizes the report and suggests some immediate actions.

Appendices are provided as background and resource information. They do not represent a consensus view of the Task Force and recommendations contained in the Appendices are not Task Force recommendations to the Department. Some of the appendices were used in part as input to the main body of this report. Other appendices are provided because they contain useful information for further discussion of matters addressed in the main body of the report.

At about the same time that the Task Force was created, the President signed a major policy directive regarding the protection of critical infrastructures such as telecommunications, electric power, and transportation. This directive resulted in the creation of a Critical Infrastructures Working Group (CIWG) to address the manner in which the directive should be implemented. The CIWG recommendations were implemented with some modification in Executive Order 13010, Critical Infrastructure Protection which was signed by the President on July 15, 1996. E.O. 13010 establishes a President's Commission to, in part,

- Assess the scope and nature of the vulnerabilities of, and threats to, critical infrastructures,
- Determine what legal and policy issues are raised by efforts to protect critical infrastructures, and

- Recommend a comprehensive national policy and implementation strategy for protecting critical infrastructures from physical and cyber threats and assuring their continued operation.

Given these parallel and closely related activities, the Task Force elected to address information warfare (defense) issues and provide conclusions from both the national and Department of Defense perspectives. However, the Task Force recommendations are specifically oriented on the Department of Defense. Department of Defense dependencies on national level activities for information warfare (defense) are provided to the Secretary of Defense for possible transmittal to the President's Commission for use in their deliberations.

TABLE OF CONTENTS

LIST OF EXHIBITS

Exhibit

EXECUTIVE SUMMARY

The Environment

The national security posture of the United States is becoming increasingly dependent on U.S. and international infrastructures. These infrastructures are highly interdependent, particularly because of the inter-netted nature of the information components and because of their reliance on the national information infrastructure. The information infrastructure depends, in turn, upon other infrastructures such as electrical power.

Protecting the infrastructures against physical and electronic attacks and ensuring the availability of the infrastructures will be complicated. These infrastructures are provided mostly (and in some cases exclusively) by the commercial sector; regulated in part by federal, state, and local governments; and significantly influenced by market forces. Commercial services from the national information infrastructure provide the vast majority of the telecommunications portion of the Defense Information Infrastructure (DII). These services are regulated by Federal and state agencies. Local government agencies regulate the cable television portion of the information infrastructure. Power generation and distribution are provided by very diverse activities—the Federal government, public utilities, cooperatives, and private companies. Interstate telecommunications are regulated by the Federal Communications Commission, intrastate telecommunications by the state public utilities commissions. Interstate power distribution is regulated by the Federal Energy Regulatory Commission, intrastate power generation and distribution by the state public utilities commissions.

Observations

Information infrastructures are vulnerable to attack. While this in itself poses a national security threat, the linkage between information systems and traditional critical infrastructures has increased the scope and potential of the information warfare threat. For economic reasons, increasing deregulation and competition create an increased reliance on information systems to operate, maintain, and monitor critical infrastructures. This in turn creates a tunnel of vulnerability previously unrealized in the history of conflict.

Information warfare offers a veil of anonymity to potential attackers. Attackers can hide in the mesh of inter-netted systems and often use previously conquered systems to launch their attacks. The lack of geographical, spatial, and political boundaries offers further anonymity and legal and regulatory arbitrage; this lack also invalidates previously established "nation-state" sanctuaries. Information warfare is also relatively cheap to wage, offering a high return on investment for resource-poor adversaries. The technology required to mount attacks is relatively simple and ubiquitous. During information warfare, demand for information will dramatically increase while the capacity of the information infrastructure will most certainly decrease. The law, particularly international law, is currently ambiguous regarding criminality in and acts of war on information infrastructures. This ambiguity, coupled with a lack of clearly designated responsibilities for electronic defense hinders the development of remedies and limits response options.

Exhibit ES-1 shows additional observations.

Exhibit ES-1. Observations

- Information warfare has been particularly troublesome for the intelligence community
- We lack a common vocabulary
- Resources are focused on classified content and systems
- It is easy to make the IW-D problem too hard
- Acquisition policy and practices pose dilemmas
- However, a lot can be done
- And DoD must start now!

What Should We Defend?

The current Administration's national security strategy for the United States suggests that the nation's "economic and security interests are increasingly inseparable" and that "we simply cannot be successful in advancing our interests-political, military and economic-without active engagement in world affairs." In the broad sense, then, the scope of national information interests to be defended by information warfare defense and deterrence capabilities are those political, military, and economic interests. These include the continuity of a democratic form of government and a free market economy, the ability to conduct effective diplomacy, a favorable balance of trade, and a military force that is ready to fight and that can be deployed where needed. These interests are supported by the delivery of goods and services that result from the conduct of functional activities such as manufacturing, governing, banking and finance, and the like. Some of these activities are critical to the nation's political, military, and economic interests. These critical functional activities, in turn, depend on information technology and critical infrastructures such as banking and finance, electric power, telecommunications, and transportation.

In general, U.S. infrastructures are extremely reliable and available because they have been designed to respond to disruptions, particularly those caused by natural phenomena. Redundancy and diverse routing are two examples of design techniques used to improve reliability and availability. However, deregulation and increased competition cause companies operating these infrastructures to rely more and more on information technology to centralize control of their operations, to support critical functions, and to deliver goods and services. Centralization and reliance on broadly networked information systems increase the vulnerabilities of the infrastructures and the likelihood of disruptions or malevolent attacks.

The information users of national interest who can be attacked through the shared elements of the national information infrastructure are those responsible for performing the critical functions necessary for the delivery of the goods and services upon which our political, military, and economic interests depend.

The Department of Defense (DoD) must preserve its ability to fulfill its basic missions. To do that, DoD must be concerned about the ensured operation of the critical functions and the availability of information necessary to fulfill those missions. The intertwined nature of the functions of national interest and supporting infrastructures add to the complexity: there are critical functions which have national security implications and which must be defended; and there are critical portions of the infrastructures which are necessary for the operation of DoD and national functions.

How Should We Defend?

- The concept for defending the information infrastructure and the information components of other critical infrastructures includes the following principles:
- Critical functions must be capable of being performed in the presence of information warfare attacks.
- Some minimum essential infrastructure capability must exist to support these critical functions.
- Point and layered defenses are preferable to area defenses.
- The infrastructure must be designed to function in the presence of failed components, systems, and networks. The risk associated with failed components, systems, and networks must be managed since it cannot be avoided.
- The infrastructure control functions should not be dependent on normal operation of the infrastructure.
- The infrastructure must be capable of being repaired.

The concept for defending is as follows. In the information age as in the nuclear age, *deter* is the first line of defense. This deterrence must include an expression of national will as expressed in law and conduct, a declaratory policy relative to consequences of an information warfare attack against the United States, and an indication of the resiliency of the information infrastructure to survive an attack. Technology to conduct information warfare is simple and ubiquitous; some form of infrastructure robustness and protection is essential. It is technically and economically impossible to *design and protect* the infrastructure to withstand any and all disruptions, intrusions, or attacks (or avoid all risk). The risk can be managed, however, by protecting selected portions of the infrastructure that support critical functions and activities necessary for maintaining political, military, and economic interests. An equally important function is to verify through independent assessments that the design principles are being followed, that protective measures are being implemented where appropriate, and that the information warfare (defense) readiness posture is as reported.

Tactical warning, damage control, attack assessment, and restoration ensures the continuance of these critical functions and activities in the presence of disruptions or attacks. The essence of tactical warning is monitoring, detection of incidents, and reporting of the incidents. Monitoring and detection of infrastructure disruptions, intrusions, and attacks are also an integral part of the defense against information warfare. Providing an effective monitoring and detection capability will require some policy initiatives, some legal clarification, and an ambitious research and development program. The telecommunications infrastructure will be subject to some form of attack and we should have some capability to limit the damage that results and to restore the infrastructure. Little research has been devoted to the basic procedures necessary to contain "battle" damage, let alone the tools which might provide some automated form of *damage control*. Some form of *attack assessment* is essential to determine the impact of an attack on critical functions and the appropriate response to an attack. Restoration of the infrastructure implies some capability to repair the damage and the availability of resources such as personnel, standby services contracts, and the like. The basic functions of monitoring, detection, damage control, and restoration must begin at the lowest possible operating level. Reports of the activity must be passed to regional, DoD, and national-level organizations to establish patterns of activity and to request assistance as needed in damage control and restoration. Finally, some form of *response* to the intrusions or attacks may be necessary to deter future intrusions or attacks. The response could entail civil or criminal prosecution, use of military force, perception management, diplomatic initiatives, or economic mandates. Because response might also involve offensive information warfare, this report does not address it in detail.

Recommendations

The Task Force makes 13 key recommendations as shown in Exhibit ES-2. The Task Force considers these recommendations as imperatives.

Exhibit ES-2. Recommendations

Bottom Line—DoD has an urgent need to:

1. **Designate an accountable IW focal point**
2. **Organize for IW-D**
3. **Increase awareness**
4. **Assess infrastructure dependencies and vulnerabilities**
5. **Define threat conditions and responses**
6. **Assess IW-D readiness**
7. **"Raise the bar" (with high-payoff, low-cost items)**
8. **Establish a minimum essential information infrastructure**
9. **Focus the R&D**
10. **Staff for success**
11. **Resolve the legal issues**
12. **Participate fully in critical infrastructure protection**
13. **Provide the resources**

DSB has been urging action on this problem for 3 years!

In addition, the Task Force made over 50 additional recommendations, which are categorized under these key recommendations. (Note that the first recommendation addresses all of information warfare, not just defensive information warfare.) The Task Force attempted to prioritize these "key recommendations," but in the end decided that portions of all of these key recommendations should be implemented immediately.

The following discussions provide all of the recommendations made by the Task Force. The parenthetical entry following each of the key recommendations identifies the section of the report in which the recommendations are discussed in detail.

1. Designate an accountable IW focal point (6.1). This is the most important recommendation the Task Force offers. The Task Force believes that the Secretary of Defense needs a single focal point charged to provide staff supervision of the complex activities and interrelationships that are involved in this new warfare area. This includes oversight of both offensive and defensive information warfare planning, technology development and resources. The SECDEF should:

1a. Designate ASD(C31) as the accountable focal point for all IW issues.

1a(1). Develop a plan and associated budget beginning in FY 97 to obtain the needed IW-D capability.

1a(2). Authorize ASD(C31) to issue IW instructions.

1a(3). Consider establishing a USD(Information).

1b. Establish a DASD(IW) and supporting staff to bring together as many IW functions as possible.

2. Organize for IW-D (6.2). This key recommendation identifies the need for specific IW-D related capabilities and organizations to provide or support the capabilities. While not specifically addressed by the Task Force, virtual organizations that draw on existing assets and capabilities can be established.

2a. Establish a center to provide strategic indications and warning, current intelligence, and threat assessments. The SECDEF should request the DCI to:

2a(1). Establish an I&W/TA center at NSA with CIA and DIA support.

2a(2). Task and resource the Intelligence Community to develop the processes for Current Intelligence, Indications and Warning, and Threat Assessments for IW-D.

2a(3). Encourage the Intelligence Community to develop information-age trade craft, staff with the right skills, and train for the information age.

2a(4). Conduct comprehensive case studies of U.S. offensive programs and a former foreign program to identify potential indicator collection, funding, training, etc,

2a(5). Establish an organization to examine and analyze probable causes of all security breaches.

2a(6). Develop and implement an integrated National Intelligence Exploitation Architecture to support the organization and processes.

In addition, the SECDEF should:

2a(7). Direct the development of IW Essential Elements of Information.

2b. Establish a center for IW-D operations to provide tactical warning, attack assessment, emergency response, and infrastructure restoration capabilities. The SECDEF should:

2b(1). Establish a DoD IW-D operations center at DISA with NCS, NSA, and DIA support.

2b(2). Develop and implement distributed tactical warning, attack assessment, emergency response, and infrastructure restoration procedures.

2b(3). Interface the operations center with Service and Agency capabilities and I&W/TA support.

2b(4). Establish necessary liaison (e.g., with military and government operations centers, service providers, intelligence agencies, and computer emergency response centers).

2c. The SECDEF should establish an IW-D planning and coordination center reporting to the ASD(C3I) with interfaces to the intelligence community, the Joint Staff, the law enforcement community, and the operations center. This center will: develop an IW planning framework; assess IW policy, plans, intelligence support, allocation of resources, and IW incidents; develop procedures and metrics for assessing infrastructure and information dependencies; and facilitate sharing of sensitive information such as threats, vulnerabilities, fixes, tools, and techniques within DoD and among government agencies, the private sector, and professional associations.

2d. Establish a joint office for system, network and infrastructure design. This office will: develop and promulgate IW-D policies, architectures, and standards; design the information infrastructure for utility, resiliency, repairability, and security; develop and implement an IW-D configuration management process; and conduct independent verification of design and procurement specifications to ensure compliance with the design. The SECDEF should:

2d(1). Establish a joint security architecture/design office within DISA to shape the design of the DoD information infrastructure.

2d(2). Establish a process to verify independently and enforce adherence to these design principles.

2e. Establish a Red Team for independent assessments. The Red Team would assess the vulnerabilities of new systems and services and would conduct "IW-like" attacks to verify the readiness posture and preparedness of the fighting forces and supporting activities. The SECDEF should:

2e(1). Establish a Red Team which is accountable to SECDEF/DEP-SECDEF and independent of design, acquisition, and operations activities.

2e(2). Develop procedures for employment of the Red Team.

3. Increase awareness (6.3). The Task Force strongly suggests the need to make senior-level government and industry leaders aware of the vulnerabilities and of the implications. To that end, the SECDEF should:

3a. Establish an internal and external IW-D awareness campaign for the public, industry, CINCs, Services, and Agencies.

3b. Expand the IW Net Assessment recommended by the 1994 Summer Study to include assessing the vulnerabilities of the DII and NII.

3c. Review joint doctrine for needed IW-D emphasis.

3d. Explore possibility of large-scale IW-D demonstrations for the purpose of understanding cascading effects and collecting data for simulations.

3e. Develop and implement simulations to demonstrate and play IW-D effects (USD(A&T) lead).

3f. Implement policy to include IW-D realism in exercises.

3g. Conduct IW-D experiments.

4. Assess infrastructure dependencies and vulnerabilities (6.4). Various infrastructures are vitally needed to support mobilization, deployment, and employment of forces and to control and sustain those forces. Some of these interconnected infrastructures are known to have single points of failure. Therefore, the SECDEF should:

4a. Develop a process and metrics for assessing infrastructure dependency.

4b. Assess/document operations plans infrastructure dependencies.

4c. Assess/document functional infrastructure dependencies.

4d. Assess infrastructure vulnerabilities.

4e. Develop a list of essential infrastructure protection needs,

4f. Develop and report to the SECDEF the resource estimates for essential infrastructure protection.

4g. Review vulnerabilities of hardware and software embedded in weapons systems,

5. Define threat conditions and responses (6.5). Conditions analogous to DEFCON should be developed to provide a common understanding of IW threat conditions. Appropriate responses to these conditions should also be developed using the Task Force suggestions outlined in the report as a starting point. The SECDEF should:

5a. Define and promulgate a useful set of IW-D threat conditions which is coordinated with current intelligence community threat condition definitions.

5b. Define and implement responses to IW-D threat conditions.

5c. Explore legislative and regulatory implications.

6. Assess IW-D readiness (6.6). A standardized process is necessary to enable commanders to assess and report their operational readiness status as it relates to their specific dependency on information and information services. Using the standard vocabulary suggested by the Task Force, the SECDEF should:

6a. Establish a standardized IW-D assessment system for use by CINCs, MilDeps, Services, and Combat Support Agencies.

6b. Incorporate IW preparedness assessments in Joint Reporting System and Joint Doctrine, for example.

7. "Raise the bar" with high-payoff, low-cost items (6.7). There are a number of low-cost activities the Department can undertake to "raise the bar" significantly for potential systems and network intruders. Three specific Task Force recommendations are that the SECDEF should:

7a. Direct the immediate use of approved products for access control as an interim until a MISSI solution is implemented and for those users not programmed to receive MISSI products.

7b. Examine the feasibility of using approved products for identification and authentication.

7c. Require use of escrowed encryption for critical assets such as databases, program libraries, applications, and transaction logs to preclude rogue employees from locking up systems and networks.

8. Establish and maintain a minimum essential information infrastructure (6.8). A strategy and an overall architecture concept employing existing core capabilities such as Milstar must be developed to serve as a means for restoring services for critical functions and adapting to large- scale outages. The SECDEF should:

8a. Define options with associated costs and schedules.

8b. Identify minimum essential conventional force structure and supporting information infrastructure needs.

8c. Prioritize critical functions and infrastructure dependencies.

8d. Design a Defense MEII and a failsafe restoration capability.

8e. Issue direction to the Defense Components to fence funds for a Defense MEII and failsafe restoration capability.

9. Focus the R&D (6.9). While many commercial and approved security products are available to meet some of the Department's needs, these products generally do not meet the Department's needs in large-scale distributed computing environments and generally do not protect against denial of service attacks. Therefore, the SECDEF should focus the DoD R&D program on the following areas.

9a. Develop robust survivable system architectures.

9b. Develop techniques and tools for modeling, monitoring, and management of large-scale distributed/networked systems.

9c. Develop tools and techniques for automated detection and analysis of localized or coordinated large-scale attacks.

9d. Develop tools for synthesizing and projecting the anticipated performance of survivable distributed systems.

9e. Develop tools and environments for IW-D oriented operational training.

9f. Develop testbeds and simulation-based mechanisms for evaluating emerging IW-D technology and tactics.

In addition, the SECDEF should work with the National Science Foundation to:

9g. Develop research in U.S. computer science and computer engineering programs.

9h. Develop educational programs for curriculum development at the undergraduate and graduate levels in resilient system design practices.

10. Staff for success (6.10). A cadre of high-quality, trained professionals with recognized career paths is an essential ingredient for defending present and future information systems. The Task Force recommends that the SECDEF:

10a. Establish a career path and mandate training and certification of systems and network administrators.

10b. Establish a military skill specialty for IW-D.

10c. Develop specific IW awareness courses with strong focus on operational preparedness in DoD's professional schools.

11. Resolve the legal issues (6.11). The advent of distributed computing has and will continue to further blur the boundaries of the systems and networks that the Department uses. Confusion also stems from uncertainty over when or whether a wiretap approval is needed. Government-wide guidance, and perhaps legislation as well, are needed in the areas of Department assistance to the private sector (e.g., Computer Security Act), tracing attackers of unknown nationality (intelligence versus U.S. persons), tracking attackers through multiple systems, and obtaining/requiring reports of computer-related incidents from the private sector owners and operators of critical infrastructures. The SECDEF should:

11a. Promulgate for Department of Defense systems:

- **Guidance and unequivocal authority for Department users to monitor, record data, and repel intruders in computer systems for self protection,**
- **Direction to use banners that make it clear the Department's presumption that intruders have hostile intent and warn that the Department will take the appropriate response.**
- **IW-D rules of engagement for self-protection (including active response) and civil infrastructure support,**

11b. Provide to the Presidential Commission on Critical Infrastructure Protection proposed legislation, regulation, or executive orders for defending other systems.

12. Participate fully in critical infrastructure protection (6.12). The Task Force makes the following recommendations to the SECDEF regarding the activities of the President's Commission on Critical Infrastructure Protection. Detailed suggestions for each of the below recommendations are outlined in Section 6.12.

12a. Offer specific Department capabilities to the President's Commission.

12b. Advocate the Department's interests to the President's Commission.

12c. Request the Commission provide certain national-level capabilities for the Department,

12d. Suggest IW-D roles for government and the private sector.

13. Provide the resources (6.13). The Task Force reviewed all of the individual recommendations categorized under the key recommendations and estimated to $5 million granularity what the implementation costs might be. The cost estimate is $3.01 billion over fiscal years 1997 through 2001. However, the Department should make a detailed estimate.

Credits

Thanks to AR of the Office of Assistant Secretary of Defense (Public Affairs), Department of Defense, for promptly sending this report. For 200-page paper copy telephone: 1-703-697-5737.
Thanks to the IW-D Task Force and contributors.
Digitized and hypertexted by JYA/Urban Deadline; with special thanks to DN.

Published January 8, 1997.
Corrections welcome; send to @jya.com.
Access Log

APPENDIX A

THREAT ASSESSMENT

A.1 THE REALITY OF THE PROBLEM

Advances in the information infrastructure and the growing dependence of the economy and government itself on that infrastructure raise questions about its security. These questions are not new. In 1990 National Academy of Sciences, Computer Science and Telecommunications Board's (CSTB) report, *Computers at Risk: Safe Computing in the Information Age,* began by observing: "We are at risk. Increasingly, America depends on computers. They control power delivery, communications, aviation, and financial services. They are used to store vital information, from medical records to business plans to criminal records. Although we trust them, they are vulnerable to the effects of poor design and insufficient quality control, to accident, and perhaps most alarmingly, to deliberate attack. The modern thief can steal more with a computer than with a gun. Tomorrow's terrorist may be able to do more damage with a keyboard than with a bomb."

In 1989, another CSTB report, *Growing Vulnerability of the Public Switched Network,* sponsored by the National Communications System, cautioned that: "Virtually every segment of the nation depends on reliable communications. . . . The committee, after careful study, has concluded that a serious threat to communications infrastructure is developing. Public communications networks are becoming increasingly vulnerable to widespread damage from natural, accidental, capricious, or hostile agents."

Since those reports were written, use of networks and network-related systems has grown in the economy at large and in the government in particular. Within the government, Department of Defense (DoD) dependence on information systems and infrastructure has grown. This growing dependence is giving rise to heightened concern about the vulnerability to electronic threats of the Defense Information Infrastructure (DII) as well as the national and global information infrastructures (NII & GII) to which it is inextricably linked (notwithstanding intentionally separate components). Additional government computer and communications network vulnerability may come from the growing use of commercial off-the-shelf (COTS) systems. For example, COTS constitutes over 90 percent of the information systems procured by DoD. Additionally, government procures over 95 percent of its domestic telecommunications network services from U.S. commercial carriers. These numbers are at levels that underscore the inherent linkage between defense, commercial, and civilian security concerns. Consider the following examples as additional input:

US Dependence on Information Systems

- Industry increasingly reliant on communications infrastructures
  - Internet presence as of May 1994 (internet info as quoted in the Computer Security Journal, Fall 1995)
  - As a sample: Exxon had 261 registered networks; GTE had 228 registered networks; Boeing had 139 registered networks; Motorola had 137 registered networks; Martin Marietta had 62 registered networks, Lockheed had 62 registered networks
  - "The number of users who have access to the Internet within companies is growing at a rate of 10% every six months." EDP Weekly, by Computer Age, 6 Nov 95, p. 4
- Governmental Structure of the US dependent on a tenuously secured communications infrastructure
  - One switch handles all federal funds transfers and transactions
- DoD information infrastructure is enmeshed
  - with other Governmental structures and industry and private citizens through shared resources of the electrical grid, telecommunications, and the Internet

Trends

- On line services are a $9.6B industry growing at 100% CGR
  - Address by Michael A. Braun, President and CEO of Kaleida Labs, Multimedia 94, 30 July 1994
- US Financial Institutions
  - transfer more than $1 trillion every day via computer
  - Federal Reserve System handles more than 24,000 wire transfers per day
    - Pittsburg City Paper, Vol 4, No.34, August 24-30, 1994, pp 8-9
- Intel Chairman Andy Grove predicts that by the end of this decade, PC sales will surpass 100 million units worldwide—more than sales of cars or TVs.
  - Egii Juliusen, "Small Computers:' IEEE Spectrum, January 1995, p. 44

- By 1993, 32.7% of US households had a personal computer
  - Marvin Sirbu, CMU
- 12 million copies of Microsoft Office have been distributed worldwide as of December 1995
  - Microsoft Corporation Annual Report 1995

A.2 ASSESSMENT OF THE THREAT

In today's information intensive environment, the information warfare threat can come in many forms. The challenge in evaluating that threat, and the appropriate level of protection or response, has been in sorting out the actual from the perceived, and determining the potential for future developments. In order to adequately assess this threat, the Task Force divided the subject into three categories:

- What is known—the validated threat.
- What is suspected—trends, indications, and the assessment process.
- What is unknown—potential events based on existing capabilities.

These threats to the National and Defense Information Infrastructures vary greatly in terms of intent, sophistication, technical means, and potential impact. The threats can be categorized into the following groups:

- Incompetent, inquisitive or unintentional blunderer; mischief makers and pranksters. 9 Hackers driven by technical challenge.
- Disgruntled employee, unhappy customer intent on seeking revenge for some perceived wrong.
- A crook interested in personal financial gain or stealing services.
- Major organized crime operation interested in financial gain or in covering their crimes.
- Individual political dissident attempting to draw attention to a cause.
- Organized terrorist group or nation state trying to influence U.S. policy by isolated attacks.
- Foreign espionage agents seeking to exploit information for economic, political, or military intelligence purposes.
- Tactical countermeasure intended to disrupt specific U.S. military weapon or command system.
- Multi-faceted tactical IW capability applied in a broad orchestrated manner to disrupt a major U.S. military mission.
- Large organized group or major nation-state intent on overthrowing the U.S. by crippling the National Information Infrastructure.

Based on validated incidents, some of these threats clearly exist today. Others are less certain, but can be estimated based on available technology and analysis of continuing trends in development. An estimate of the likelihood for each of these threat categories is shown below.

IW Threat Estimate

	Validated Existence	Existence Likely but not Validated	Likely by 2005	Beyond 2005
Incompetent	W			
Hacker	W			
Disgruntled Employee	W			
Crook	W			
Organized Crime	L		W	
Political Dissident		W		
Terrorist Group		L	W	
Foreign Espionage	L			
Tactical Countermeasures		W		
Orchestrated Tactical IW			L	W
Major Strategic Disruption of U.S.				L

W = Widespread; L = Limited

The information throughout this Appendix was compiled from unclassified sources and briefings received by the DSB from subject matter experts within the Department of Defense, and throughout the civilian sector.

A.3 THE VALIDATED THREAT

IW-related incidents date back to the mid 1980s with the growth of personal computers on a worldwide scale.

IW-Related Incidents

There Really Is A Smoking Gun

- Hanover Hackers — late 1980s
  - Software time bombs in Public Network switches in Denver, Atlanta, and New Jersey — mid-1989
  - Dutch teenagers intrusion into Pentagon computers during the Gulf War. — Nov 1991
- Rome Labs INTERNET intrusions. — Apr 1994

IW-Related Incidents
There Really Is A Smoking Gun
(continued)

- Organized crime attack on Citibank. Aug 1994
  - INTERNET Liberation Front: 22-man group; 4 currently indicted. Dec 1994
  - Numerous other hackers apprehended and awaiting prosecution (e.g. Mitnick, Poulsen). Ongoing
- Sniffer programs found on all major INTERNET providers.
  - MCI Communications switch penetrations.
- USAF Captain hacks into U.S. Atlantic Fleet ship computers as a test of system vulnerability. Sep 1995

The well known case involving the Hanover Hackers is one of the first recorded incidents and is considered to be an example of hacker activity performed for the challenge of gaining entry into someone else's system—without malicious intent.

Although most Public Network (PN) attacks are aimed at accessing other systems, or avoiding toll charges, the software time bomb attacks indicate that denial of service was the objective.[1] (Note: References are at Attachment 1 to this Appendix). In the case involving Dutch teenagers, sensitive information related to U.S. war operations during Desert Storm was modified or copied. Access techniques used in this case included INTERNET and other networks.[2] The Rome Labs incident is another highly publicized case which eventually revealed that over 150 INTERNET intrusions had occurred between 23 March and 16 April 1994. The intrusions were accomplished by a 16-year old British hacker and an unknown accomplice. Several research programs and systems were compromised through the use of Trojan Horses and Network Sniffers. The individual was eventually apprehended by Scotland Yard, and is awaiting prosecution.[3]

In the 1994 attack on Citibank, an international crime group used the electronic transfer system and the international phone network to gain access and transfer approximately $12M to their own accounts. Prosecution of individuals apprehended in Russia and several European countries is pending at this time.[4] In December 1994, a group known as the INTERNET Liberation Front was charged with stealing phone net data, performing INTERNET attacks for money, and development of highly sophisticated attack tools. Numerous phone, information service, and INTERNET providers were attacked, including some government systems. There was also a substantial international component to their activity based on membership involving at least eight countries.[5] The MCI incident involved an engineer who electronically collected 60,000 calling card numbers and sold them to an international crime ring. To accomplish this task, the individual penetrated several barriers which could have shut down the switch for a prolonged period.[6]

A final example is a case involving a programmed test of electronic systems vulnerabilities. An Air Force hacker remotely entered the command and control system of a ship at sea, through use of a standard computer, INTERNET connection, and the E-mail system onboard the ship. Access included ship navigational control systems which could have effected ship performance or response to guidance commands.[7]

The cases listed here are certainly not an all-inclusive list. They do support an alarming trend toward widespread vulnerability on a case by case basis. The major concern involves what the potential outcome would be if these types of attacks were coordinated to occur simultaneously, or if the tools and techniques used were applied with a more subversive intent.

A.4 THE SUSPECTED THREAT—AND THE ASSESSMENT PROCESS

In order to more clearly identify the suspected threat, the Task Force considered a variety of sources for analytical support, and paid particular attention to some of the more detailed threat and vulnerability assessments accomplished within the last year.

The Defense Information Systems Agency (DISA) conducted an extensive vulnerability assessment of government network systems in 1994 and 1995. A summary of the DISA focus, and findings is shown below [8]:

IW Assessments—DISA Report
(Developing the Information Warfare Defense: A DISA Perspective, Dec 1995)

Focus:

- DISA ability to support defensive information warfare (DIW) initiatives.
- Assessment of vulnerabilities.

Findings:

- DISA is organized to effectively support DIW initiatives, but lacks personnel and funding in many key areas.

IW Assessments—DISA Report
(Developing the Information Warfare Defense: A DISA Perspective, Dec 1995)
(continued)

- It is estimated that DoD is attacked about 250,000 times per year, but only 1 in 500 attacks are detected and reported.
  - DISA assessment verified that less than 5% of all attacks are ever detected, and of those, less than 3% are reported.
  - Most damaging attacks come from insiders, but hacker tools commonly available on the Internet are capable of intruding on a majority of DoD systems.

The result of this report was an increased awareness of a growing problem, but the initial actions were primarily focused on security awareness training, and increased training for Local Area Network (LAN) managers. Indications from DISA are that numbers of reported attacks remain at single digit percentage levels, and the problem continues to grow.

At the request of Congress, the General Accounting Office (GAO) conducted an assessment, with the report published in June, 1996. A summary of the GAO focus, findings, and recommendations is shown below [9]:

IW Assessments—GAO Report
(Information Security: Computer Attacks at Department of Defense Pose Increasing Risks, 22 June '96)

Focus:

- Potential for further damage to DoD computer systems.
- Challenges DoD faces in securing sensitive information on its computer systems.

Findings:

- DoD relies on a complex information infrastructure to design weapons, identity and track enemy targets, pay soldiers, mobilize reservists, and manage supplies.
- Use of the Internet to enhance communication and information sharing has increased DoD exposure to attack.
- DoD information is unclassified, but it is sensitive, and should be protected.
- DISA estimates that DoD is attacked about 250,000 times per year, but only 1 in 500 attacks are detected and reported.

IW Assessments—GAO Report
(Information Security: Computer Attacks at Department of Defense Pose Increasing Risks, 22 June '96)
(continued)

- Attackers have stolen, modified, and destroyed data and software, disabled protection systems, and shut down entire systems and networks.
- Security breaches cost DoD hundreds of millions of dollars annually, and pose a risk to national security, yet CERT teams are inadequately staffed, limiting response capability.
- Policy and training regarding computer security and network management are greatly outdated. There is no uniform policy for assessing risks, protecting systems, responding to incidents, or assessing damage.

Recommendations:

- Develop departmentwide policies for preventing, detecting, and responding to attacks, mandating the following:
  - Report all security incidents within the Department.
  - Perform risk assessments routinely.
  - Correct vulnerabilities and deficiencies expeditiously.
  - Expeditiously assess damage from intrusions to insure integrity of data and systems compromised.
- Require military services and Defense agencies to use training and other mechanisms to increase awareness and accountability.
- Require trained information system security officers at all installations.
- Continue developing and cost-effectively using departmentwide network monitoring and protection technologies.
- Evaluate the incident response capabilities within DISA, the military services, and the Defense agencies to ensure that they are sufficient to handle the projected threat.
- The Secretary should assign clear responsibility and accountability within OSD, the military services, and Defense agencies for ensuring the successful implementation of this computer security program.

Results of this report have been forwarded to the Senate Armed Services Committee and House Committee on National Security; the Senate Committee on Appropriations, Subcommittee on Defense, and the House Committee on Appropriations, Subcommittee on National Security; the Senate Select Committee on Intelligence and the Permanent Select Committee on

Intelligence; the Secretary of Defense; the secretaries of the military services; and the Director, Defense Information Systems Agency.

The report concludes that there are significant risks based on these findings:

- Defense cannot locate or deliver supplies promptly without properly functioning inventory and logistics systems.
- Defense relies heavily on computer technology—especially a network of simulators that emulate complex battle situation—to train staff.
- It is impossible to pay, assign, move, or track people without globally networked information systems.
- Defense cannot control costs, pay vendors, let or track contracts, allocate or release funds, or report on activities without automation.
- Defense systems handle billions of dollars in financial transactions for pay, contract reimbursement, and economic commerce.

According to the FBI and Defense Investigative Service (DIS), high technology and defense- related industries remain the primary targets of foreign economic intelligence collection operations. This finding continues a trend reported in the *1995 Annual Report.* The most likely industry targets of economic espionage and other collection activities during the past year include the following areas, most of which are included on the 1996 Military Critical Technology List (MCTL):[10]

- Advanced materials and coatings
- Advanced transportation and engine technology
- Aeronautics systems
- Aerospace
- Armaments and energetic materials
- Biotechnology
- Chemical and biological systems
- Computer software and hardware
- Defense and armaments technology
- Directed and kinetic energy systems
- Electronics
- Energy research
- Guidance, navigation, and vehicle control
- Information systems
- Information warfare
- Manufacturing processes

- Marine systems
- Materials
- Nuclear systems
- Semiconductors
- Sensors and lasers
- Signature control
- Space systems
- Telecommunications
- Weapons effects and countermeasures.

According to a DIS summary of suspicious contacts reported in FY95, entities associated with 26 foreign countries displayed an interest in 16 of 18 technology categories listed in the new MCTL. The U.S. considers all of the above industries to be strategically important because they produce classified products for the government, produce dual-use technology used in both the public and private sectors, or are responsible for the leading-edge technologies required to maintain U.S. economic security.[10]

FBI Director Freeh provided the following five examples of foreign targeting activities in his 28 February 1996 statement before the Senate Judiciary and Intelligence Committees:

- One foreign government controlled corporation targeted U.S. proprietary business documents and information from U.S. telecommunications competitors.
- Another foreign competitor acquired the technical specifications from a U.S. automotive manufacturer.
- In violation of U.S. export laws, a foreign company attempted to acquire a U.S. company's restricted radar technology.
- Several U.S. companies reported the targeting and acquisition of proprietary biotechnology information.
- One U.S. company reported the foreign theft of its manufacturing technology regarding its microprocessors.

Types of U.S. government economic information—pre-publication or unpublished "insider" data—of special interest to governments and intelligence services include:[10]

- Bid proposals
- Economic, trade, and financial agreements
- Energy policies
- Marketing plans

- Price structuring
- Proposed legislation affecting the profitability of foreign firms operating in the U.S.
- Tax and other monetary policies
- Technology transfer and munitions control regulations
- Trade developments.

Three additional case studies were reviewed by the Task Force involving a southeast U.S. port city, a rail traffic control center, and a 1996 Federal Aviation Administration (FAA) vulnerability assessment. A summary of the findings:

- Port City Assessment:
  - Identified single point of failure for infrastructures supporting military mobilization and deployment
- Rail Traffic Control Center Assessment:
  - Central control switching facility for east coast rail traffic.
  - Potential contributor to problems resulting in fatal Maryland rail collision of AMTRAC and MARC trains in fall of 1995.
- FAA Assessment:
  - Not vulnerable today due to antiquated systems, limited networking, and proprietary software.
  - Upgrades will lead to vulnerabilities due to widespread use of COTS technologies and increased networking.

Details of the assessment which could impact deployment of units and follow-on forces which rely on transport out of the port terminal region are provided in Reference 13. Investigation of the AMTRAK—MARC collision indicated human error, but vulnerabilities were detected in the control center, making it a potential single point of failure for exploitation. The FAA assessment, provided in briefing form to the Task Force in June, 1996, concluded that even though vulnerabilities were likely to grow, financial realities restricted the ability to plan protective measures into proposed upgrades until mandated, or in worst case, following a major incident.[11]

A.5 ARE WE AWAITING AN ELECTRONIC PEARL HARBOR?

The trends seen in development of intrusive tools on the INTERNET, growth in hacker activity, and related incidents cause further concern. A summary of recent trends is given below:

IW Trends

- Open availability of intrusion tools.
  - SATAN made available to the public, April 1995.
  - Rootkit: Recently available, used to mask intrusions.
- Continued growth of hacker activity:
  - Masters of Deception: Programmed attacks on phone companies.
  - Legion of Doom: Phone switching/billing, and credit card abuses.
  - Poulsen/Mitnick/Shadowhawk: Phone, system access, computer code abuses.
  - 5 hacker group break-in of computers at University of Washington, Bank of America, ITT, and Martin Marietta, (1993).
  - Operation Moon Angel: Federal agents arrest 74 hackers nationwide for unauthorized entry into business and government computers (April 1995).
- Continued growth in reported computer crimes:
  - Academy of Criminal Justice Sciences Study indicates that 98.5% of participating businesses had been victims of computer theft or attempted theft.
- Cell phone cloning.
- Terrorist acts: World Trade Center Bombing.

Tools: The NSTAC *Assessment of Risk to Security of Public Networks* reported in February, 1996 that SATAN, the Security Administrator Tool for Analyzing Networks, scans and reports system vulnerabilities, which if improperly used, could enable system attacks. It was made openly available on the INTERNET in April, 1995. The report also identifies Rootkit as a tool which falsifies data, making detection of intrusion difficult even with state-of-the-art technology. Rootkit is also openly available on the Internet

Hacker growth: Additional case study information is provided at Attachment 1 for first three listings. In the case of the 5-hacker group, one raid wiped out data on the Learning Link, a NYC public television station computer serving hundreds of schools.[2] The Moon Angel offenses included breaking into NASA computers controlling the Hubble telescope, and rerouting calls from the White House.

In October, 1995 New York officials made arrests in what was declared the largest cell phone cloning operation in the country. Estimates are that over 27,000 phones were cloned within a seven month period at an estimated loss of $1.5M per day in cell phone revenue nationwide.[2]

Finally, consider the World Trade Center bombing as a case which might be a good example of physical versus virtual attack: Twin tower, 110 story building; 50,000 workers and 80,000 visitors daily vs. Global marketplace nerve center, many City/State/Federal offices, several international office, $3M phone switch station, telecom for Wall Street to the World.[12]

These trends are cause for a growing concern—the unknown threat, and the potential for an attack having strategic significance.

A.6 THE UNKNOWN THREAT—POTENTIAL EVENTS BASED ON EXISTING CAPABILITIES (THE DEVELOPMENT OF A STRATEGIC THREAT)

Existing, easily acquired capabilities make the potential for an attack having strategic significance a reality. The most common capabilities for IW-related attacks are, by themselves, often seen as more of a localized nuisance, rather than a strategic threat. When applied in a coordinated attack however, the results are far more widespread. Consider the Nth order effects in the following example from Col Charles Dunlap's essay, "How We Lost the High-Tech War of 2007", published in *The Weekly Standard,* January 29, 1996:

The Setting: (The year 2007):

- Downsizing and cuts in military infrastructure are "off-set" by information technology.
- COTS technology used widely by U.S. and her adversaries.
- Open architecture provides information equality—not information dominance.
- U.S. insistence on open architecture leaves sources of information readily available to opponents—News media is a particularly valuable source.
- Warfare has become even more savage—not cleaner, more high-tech. Televised atrocities and deaths of U.S. troops become a tool of adversaries to sway public opinion.

The Indirect Attack—U.S. C2-Protect efforts are successful in countering direct attack—leading adversary to indirect attack with many Nth order effects:

- Mexican economy attacked—computers corrupted on a massive scale

- Counterfeit electronic pesos flood Mexican bank accounts
- Hyperinflation; economy collapses
- Refugees flood into U.S.
- Call for troops to be brought home to face domestic situation.

The technologies required to perform this these types of attack are available today. The issue of whether or nor they comprise a strategic threat is more a matter coordinated timing. Some may come in the form of a simple attack on a target identified as a single point of failure:

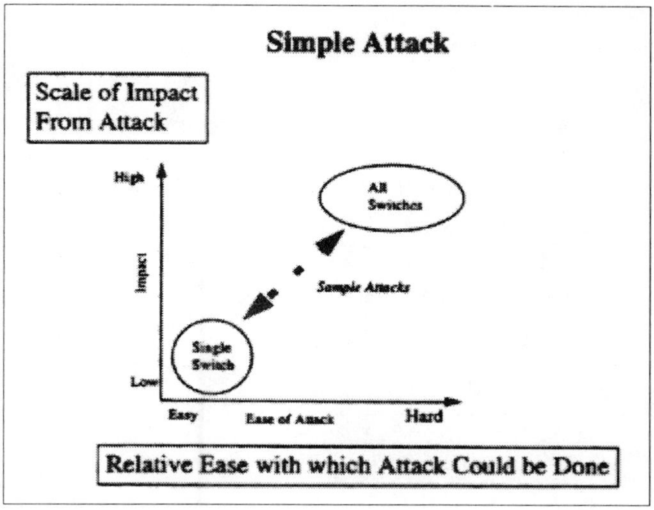

A more complex, coordinated attack takes on a multi-dimensional nature:

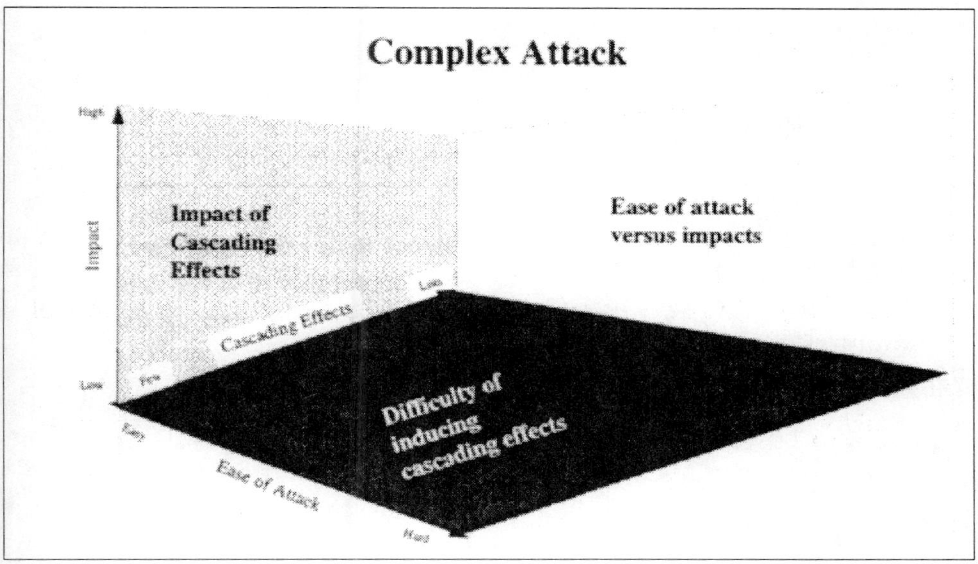

In either of these cases, the timing of the attack is what in fact may have made it strategic in nature. Consider the port city example:[13]

- A power outage, communications failure, or road/rail disruption would be an inconvenience to citizens on an average day.
- However, these same incidents coordinated to occur at the peak of Desert Storm deployment could easily have constituted a strategic threat which would have altered arrival of troops and equipment which played a critical part in the outcome of the war.
- Combine these with the previous examples of attacks on Pentagon computers. Rome Lab, Citibank, and the MCI switch, and the result is widespread loss of confidence in the government's ability to respond to problems both at home and abroad.

To demonstrate the relative ease of achieving an IW capability, the Threat Panel prepared the following table:

Enabling Technologies / Country	C2W														
	OPSEC	PSYOPS	EW	Deception	Destruction Lethal	Destruction Non-Lethal	Encryption	S/W Eng	Network Eng	Computer Security	Info Security	Comms Technologies	HPMW	Physical Security	Intelligence
Russia	↗ D	= D	↗ D	↗ D	↗ D	= D	= D	↘ A	↘ A	↘ D	= D	↘ D	↗ D	↗ D	↗ D
China	↗ D	= D	= A	= D	↘ D	↘ D	= D	↘ A	↘ A	↗ A	= D	↘ D	= D	↗ D	= D
North Korea	↗ D	↗ D	↘ A	= D	↘ D	↘ A	↘ D	↘ A	↘ A	↘ A	= D	↘ A	↘ A	↗ D	= A
Iraq	↘ D	↘ A	↘ A	↘ A	↘ A	↘ A	↘ A	↘ A	↘ A	↘ A	↘ A	↘ A	↘ A	↗ D	= A
Iran	↘ D	↘ A	↘ A	↘ A	↘ A	↘ A	↘ A	↘ A	↘ A	↘ A	↘ A	↘ A	↘ A	= D	= A
India	= D	↘ A	↘ A	↘ A	↘ D	↘ A	↘ D	↗ D	↗ D	↘ D	↘ A	= D	↘ A	= D	= A
Egypt	↘ D	↘ A	↘ A	↘ A	↘ A	↘ A	↘ A	↘ A	↘ A	↘ A	↘ A	↘ A	↘ A	= D	= A
Cuba	↘ D	↘ A	↘ A	↘ A	↘ A	↘ A	↘ A	↘ A	↘ A	↘ A	↘ A	↘ A	↘ A	= D	= D
Libya	↘ D	↘ A	↘ A	↘ A	↘ A	↘ A	↘ A	↘ A	↘ A	↘ A	↘ A	↘ A	↘ A	= D	= A
Syria	↘ D	↘ A	↘ A	↘ A	↘ A	↘ A	↘ A	↘ A	↘ A	↘ A	↘ A	↘ A	↘ A	= D	= A

Legends	
↗	significant, equal to our best effort
=	average or good capabilities
↘	minor or nascent capabilities
D	Developer - making and/or exporting
A	Acquiring from external sources

As an example of a country heavily involved with developing their own capability, consider Russia. Of the 15 categories listed, Russia has a significant capability in seven categories. and a good capability in four (total: 11 of 15). These developments continue, even in the face of widespread economic difficulties. More importantly, almost any nation is capable of developing significant Information Warfare capabilities. Unlike nuclear capabilities, however, IW is relatively inexpensive, and quick to obtain, given the volume of available markets. Thus, a country such as Iran could acquire a strategic capability to threaten the United States without requiring a significant investment, or a long-term development cycle.

A.7 THREAT CONCLUSION

In order to best understand the significance of a potential IW threat, we must consider the often opposing views of information security between the private/commercial sector, and the national security view:

Merging Two Views On Information Security Into One
Strategic Sanctuary Is At Risk

(Concepts expressed in NSA briefing "Ensuring Information Superiority for the 21st Century", presented by LtGen Minihan at NSTAC session, May 1996)

National Security View:

- Protection of information has intrinsic value—National interest.
- Cost of compromise difficult—can be life threatening.
- Risk avoidance approach is traditional response.

Private Sector/Commercial View:

- Cost of doing business—pass the expense on to the customer.
- Countermeasures have a definite expected value.
- "Insurance" approach is the traditional response.

National and Private Sector Information Security Are Now Inexorably Intertwined:

- Zone of cooperation is emerging.
- Risk management approach is needed.

The private sector has viewed IW as a cost of doing business that was often passed on to the customer. The national focus still struggles with the concept of what constitutes a strategic threat. The response has been to avoid risk rather than manage and anticipate it. A zone of cooperation is now emerging which must be better defined:

- Where do protection, detection, and response responsibilities lie?
- Risk management rather than risk avoidance is a critical step.

These issues are at the heart of the defensive information warfare issues.

ATTACHMENT 1

REFERENCES

[1] NSTAC *Assessment of Risk to Security of Public Networks*, Feb 1996.

[2] *Trends and Experiences in Computer-Related Crime*, Academy of Criminal Justice Sciences, 1996.

[3] Rome Lab Attacks Final Report, 20 Jan 1995.

[4] Senate hearing on Security in Cyberspace, 5 June 1996.

[5] *Trends and Experiences in Computer-Related Crime*, Academy of Criminal Justice Sciences, 1996.

[6] 1995 DSB Report.

[7] "Hacker Exposes U.S. Vulnerability," *Defense News*, Oct 9-15, 1995.

[8] DIS Briefing, "Developing the Information Warfare Defense: A DISA Perspective," given by Mr. Bob Ayers, March 1996.

[9] GAO Report, *Information Security: Computer Attacks at Department of Defense Pose Increasing Risks*, 22 June 1996.

[10] *1996 CSI/FBI Computer and Security Survey*, as reported in Computer Security Issues and Trends, Volume II, Number 2, Spring 1996.

[11] FAA Briefing, "Security of the Air Traffic Control System," given by FAA representatives, Mr. Dennis Hupp and Ms. Trish Hammer, June 1996.

[12] NSA Briefing, "Ensuring Information Superiority for the 21st Century," Driven by LtGen Minihan, May 1996.

[13] Joint Program Office (JPO) Briefing, "Infrastructure Assurance Supporting Military Operations," given by the Joint Program Office for Special Technology Countermeasures, Ms. Susan Hudson and Mr. Bob Podlesney, July 1996.

APPENDIX B

NATIONAL INTELLIGENCE EXPLOITATION ARCHITECTURE

The Task Force was briefed by a wide variety of officials and members of the Defense and Intelligence Communities. Several consistent themes became apparent. The "Changed World" in which we live, the changes in threats to the United States, the impact of the "Peace Dividend," and our concomitant expanded "global vision" all drive us to realize that information, per se, has become a precious commodity to the U.S. Further, our existing intelligence structure, collection, analytic and information integration capabilities are optimized for yesteryear. Even though we clearly need specific intelligence collection in a number of areas, such as from networked systems such as the INTERNET and other open sources, it became readily apparent that we do not effectively exploit all of the data that we already collect! The decrements, often horizontal, in analytic resources increase the need for effective integration of the analysis processes across the Intelligence Community (IC). Neither our current structure nor our information processing systems are optimized for the new world set of problems.

Issues relating to Intelligence Community resources, IC support to the military acquisition processes, strengthened IC Issue Managers, review of intelligence collection investment strategies, and developing areas of IC business excellence, all point to a critical need for improved intelligence information integration. We simply will not return to the larger number of analytic resources of the cold war; thus a new paradigm is urgently needed. A "New Vision" is proposed.

The Director of Central Intelligence, in concert with the Secretary of Defense, should create "A New Vision" for intelligence exploitation in the U.S. This "New Vision" would lead to the mission of an integrated National Intelligence Exploitation Architecture which, over time, would develop the meta core system (integrated system of systems) for the National Intelligence Community.

Why do we need this New Vision? Because our existing exploitation and analytic systems were created and built during a period in which virtually all such systems were custom designed and implemented; and thus "stove piped" for specific tasks without particular regard for interfaces and compatibility with other systems. These existing systems were built for what became relatively well understood problem sets with characteristics such as:

- known geography and political boundaries
- known jargon and syntax
- known major nations, entities, forces, units, etc.
- known doctrine and tactics
- known, or at least slowly evolving, military capabilities

- known goals and objectives
- known cultures and ethics
- parametric and other signature characteristics known to and largely exploitable by our collection systems
- well-defined and established data bases, data dictionaries and processing techniques to exploit and analyze observables.

In relative terms, this was a fairly static target set for many years.

Our world has changed to the extent that we now do not know who all of our potential adversaries are, or might be. Who would have predicted two years ago that Rwanda, Somalia and Haiti would demand so much from our intelligence communities. Our requirements now are for globally based intelligence, dynamic in nature and rich in detail, which include increased exploitation of open sources and networks in a totally new information age. The INTERNET provides potential for access to rich repositories of open source information. However, IC access to the INTERNET raises difficult questions and serious concerns about conflicts between law enforcement, intelligence activities and constitutional guarantees. These issues will have to be addressed as part of the NIEA.

Given this shift, we should not be surprised that our existing exploitation and analyses systems do not provide the level of capability required in the new world in which we live.

It is estimated that the National Intelligence Community invests between one and two billion dollars a year in new information management technology (not signal processing, although substantial sums are invested there as well). Additionally, DARPA has over $300 million in its budget for advanced computational technologies alone.

If we focused one-quarter of these moneys towards a coherent, integrated and distributed exploitation and analysis system(s), the U.S. could invest the necessary resources to realize this architecture over the next decade. Why would we wish to do so?

First, our dramatically changed world has been exacerbated by our "Peace Dividend" in the sense that we have taken substantial billet reductions in analytic resources. Thus, we can no longer afford the redundancies of the past and at the same time, we must be able to adapt our exploitation and analysis by sharing (or in many cases shifting) the analytic processes in real-time.

Second, hardware and software technology permit us now to design and implement relatively open systems that have a high degree of interoperability and can use a great deal of COTS. We no longer require the vast majority of our systems to be custom and monolithic! We can have systems that can be modified quickly to address changing requirements and to take maximum advantage of rapidly developing technologies.

Third, the NFIP is going to spend (has budgeted) a larger amount on information technology per year, but in a largely diffused manner. Today, there is neither a coherent architecture, nor focused activity to distribute our development activities in such a way to bring together the best IT resources of the entire Intelligence Community into a highly distributed and interoperable meta system. Surely there are some excellent IT development activities occurring in the IC. These should continue, but in the broader framework of where we need to be in the 21st century. If the past ten years are an example and absent a coordinated development activity as proposed here, we will spend billions of dollars on information technology in the NFIP alone without achieving the kind of integrated capabilities we must have for the 21st century.

How would this work? The "New Vision" would drive the creation of an architecture, supported by enforced building codes (standards, interfaces, security protocols, etc.) in which development activities would be parsed to specific agencies of the IC and supported by the experience of, for example, DoD's experience in promoting interoperability in the JROC process and the development of the GCCS. Over time, each agency would become a center of excellence for a particular area of technology. All of this would fit within the framework of the architecture to permit modules of capability to be developed by one agency for the benefit of any or all other agencies. Let's take but a few examples:

- Text processing is crucial to us all. CIA and NSA arguably have the preponderant IC capability and interest. Each of them should be charged with developing several tiers of text processing capabilities which would build first the foundation and thence the 21st century capabilities we need to exploit open source and covertly collected information. CIA, for instance, might be asked to concentrate on data vaults (archiving large amounts of unformatted data), context based retrieval and query by example. NSA might be asked to concentrate on automated exploitation of facsimile, automated translation and understanding of foreign languages and natural language interfaces.
- Image processing in all of its forms will explode in importance to the IC. While we have no processes today to automatically index, search, retrieve or exploit video, such capabilities will be crucial to our ability to deal with intelligence requirements of the 21st century. Thus the new National Imagery and Mapping Agency would be charged with developing the technology, tools and techniques to exploit video; and to handle the huge volume of SAR & EO/IR imagery from both airborne and spaceborne sensors. This is needed even though we do not have today robust collectors of video. But we will, and we must lay the groundwork today for that future.
- DARPA might logically be asked to develop the tools needed for 21st century automated integration of systems and a number of other tech-

nologies which potentially span and integrate with other IT developments. Examples of these include decision aids, correlation and fusion aids and visualization tools that work across disparate sources of intelligence and support a collaborative community of effort.

In each of these cases, there would be strong emphasis on exploiting and integrating technology from the private sector as well. For example, Time-Warner and Salomon Brothers are but two firms investing hundreds of millions of dollars in advanced information technology. We need to assess where the private sector will be providing the research and development so as to leverage those developments and conserve our scarce resources for R&D specific to IC/DoD needs and for integration of commercial technologies.

Over time, the IC could develop a series of interoperable systems which would be less expensive and more powerful by several orders of magnitude for the 21st century, than if we proceed during the next decade in the same mode as the last decade.

The Intelligence Community will have to change in response to the NIEA. Carrying out the initiative may require a more highly centralized focus on information systems that are both within individual organizations and across the IC as a whole. Long-debated plans for Central Information Services Offices may have to be implemented to create the budgetary resources and organizational authority needed to guide an internal information revolution. A Central Information Council may be needed at the SECDEF/DCI level (perhaps to include other government agencies) to establish policy and to guide the IC to focus on common interests. Whatever organizational reforms are taken to ensure the success of the initiative, change will be needed to break down resistance to change, shift established patterns of investment, and enforce a high degree of cooperation and interoperability.

- Investments in information systems must be shifted from operations and maintenance of existing, slowly-evolving systems to the development of more powerful and adaptable systems that are the focus of the initiative.
- Higher levels of cooperation and coordination between the collection, exploitation and analytical communities are needed to support the dynamic, uninhibited research environment envisioned in the initiative.
- An unconstrained research environment will break down the isolation of analysts from each other and the policymakers and encourage the integration of military, economic, ethnic, political and technological factors in analysis. Analysis standards must be established and enforced throughout the IC to avoid "tabloid" intelligence reporting and to ensure the presentation of sound, but divergent views.

The most difficult part is to make the decision. To commit to a road map of information technology which will become the exploitation and analysis meta system (or system of systems) of the 21st century. The target environment is an integrated, yet highly distributed, heterogeneous IT infrastructure which—over time—will permit an individual in the Intelligence or Defense Communities to query this information environment (much as can be done today on INTERNET). The responses would be relevance ranked and presented in a contextual framework pertinent to that particular user. Thus, military commanders/CINCs, acquisition managers, intelligence analysts and a myriad of other users could gain access to the most comprehensive and broadly based information and intelligence available. Sure security is both an issue and a potential problem. Keep in mind that we are talking about the system for the 21st century and many of the security issues of today will be resolved either procedurally (a la the Joint Security Commission Report) or technically through protective hardware and software. Figure 1 illustrates the concept whereby an individual seeking information "goes fishing in the sea of data." The system would understand some of the context from which the query was made. and as the user asked additional and clarifying questions, the system builds more and more relationships pertinent to the user's needs. Additionally, the system is capable of making queries by example. Thus, the user could ask, "tell me if you know of any more of those." Alerting tools would be embedded in this as well. Both AT&T and American Express have profiles of millions of users which, when violated or exceeded, alert the service provider of anomalies in spending, geography or activity norms. Some of these norms are dynamically adjusted automatically as people use their phone or credit cards.

What the User Needs

An easy capability to extract information related to his problem from the mass of data (national and otherwise) available on distributed problem solving networks

Figure 1.

Figure 2 illustrates the distributed nature of the component systems. They could be spread across Washington, the country, or the world. The key is that, like the INTERNET, the user does not have to know where information is stored in order to retrieve it! His query will seek data through the network of servers/routers/switches that dynamically interface the systems. Although today's INTELINK is a significant improvement over a couple of years ago in accessing intelligence from remote, distributed agencies and commands, it relies on pre-identified and indexed intelligence. What we need for the future is a system that aids the user in finding "unknown" information.

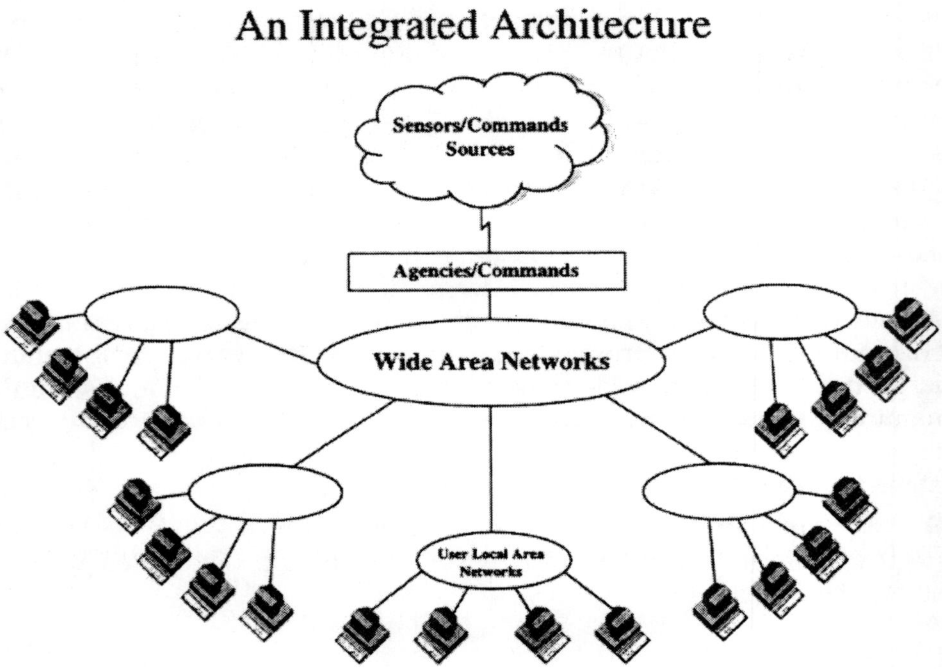

Figure 2.

Figure 3 shows how an integrated interactive multi-media workstation would have (or access) decision aids, correlation and fusion aids and visualization tools to provide the user the most pertinent and timely information. There is no intent to create and keep current monolithic data bases from which searches would be made. Data bases, as we currently know them, are necessary but hardly sufficient for our 21st century purposes. More about that shortly.

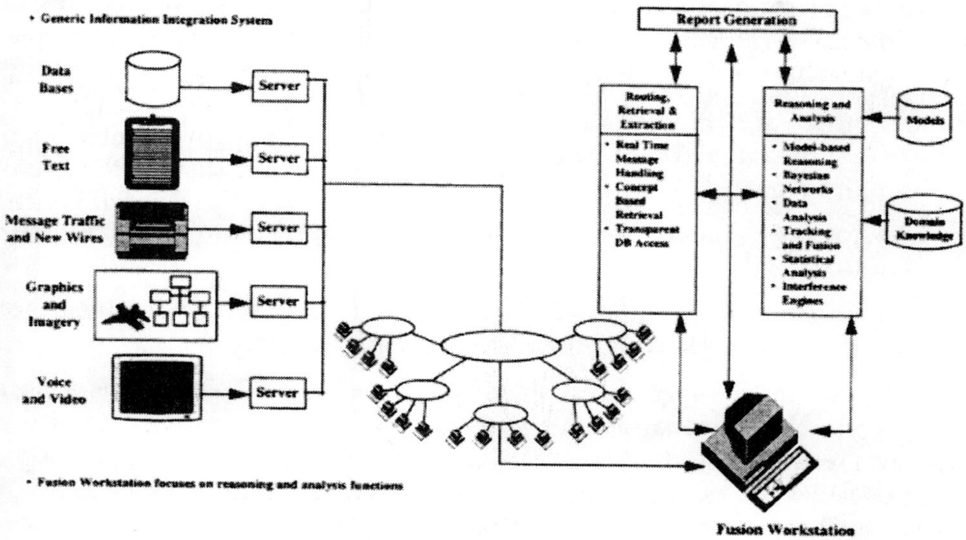

Figure 3.

A powerful aspect of this proposed National Intelligence Exploitation Architecture is that this identical infrastructure could support all of DoD, or all of the government. The tools, techniques, technology and integration required to build and implement this system, need only to provide access to the data sources others might need to serve all of DoD or all of Government. Surely there would be requirements for domain specific tools, decision aids and presentation unique or nearly unique to particular user communities. But the underlying infrastructure would be as widely applicable and robust for all, as is the INTERNET today and tomorrow. There is within the DoD a Common Operating Environment (COE), used principally as the core of the GCCS and some other C2 systems. This may offer a starting point—a building block—from which design the NEA.

Surely many of the issues associated with the successful implementation of this architecture seem intractable today. A great deal of technology R&D and technical development must be accomplished and integrated over time to achieve these goals. This is a journey we believe is absolutely essential. Our existing "stovepipe" systems were built with old technologies under different paradigms. We have a new world, and a new paradigm for sharing information—most of which will now be unformatted, in contrast to most information in existing intelligence data bases.

Most of the existing analytical support systems in use today deal with three major types of information in various storage sizes (e.g., megabytes to terabytes). The information types are:

- Fixed-format database—file, record, field with predefined field sizes and attribute names. Collected data which cannot be fitted within the existing data definition must be discarded, since there is no way to store and retrieve it.
- Free-form text databases—unformatted messages, open source materials, etc.
- Pictorial or graphic databases—graphics, imagery, etc. (Note: these are largely still images, with limited animation or video.)

Enormous effort has gone into developing automated systems to support filing collected information into one of these types such that it can be queried, retrieved, and disseminated using existing (circa 1980) indexing and database technology The "New World Order" and the emergence of new database types such as analog and digital video, voice, and new National collection capabilities are generating a need for tools and techniques for dealing with extremely large data vaults.

The term data vault describes a repository of data of information in a multiplicity of formats Boolean, single character, character string, or numeric fields; free-form text; and "blobs" (Block of Bytes). Blobs can contain images, digitized audio, video, etc. Dealing with data vaults of the magnitude which we can now collect will require substantial innovation in relational and/object oriented database storage, indexing, and retrieval that are needed are:

- High speed, high volume storage and retrieval, including full automated stuffing of formatted databases from text messages and packed parameterized data streams.
- Automated means of storing, indexing and accessing blobs of non-textual materials (graphics, imagery, video, etc.) by content.
- High speed data transmission of the contents of entire data vaults or subsets thereof.
- Super high performance object database systems. Automatic format recognition and transformation. (Simple example: PICT to TARGA and the reverse. More complex examples: Rembrandt or PictureTel to Fractal or DVI and reverse, Model 204 to SQL and Back.)
- "Profiling" of non-textual materials better than the way we now do text and messages.
- Fully automated formation of hyperlinks.
- Case Frame of Concept-based retrieval.

- Intelligent User Assistance Agents ("knowbots").
- Self Organizing databases (especially text, imagery, video, etc.).
- Superior query techniques for sporadic users who are not (and do not have time to become) data retrieval specialists (see next section).
- All of this within a secure environment (classified and unclassified).

There are a number of systems under development which may attack some of these issues. For example, EOSDIS will collect, store and make accessible on the order of terabytes a day.

Refer back to Figure 1 which illustrates the capability needed for the user in response to the explosion of dissimilar information to which we have and need access and understanding. The technologies cited above can be referred to as those necessary to provide Distributed Problem Solving (DPS) capabilities to intelligence analysts and others.

While we attempt to attack the multi-source correlation and fusion problem with the automation, we often overlook the finest and fastest correlation system available—the human eye, ear, and brain. Further, almost since the beginning of application of ADP technology to intelligence problems, analysts have asked for a "smart map."

The third fundamental piece of the puzzle is finding ways of displaying complex and voluminous disparate data streams such that our premier correlation tool can visualize them. A true smart map is one presentation approach, to which almost any analyst can relate. Some of the features included in smart maps would be: pan and zoom, movable viewpoints, active regions, alerts and alarms, validity representation and so on.

In addition to these smart map capabilities, we need better ways to visualize dynamic phenomena such as occurrences of scenario events with respect to time, and to integrate temporal and spatial relationships in displays, operations clocks, etc. These need to be integrated with the smart map display, with corresponding active regions on the timeline displays such that the analyst can access the same information from either place. Display techniques are needed to allow visualization of problems with dimensionalities higher than four (three-space + time).

Additionally, efforts in voice recognition technology could minimize keyboard entry of database and knowledge base input and queries.

The technologies cited above (and a number of others such as imagery processing, compression techniques, interactive multi-media, etc.) represent a panoply of capabilities, some of which are far more attainable or cost effective than others. Some are more likely at the end of a decade, others reasonably soon. Much of the needed technology is being developed, or will be developed in the private sector. Systems for voice recognition and understanding are already replacing commercial telephone operators; office work stations are already taking dictation; personal computers are translating scientific journals from Japanese into English. Image understanding systems are being used to read x-ray mammograms and inspecting cell cultures. Advanced computer systems are being used by commercial airlines for resource allocation and logistics planning beyond human capabilities. Other applicable commercial developments include worldwide, point-to-point voice, compressed data, and even encrypted communications for cellular phones and the INTERNET. The entertainment industry is investing huge sums to develop new wideband data distribution systems (*i.e.*, high definition television) and direct, digital broadcast satellites.) These are all technologies which are directly applicable and will be developed far faster by commercial industry than by the government.

These technologies need not be developed twice. The trick, or course, is to pick the right ones; fit them into a critical path, and integrate them into the National Intelligence Exploitation Architecture. This drives us to realize that integration, per se, is becoming and must become a technology in its own right. Advanced integration tools, techniques and testing require significant development. DARPA, in concert with the private sector, is the obvious candidate to tackle these issues.

The challenge then for the Intelligence Community (the DCI and SECDEF) is to:

- Develop the "New Vision " This should be accomplished working with the customer base to derive a set of design objectives for the National Intelligence Exploitation Architecture. Next.
- Develop the basic system architecture road map; evaluating various technologies and approaches, and then
- Create a detailed program plan to implement the infrastructure, and
- Make needed organizational adjustments to ensure the program is carried out.

It is believed that adequate funds are present in existing NFIP (with partial DARPA support) budgets to support this architecture. Success would take commitment to a coherent road map and parsing varied development activities to agencies which would essentially become centers of excellence for varied components of this architecture. This program provides the framework for our 21st century intelligence exploitation and analysis support to government.

APPENDIX C

A TAXONOMY FOR INFORMATION WARFARE?

Taxonomy:

1. The classification of organisms an ordered system that indicates natural relationships.

2. The science, laws, or principles of classification; systematics.

3. Division into ordered groups or categories: "Scholars have been laboring to develop a taxonomy of young killers" (Aric Press)

[French *taxonomie*: Greek *taxis*, arrangement; see *TAXIS* + *-nomie*, method (from Greek *nomia*; see *-NOMY*).] American Heritage Dictionary

Summary: A taxonomy of information warfare that describes information warfare was derived by the Defense Science Board Task Force on Information Warfare Defense. Unfortunately, as in most cases where both objects and processes are present, this taxonomy would not scale in a linear manner beyond three levels. This is the result of the number of permutations and combinations by which the attacks could be mounted against a particular process, over variable time periods. The derivation of the taxonomy is discussed latter in this Appendix.

However, by adopting concepts from Joint Pub sources and inputs of the Threat and Policy Panels of the Task Force on Information Warfare (Defense), a standard vocabulary for use in threat alerting and for the assessment and reporting of defensive preparedness, tied to specific information dependent processes, was developed for information warfare defense.

Such a tailored warning, assessment and reporting system can and should be developed for use in each civil agencies and in various domains of commercial sector such as electrical power and financial services. A caution: Whatever schema is used to evaluate the operational readiness of information dependent processes and activities, it must be timely and reflect the current state of the security policy being implemented, the supporting infrastructures (computers, communications, electricity and other supporting utilities) and the training status of the personnel, both systems administrators and users of information and information systems.

A range of standardized scenarios should be promulgated for use by the components of the Department of Defense in conducting preparedness surveys and for use in military planning. A proposed partitioning of increasingly robust assessment scenarios for use in planning and assessments follows:

1) accident. (The inclusion of accidental failure is important because in many cases the cause of failure may never be determined but it is still

important to know the range of potential effect on the information dependent process.),

2) amateur hackers,

3) experienced hackers,

4) well-funded non-state group or actor able to purchase or hire advanced information warfare capabilities,

5) state-sponsored information warfare, and

6) state-sponsored information warfare with the active collusion of an authorized insider (worst case).

A standardized set of methods *for assessing information dependent processes* should be used so that reporting is consistent across a wide range of information dependent activities. A proposed partitioning of *assessment methods* follows:

a) an unknown information assurance capability for a specified assessment scenario,

b) an engineering estimate of information assurance, based on a review of design and recovery plans, but no physical testing for a specified assessment scenario,

c) an engineering estimate of information assurance, based on design parameters, simulation exercises, and the review of detection capabilities and recovery plans, but no physical testing for a specified assessment scenario,

d) an internal information assurance audit by an internal but independent organization, based examination of the written record of security and accidental incidents and responses from a live contingency plan exercises designed to simulate a specified assessment level defined above,

e) an internal information assurance audit by an internal but independent organization, based on testing and examination of security and accidental incidents and responses from a live contingency plan exercise designed to simulate a specified assessment scenario defined above. and

f) an information assurance audit by a totally independent security assessment organization, based on testing and examination of security and accidental incidents and responses from a live contingency plan exercise designed to simulate a specified assessment scenario defined above (most stringent test case).

Note that all organizations would not be expected to meet the most stringent assessment scenario. The application of an evaluation level would be determined by the criticality of the information dependent process to the overall activity.

In such an information assurance, planning, testing and evaluation construct, the most robust and resilient organization would have demonstrated a 6-f capability of information assurance.

Although not a taxonomy of information warfare, this approach provides a standard vocabulary for assessing and reporting operational readiness of organizations to carry out information dependent processes in an information warfare environment. This construct also provides a basis for developing an *information warfare readiness reporting* process.

Within the Department of Defense, suitable information assurance reporting criteria along the above lines should be added to the Status of Resources and Training System (SORTS) (or a SORTS-like report); Communications Spot Report (COMSPOT) and daily Communications Status Report (COMSTAT); annual CINCs Preparedness Assessment Report (CSPAR); Combat Support Agency Assessment System (CSAAS); and the Base Defense and Operations Security evaluation schemes.

In addition to preparedness assessments, which address specific information dependent processes, a generalized threat warning system is needed to communicate a heightened level of alert to numerous interconnected information dependent activities.

Design of a warning system is complicated by the interconnectivity of the national (and global) information infrastructure. A heightened state of alert must extend to all connected systems but at higher threat levels appropriate actions could include disconnecting from the infrastructure so a warning method is needed that does not fully depend upon the interconnected infrastructure. Conceivably, preparation could include "war modes" that extend across lower levels of network protocols (physical level through transport layer protocols). In addition, a workable information warfare alert and response process will require a comprehensive legal. regulatory and operational infrastructure.

Detection of information warfare attacks will likely not come directly from intelligence or the managers of individual systems. "Warlike" attacks may have many diverse targets but probably will not follow the pattern of normal thefts or disruptions caused by amateur intruders, except as cover, concealment or deception.

Reporting of incidents, particularly of attacks on civil information users of national interest, will neither be automatic nor directed to a common point unless a distributed structure is created now, like the Center for Disease Control. Creation of a distributed reporting structure that filters upward with a focus on finding broader and broader patterns through indirect measurement and iterative analysis is essential as most "problem" detection will take place locally in a very decentralized fashion without the necessary visibility to detect the linkages between apparently unconnected events.

The Tactical Warning/Attack Assessment functions will require the synthesis of diverse and apparently unrelated information. Specialists in offensive information warfare should be included in the make-up of Department of Defense and national TW/AA centers to ensure suitable tradecraft is applied to the IW/AA process.

On receipt of an information warfare alert message or threat condition, the individual managers of information dependent processes could initiate appropriate defensive actions to include disconnecting from the shared infrastructure. Although Alert Conditions could be issued as a result of strategic warning, most would be triggered by an aggregation of tactical warning reports of individual incidents which will show a pattern of an attack rather than isolated incidents.

A set of proposed information warfare (IW) Alert Conditions and Responses for use by the Federal government, in both civil and national security activities, follow:

IW Alert Condition I

Situation—Normal

Normal level of threat from accident, crime and amateurs

Normal level of unexplained activities in all sectors of the nation

Response Required:

Normal protective actions to include:

- Due diligence in protecting information systems and assets
- Reasonable level of maintenance activities
- Compliance with IRS transaction auditing requirements
- Compliance with all applicable rules, regulations and laws

Normal level of unexplained activities in all sectors of the nation

IW Alert Condition II

Situation—Perturbation

a) 10% increase in incidence reports, either regional or within a functional information dependent activity of national interest

- Regional would include a large communally served geographic area
- Functional would include sectors of the infrastructure, including but not limited to
  - Sector systems, such as medical systems or financial systems
  - Telecommunications service providers — Public utilities

b) 15% increase in all incidents

- Not limited to obvious infrastructure connections

Response:

Increase incident monitoring and cooperative analysis Look for patterns across a wide range of variables

- Including source, users, time, connection, and type of equipment

Alert all agencies to increase awareness of activities

- Including Federal agencies, regulatory bodies, trade groups, professional organization, and corporate entities

Begin selective monitoring of critical information services

- Initiate expanded audit and tracking capabilities with increased reporting to central manager

IW Alert Condition III

Situation—Heightened Defense Posture

a) 20% increase in incidence reports across the board, even with no apparent connection

b) Condition II with special contexts

- Contextual sensitivity subject to integration with all other operations and activities of the U.S.

Response:

Disconnect all unnecessary connections

- Advisory notices broadcast over diverse media to all elements of infrastructure: an IW emergency broadcast warning

- Limiting connections should force a channeling of hostile activity and reduce the number of backdoors that can be exploited

Turn on real time audit for critical information systems

- Augment audit analysis teams to handle the increased loads

Begin mandatory reporting to central manager

- Support forensic investigations and help determine the identity of the aggressors

IW Alert Condition IV

Situation—Serious Situation

a) Major regional or functional events that seriously undermine U.S. interests

b) Conditions II or III with special contexts

- Contextual sensitivity subject to integration with all other operations and activities of the U.S.

Response:

Implement alternate routing

- Example: replace a beleaguered switch with an ACTS satellite until the system can be rebuilt

Limiting interconnectivity to minimal states

- Begin triage to protect the main body

Begin "aggressive" forensics investigations

- Require legal back-up to allow active tracing of activities independent of identity or citizenship constraints
- Includes proactive defensive measures
- Includes intent to prosecute or exact retribution

IW Alert Condition V

Situation—Brink of War

a) Widespread incidents that undermine U.S. ability to function

b) Conditions III or IV with special contexts

- Contextual sensitivity subject to integration with all other operations and activities of the U.S.

Response:

Disconnect critical elements from the public infrastructure

- Deploy the Minimum Essential Information Infrastructure and temporary systems as required

Implement WARM protocols

- For critical systems, implement alternate protocols for network to transport layers of systems

Declare state of emergency

Prepare for warfare, including retribution against aggressors using the full force of the U.S.

Consideration of A Taxonomy for Information Warfare

Many of the definitions, concepts and words that follow are drawn from the Joint Publication System, and in particularly from the Joint Doctrine for Command and Control Warfare and the Joint Reporting Structure.

The central concept of information warfare is straightforward: The ultimate target of information warfare is an information dependent process, whether human or automated. The use of the word "warfare" should not be construed as limiting information warfare to a military conflict, declared or otherwise.

The root concept of information warfare is offensive in nature. In turn, the concept of information warfare defense flows from the offense. This is not surprising as most defensive actions (counter-air, anti-submarine warfare, counter-mine, anti-crime, anti-drug) only have meaning within the context of action-reaction. Offensive information warfare targets information or information systems in order to affect the information dependent process, whether human or automated. Defensive information warfare protects the information dependent process, whether human or automated.

The question of interest is whether a useful taxonomy information warfare can be derived.

In Joint Pub 3-13.1, Joint Doctrine for Command and Control Warfare, an "information system" is defined as the organized collection, processing, transmission, and dissemination of information, in accordance with defined procedures, whether automated or manual. This includes the entire infrastructure, organization, and components that collect, process, store, transmit, display, and disseminate information. It includes everything and everyone that performs these functions — from a laptop computer to local and wide-area voice and data networks, broadcast facilities, buried cable and, most importantly, the people involved in transmitting, receiving, processing, and using the information. People, decisionmakers at all levels, are the most important part of the information system.

However, information systems themselves are part of larger information infrastructures. These infrastructures link individual information systems in a myriad of direct and indirect paths. The growing information infrastructures of today transcend industry, media, and the military and includes both government and non-government entities. The collection, processing, and dissemination of information by individuals and organizations comprise an important human dynamic, which is an integral part of the information infrastructure. A news broadcast on CNN, a diplomatic communique, and a military message ordering the execution of an operation all depend on the global information infrastructure. The information infrastructure has been assigned three categories global information infrastructure (GII), national information infrastructure (NII), and defense information infrastructure (DII).

- The GII is the worldwide interconnection of communications networks, computers, data bases, and consumer electronics that make vast amounts of information available to users. It encompasses a wide range of equipment, including cameras, scanners, keyboards, fax machines, computers, switches, compact disks, video and audio tape, cable, wire, satellites, fiber-optic transmission lines, microwave, nets, switches, televisions, monitors, printers and much more. The GII, however, includes more than just the physical facilities used to store, process, and display voice data. The personnel who operate and consume the transmitted data constitute a critical component of the GII.
- The NII is the subset of the GII within the U.S. used for social, economic and national security activities.
- The DII is the shared or interconnected system of computers, communications, data applications, security, people, training, and other support structures serving DoD's local, national and worldwide information needs. The DII connects DoD mission support, command and control (C2), and intelligence computers through voice, telecommunications, imagery, video and multimedia services. It provides information processing and services to subscribers over the Defense Information Systems Network. It includes C2, tactical, intelligence and commercial communications systems used to transmit DoD data

In actuality the GII, NII and DII labels are misleading as there are few distinct boundaries in the information environment. The DII, NII, and GII are inextricably intertwined, a trend that will only intensify with the continuous application of rapidly advancing technology. Again, no ordered structure is readily apparent on which to base a taxonomy.

If information warfare targeting and information warfare defense are shaped by particular information dependent processes then perhaps ordering information dependent processes will lead to a structure. However, only a little reflection leads to the conclusion that there are an infinite variety and scope of information dependent processes. Clearly, there is no "ordered system" that will tie these potential processes together, other than the shared characteristic of depending on information. Enumerating information dependent processes will not yield a taxonomy.

What of the methods of information warfare? Consider that attacks and defenses may involve:

- Physical attacks the components of the information infrastructure, e.g., computers, communications devices, software, cables, control devices, etc.
- Physical attacks on the components containing or supporting the information infrastructure such as buildings, power systems, environmental services.
- Physical attacks on or the subversion of the people (witting or unwitting) who operate elements of the information infrastructure.
- Physical destruction of information (erasure or over-write) without harming the infrastructure components.
- Logic (malicious code) attacks on the components of the information infrastructure, e.g., computers, communications devices, software, control devices, etc.
- Logic attacks on computer-controlled components supporting the information infrastructure. These may include air conditioners, air handlers, power distribution, and cooling water.
- Attacks on information provided via the information infrastructure that is used by a specific function(s) (e.g., deception operations, and insertion of false information).
- Corruption of information using logic or digital attacks without harming the components of information infrastructure. (The greatest harm may result from an attack which corrupts or injects false information in a manner that cannot be detected by the users of that information who subsequently take actions based on the corrupted or false information.)
- Combined attacks where both physical and logical attacks on the information infrastructure or supporting elements are undertaken in combination to either mask one or the other types of attack or to obtain the benefits of a combined attack.

From the above it follows that at the highest level information dependency can be partitioned into two elements: one, the availability of information needed by the process; and two, the integrity of information used in the process. Some would add a third element, the confidentiality of information, as it is an important factor in many civil and military information dependent processes. In the following derivation all three are addressed. Note that this trial taxonomy is irrespective of the offensive or defensive actions that may be undertaken to achieve or defend against these conditions it is just a structure for information warfare.

A top-level taxonomy for information warfare

Availability of information or information services

Loss of information

- Detected on occurrence
- Detected after n* units of time
- Undetected

Delay in receipt of information

- Detected on occurrence
- Detected after n units of time
- Undetected

Loss of an information service

- Detected on occurrence
- Detected after n units of time
- Undetected

Delay in an information service

- Detected on occurrence
- Detected after n units of time
- Undetected

Integrity of information

Unauthorized change in data

- Detected on occurrence
- Detected after n units of time
- Undetected

Insertion of false data

From a correct source

- Detected on occurrence
- Detected after n units of time
- Undetected

From an incorrect source

- Detected on occurrence
- Detected after n units of time
- Undetected

Confidentiality of information

Compromise detected on occurrence

Compromise detected after n units of time

Compromise undetected

*The unit of time can vary from microseconds to years. The criticality of n is determined by the information dependent process in each particular case

Although only at three levels of complexity this sample taxonomy rapidly becomes unwieldy. Complexity grows at the next level as each of these conditions can be the result of accident or caused by deliberate intent. In many cases it may be impossible to determine which led to the condition.

At the next level deliberate intent can be carried out by an exterior actor, an insider with authorized access to the information or information services use in an information dependent process, or by both internal and external actors may be working in concert. Then there is the factor of time. If the failure was detected only after n units of time had elapsed, the affects that matter cannot be generalized but rather are unique to a specific information dependent process. The introduction of process-dependent timing takes us back to the earlier infinite variety of processes which has already been rejected as a basis for a taxonomy.

But to press on with this sample taxonomy, we recognize that all of these events can be arrayed in multiple sequences and combinations. There are an infinite combination and permutation of such attack methods and countering defenses available for application within the intertwined DII/NII/GII environment. Thus, an attempt to add successive layers to the taxonomy sketched out above would explode into incomprehensible complexity. Each element of data; each bit and byte of software; each device, whether in a computer at an end-node or along a communication path; each waveform; and each person with access to any of the components would have to be mapped onto the structure.

It is just this complexity that is large part of the challenge facing the defender: he cannot know or protect against all the possible means of attack to succeed, the attacker needs only to know one weakness that the defender has left unprotected or have a weapon that can breech one point in defense. This is the imperative for risk management, resilient systems, and robust recovery capabilities. Again, although a top-level information warfare taxonomy can be sketched, it does not scale to a useful construct. (See the last page of this Appendix for a footnote on complexity.)

Now the principle reason an information warfare taxonomy is a desired objective is that it adds precision to communication. Although the simple taxonomy sketched above does not meet that goal, a workable alternative is proposed that can be inserted into existing reporting structures. The development of this alternative to a taxonomy has the benefit that it builds on existing models from the Joint Publications System.

Joint Publication 1-03, **"Joint Reporting Structure (JRS),"** establishes a standard reporting vocabulary for the Department of Defense. Joint Publication 1-03.3 establishes the **"Status of Resources and Training System (SORTS)"**, and provides the general provisions and detailed instructions for collecting and preparing data on units of the U.S. Armed Forces and selected foreign and international organizations. In practice, the utility of SORTS is not optimum because of the timeliness and quality of data submitted. Whether incorporated in SORTS or a stand-alone method, an information warfare SORTS-like reporting scheme is needed.

SORTS functions as the following:

a. Central Registry of All Operational Units in the U.S. Armed Forces. SORTS is the single, automated reporting system within the Department of Defense that provides the National Command Authorities (NCA) and the Chairman of the Joint Chiefs of Staff with authoritative identification, location, assignment, personnel, and equipment data for the registered units and organizations of the U.S. Armed Forces, Defense agencies, and certain foreign and international organizations involved in operations with U.S. Armed Forces. The composite registry of all units is maintained by the Joint Staff. After initial registration, SORTS is designed to receive reports by exception when changes occur.

b. Repository of Resource Status of Selected Units. For selected registered units, SORTS also provides the condition and level of resources and training. This includes the unit commander's assessment of how resources and training levels will affect the unit's ability to undertake its wartime mission. Units report by exception within 24 hours of a change or as directed by the Chairman of the Joint Chiefs of Staff. If no change in unit status occurs within 30 days of report submission, units submit a validation report.

SORTS contains provisions for reporting various readiness items:

(a) Overall C-Level (OVERALL) Set. Data in this set include the overall C-Level for the unit and the codes for primary, secondary, and tertiary degradation reasons. The overall readiness showing how well the unit meets prescribed levels of personnel, equipment, and training for the wartime mission for which the unit has been organized or designed is ranked in descending order from C-1 to C-5:

C-1. The unit possesses the required resources and is trained to undertake the full wartime mission(s) for which it is organized or designed. The resource and training area status will neither limit flexibility in methods for mission accomplishment nor increase vulnerability of unit personnel and equipment. The unit does not require any compensation for deficiencies.

C-2. The unit possesses the required resources and is trained to undertake most of the wartime mission(s) for which it is organized or designed. The resource and training area status may cause isolated decreases in flexibility in methods for mission accomplishment but will not increase vulnerability of the unit under most envisioned operational scenarios. The unit would require little, if any, compensation for deficiencies.

C-3. The unit possesses the required resources and is trained to undertake many, but not all portions of the wartime mission(s) for which it is organized or designed. The resource and training area status will result in significant decreases in flexibility for mission accomplishment and will increase vulnerability of the unit under many, but not all, envisioned operational scenarios. The unit would require significant compensation for deficiencies.

C-4. The unit requires additional resources or training to undertake its wartime mission(s), but it may be directed to undertake portions of its wartime mission(s) with resources on hand.

C-5. The unit is undergoing a Service-directed resource action and is not prepared, at this time, to undertake the wartime mission(s)) for which it is organized or designed.

(b) Personnel Level (PERSONEL) Set. Data in this set include the personnel level (P-level) and a code for the primary reason for degradation in the personnel area.

(c) Equipment and Supplies On Hand Level (EQSUPPLY) Set. Data in this set include the equipment and supplies on hand level (S-level) and a code for the primary reason for degradation in the equipment and supplies on hand area.

(d) Equipment Condition Level (EQCONDN) Set. Data in this set include the equipment condition level (R-level) and a code for the primary reason for degradation in the equipment condition area.

(e) Training Level (TRAINING) Set. Data in this set include the training level (T-level) and a code for the primary reason for degradation in the training area

(f) Forecasted Category Level (FORECAST) Set. Data in this set include the forecasted C-level for the unit and the date the unit expects to attain that C-level.

(g) Category Level Limitation (CATLIMIT) Set. Data in this set include the imposed maximum C-level for the unit, if any, and the primary resource area causing the limitation.

An additional category should be added to SORTS specifying at what level of assessment scenario the unit is prepared to operate and how this preparedness was assessed using the terminology described earlier.

Joint Pub 1-03.10, **"JRS Communications Status,"** provides for the Defense Information Systems Agency to provide near-real-time status information on a serious degradation of the Defense Communication System (DCS) via a Communications Spot Report and to provide a summary of significant status information on the DCS via a daily Communications Status Report.

These reports should be expanded to include information systems and information services. Further, these reports should be used by the military departments, serv-

ices, combat support agencies and the CINCs to report the status of information systems and services.

Joint Pub 1-03.31, *"Preparedness Evaluation System,"* establishes the CINCs Preparedness Assessment Report (CSPAR). These report provide a biennial appraisal of the preparedness of the unified and specified commands to accomplish Joint Strategic Capability Plan tasks (both supporting and supported) within the constraints of the total apportioned force (Active and Reserve). In the CSPAR, each CINC identifies overall strengths and significant deficiencies affecting the command's ability to carry out assigned missions and execute the plans produced during the most recent planning cycle. In submitting the CSPAR, CINCs are reporting on their ability to accomplish a specific task using available capabilities.

The CINCs should be required to include an assessment of their ability to carry out assigned missions at the appropriate assessment scenario level and indicate the process used to determine preparedness.

Joint Pub 1-03 32 1. **"Combat Agency Assessment System,"** sets forth the guidelines and procedures for operating the Combat Support Agency Assessment System (CSAAS), a uniform system for reporting to the Secretary of Defense, the commanders of the unified and specified commands (CINCs), and the Secretaries of the Military Departments concerning readiness of each combat support agency to perform with respect to a war or threat to national security.

Chairman, Joint Chief of Staff (CJCS)-sponsored exercises provide the principal means of on-site evaluation of agency responsiveness in reacting to National Command Authority decisions and CINC warfighting requirements. In the event no such exercises are scheduled during the first two quarters of even-numbered fiscal years, Joint Staff observers conduct independent site visits to each of the combat support agencies. Although the CSPAR is the principal means for the combatant commands to assess agency support, Joint Staff observers may also visit combatant command headquarters to discuss overall support, agency supporting plans, and ongoing efforts to improve shortfalls.

These reports should be modified to include an annual assessment of the preparedness of the combat support agencies, at a specified assessment level to carry out their mission. The current two year schedule currently followed in assessing the readiness of combat support agencies is not realistic in an age of information warfare. The information dependent processes of these agencies are directly tied to the ability to mobilize, deploy and sustain the forces. Currently, this is an unknown in the age of information warfare.

Joint Pub 3-10.1, **"Joint Tactics, Techniques, and Procedures for Base Defense,"** categorizes threats to bases in the rear area by the levels of defense required to counter them. Emphasis on specific base defense and security measures may depend on the anticipated threat level. (These threat levels are discussed in detail in Joint Pub 3-10.)

a. Level I threats can be defeated by base or base duster self-defense measures.

b. Level II threats are beyond base or base cluster self-defense capabilities but can be defeated by response forces, normally military police (MP) units assigned to area commands with supporting fires.

c. Level III threats necessitate the command decision to commit a Theater Contingency Force. Level III threats, in addition to major ground attacks, include major attacks by aircraft and theater missiles armed with conventional weapons or nuclear, biological and chemical (NBC) weapons.

The threat to bases in the rear area should be modified to include information warfare attacks.

Joint Pub 3-10.1 also spells out Threat Conditions and Responses and states that in combating terrorism, bases should use common terrorist threat conditions (THREATCONs), each with its specific security measures and required responses.

Threat assessments are used to determine threat levels, to implement security decisions, and to establish awareness and resident training requirements. Threat levels are determined by an assessment of the situation using the following six terrorist threat factors:

(1) Existence. A terrorist group is present, assessed to be present, or able to gain access to a given country or locale.

(2) Capability. The acquired, assessed, or demonstrated level of capability to conduct terrorist attacks.

(3) Intentions. Recent demonstrated anti-U.S. terrorist activity, or stated or assessed intent to conduct such activity.

(4) History. Demonstrated terrorist activity over time.

(5) Targeting. Current credible information on activity indicative of preparations for specific terrorist operations.

(6) Security Environment. The internal political and security considerations that impact on the capability of terrorist elements to implement their intentions.

The severity of the terrorist threat is indicated by the designated threat level, assigned through. analysis of the above threat assessment factors. Threat levels, and associated factors, are:

(1) Critical. Factors of existence, capability, and targeting must be present. History and intentions may or may not be present.

(2) High. Factors of existence, capability, history and intentions must be present

(3) Medium. Factors of existence, capability, and history must be present. Intentions may or may not be present.

(4) Low. Existence and capability must be present. History may or may not be present.

(5) Negligible. Existence and/or capability may or may not be present.

The terrorist threat level is one of several factors used in the determination of terrorist THREAT CON. Factors that enter into the decision to assign a particular THREATCON and its associated measures include threat, target vulnerability, criticality of assets, security resource availability, impact on operations and morale, damage control, recovery procedures, international regulations, and planned U.S. Government actions that could trigger a terrorist response.

The terrorist THREATCON system provides a common framework to facilitate inter-Service coordination, support U.S. military anti-terrorist activities, and enhance overall DoD implementation of U.S. Government anti-terrorist policy. THREATCONs are described below:

(1) THREATCON NORMAL. Applies when a general threat possible terrorist activity exists, but the threat warrants a routine security posture.

(2) THREATCON ALPHA. Applies when there is a general threat of terrorist activity against personnel and installations, the exact nature and extent of which are unpredictable and circumstances do not justify full implementation of THREATCON BRAVO measures. However, base defense forces may have to implement selected measures from higher THREATCONs based on intelligence received. Base defense forces must be able to maintain the measures in this THREATCON indefinitely.

(3) THREATCON BRAVO. Applies when an increased and more predictable threat of terrorist activity exists. Base defense forces must be able to maintain the measures of this THREATCON for weeks without causing undue hardship, without affecting operational capability, and without aggravating relations with local authorities.

(4) THREATCON CHARLIE. Applies when an incident occurs or when intelligence indicates an imminent terrorist action against U.S. bases and personnel. Implementation of measures in the THREATCON for more than a short period probably will create hardship and affect peacetime activities of the unit and its personnel. Sustaining

this posture for an extended period probably will require augmentation.

(5) THREATCON DELTA. Applied in the immediate area where a terrorist attack has occurred or when intelligence has been received that terrorist action against a specific location is likely. Normally, this THREATCON is declared as a localized warning.

The description of threat levels, threat assessments, severity of threat, and threat condition found in Joint Pub 3-10.1 is a good model for information warfare defense preparation, assessment, and warning.

Finally, Joint Pub 3-54, **"Joint Doctrine for Operations Security,"** Change 1, Appendix E, outlines procedures for Operations Security (OPSEC). These surveys in general:

a. Thoroughly examine an operation or activity to determine if adequate protection from adversary intelligence exploitation exists.

b. Check on how effective the OPSEC measures the operation or activity being surveyed in protecting protect its critical information.

c. Cannot be conducted until after an operation or activity has at least identified its critical information for without a basis of identified critical information, there can be no specific determination that actual OPSEC vulnerabilities exist. (This is also true in information warfare.)

Each OPSEC survey is unique. Surveys differ in the nature of the information requiring protection, the adversary collection capability, and the environment of the activity to be surveyed

a. In combat, a survey's emphasis must be on identifying operational indicators that signal friendly intentions, capabilities, and/or limitations and that will permit the adversary to counter friendly operations or reduce their effectiveness.

b. In peacetime, surveys generally seek to correct weaknesses that disclose information useful to potential adversaries in the event of future conflict. Many activities, such as operational unit tests, practice alerts, and major exercises, are of great interest to a potential adversary because they provide insight into friendly readiness, plans, crisis procedures, and C2 capabilities that enhance that adversary's long-range planning.

OPSEC Surveys are not Security Inspections:

a. OPSEC surveys are different from security evaluations or inspections. A survey attempts to produce an adversary's view of the operation or activity being surveyed. A security inspection seeks to determine if an organization is in compliance with the appropriate security directives and regulations.

b. Surveys are always planned and conducted by the organization responsible for the operation or activity that is to be surveyed. Inspections may be conducted without warning by outside organizations.

c. OPSEC surveys are not a check on the effectiveness of an organization's security programs or its adherence to security directives. In fact, survey teams will be seeking to determine if any security measures are creating OPSEC indicators.

d. Surveys are not punitive inspections, and no grades or evaluations are awarded as a result of them. Surveys are not designed to inspect individuals but are employed to evaluate operations and systems used to accomplish missions.

e. To obtain accurate information, a survey team must depend on positive cooperation and assistance from the organizations participating in the operation or activity being surveyed. If team members must question individuals, observe activities, and otherwise gather data during the course of the survey, they will inevitably appear as inspectors, unless this non-punitive objective is made clear.

f. Although reports are not provided to the surveyed unit's higher headquarters, OPSEC survey teams may forward to senior officials the lessons learned on a non-attribution basis. The senior officials responsible for the operation or activity then decide to further disseminate the survey's lessons learned.

There are two basic kinds of OPSEC surveys: command and formal.

a. A command survey is performed using only command personnel and on events within the particular command.

b. A formal survey requires a survey team composed of members from inside and outside the command and will normally cross command lines (after prior coordination) to survey supporting and related operations and activities.

c. Both types of surveys follow the same basic sequence and procedures.

Although Joint Pub 3-54 is scheduled to be rewritten, it is quoted extensively as another possible model for conducting information warfare assessments. The assessment methodology cited at the beginning of the annex should yield more rigorous conclusions.

By adopting concepts from each of the Joint Pub sources cited above a standard vocabulary of status reporting, tied to specific information dependent processes, can be developed for information warfare. Such an assessment and reporting system should be developed that stands on its own for use in civil agencies and the commercial sector. Within the Department of Defense this may be more easily achieved by making suitable modification of the several portions of the Joint Reporting System.

In the case of information warfare, as in the terrorism example above, a range of *standardized threat scenarios* should be promulgated for use in conducting preparedness surveys, as standardized *assessment conditions* for planning purposes, and a set of standardized *threat warnings or THREATCONS*, if warning is available.

Whatever schema is used to evaluate the operational readiness of information dependent processes and activities, it must be timely and reflect the current state of the security policy being implemented, the supporting infrastructures (computers, communications, electricity and other supporting utilities) and the training status of the personnel, both systems administrators and users of information and information systems.

Complexity Footnote:

A military example of how the complexity builds is found in command and control warfare (C2W). The U.S. military defines C2W as an application of information warfare in military operations.

The execution of C2W involves the integrated use of some or all of the tools of psychological operations (PSYOP), military deception, operations security (OPSEC), electronic warfare (EW), and physical destruction, mutually supported by intelligence, to deny information to, influence, degrade, or destroy adversary C2 capabilities while protecting friendly C2 capabilities against such actions. Again, these are just means to carry out information warfare in a particular military environment.

Defensive tools called out in Joint Pub 6.0, Doctrine for C4 Systems Support to Joint Operations, include:

(1) Physical security of facilities,

(2) Personnel security of individuals authorized access to systems,

(3) Operations security (OPSEC) procedures and techniques protecting operational employment of C4 system components,

(4) Deception, deceiving the adversary about specific system configuration, operational employment, and degree of component importance to mission accomplishment,

(5) Low probability of intercept (LPI) and low probability of detection (LPD) capabilities and techniques designed to defeat adversary attempts to detect and exploit transmission media

(6) Emissions control procedures designed to support OPSEC and LPI/LPD objective,

(7) Transmission security capabilities designed to support OPSEC and LPI/ LPD objectives,

(8) Communications security (COMSEC) capabilities to protect information transiting terminal devices and transmission media from adversary exploitation,

(9) Computer security capabilities to protect information at rest, being processed, and transitioning terminal devices, switches, networks, and control systems from intrusion, damage, and exploitation,

(10) System design and configuration control (e.g., protected distribution systems, protection from compromising emanation (TEMPEST)) to mitigate the impact of information technology vulnerabilities, and

(11) Identifying technological and procedural vulnerability analysis and assessment programs.

To this list can be added non-repudiation, identification and authorization, end-user use of encryption services, transmission encryption, replication, and a host of other techniques to protect various elements of the information infrastructure. As in the case of C2W, these are tools and in themselves, they are not information warfare.

APPENDIX D

ORGANIZATIONAL MODELS

The Task Force reviewed three organizational models for possible application at the Department of Defense or national levels. These included the Centers for Disease Control and Prevention, the Federal Emergency Management Agency, and the National Drug Intelligence Center. The following reviews are provided for reference only.

D.1 CENTERS FOR DISEASE CONTROL AND PREVENTION

Surveillance, Research, Prevention Efforts in The Area of Infectious Diseases:

Applicability Of CDC Experience To A National Center For Information Systems Security

D.1.1 Introduction

In the United States, the threat of infectious disease is changing rapidly in conjunction with dramatic changes in global society and environment. Worldwide, there is explosive population growth with expanding poverty and urban migration which, with rapid environmental changes, is resulting in the emergence of new and the reemergence of previously controlled infectious diseases; international travel is increasing so that infectious microbes can easily travel across borders with their human or animal hosts. Diseases that arise in other parts of the world are repeatedly introduced into the United States, where they may threaten our national health and security.

The threats to the U.S. Information Technology (IT) infrastructure bear similarities to the emerging infectious disease threat to public health. In particular, the context of Information Warfare Defense is parallel to that in public health. IT infrastructure growth, changing technology and increasing network interconnectivity correspond to global population growth, environmental change and increased travel. The U.S. Government approach to the increasing public health threat, led by the Centers for Disease Control and Prevention (CDC), can provide lessons in responding to national IT security threats.

D.1.2 Background and Legislative History

The Centers for Disease Control and Prevention (CDC) is an agency of the Public Health Service, in the Department of Health and Human Services. Its mission is to promote health and quality of life by preventing and controlling disease, injury, and disability. As the nation's prevention agency, the CDC accomplishes its mission by working with partners throughout the nation and the world.

The CDC formally came into being in a department reorganization in 1980. In 1993, the organization officially became known as the Centers for Disease Control and Prevention, but the commonly known abbreviation CDC remained.

The CDC traces its beginnings to 1946 when the Communicable Disease Center was established as a Field Station of the Bureau of State Services in the Public Health Service. It took over the offices and some responsibilities of the DoD's Office of Malaria Control which was being disestablished. The primary mission was to work with the States in tracking and controlling the spread of communicable diseases in the United States.

The Center grew out of the general authority granted to predecessor organizations of the Department of Health and Human Services (HHS) That is, no specific legislation was required for its establishment. However, it is noteworthy that in 1893 Congress mandated that state and municipal authorities report information weekly about the incidence of certain diseases to the Public Health Service. Currently, CDC general authority flows through the general authority given to the Secretary for Health and Human Services. Funding for studies on specific programs such as lead poisoning prevention, HIV, and breast cancer prevention are contained in various legislative acts.

CDC supports surveillance, research, prevention efforts, and training in the area of infectious diseases through its National Center for Infectious Diseases (NCID). Created in 1981, NCID is committed to the prevention and control of traditional, new, and re-emerging infectious diseases in the United States and around the world.

NCID accomplishes its mission of preventing illness and death from infectious diseases by focusing its resources in five areas:

- Epidemic Assistance
- Surveillance Of Infectious Diseases, In Collaboration With State And Local Health Departments
- Epidemiological And Laboratory Research
- Formulating, Disseminating, And Evaluating Prevention And Control Strategies
- Training And Consultation Programs In Cooperation With Other CDC Units And Outside Agencies And Organizations

D.1.3 Concept of Operations: The CDC Approach to the Global Threat of Infectious Disease

NCID Surveillance Activities

NCID collects, analyzes, and interprets reports of nationally notifiable infectious diseases and outbreaks submitted by state and local public health agencies and disseminates the findings. In addition to this traditional form of surveillance, the center uses supplemental, non-traditional systems to monitor trends in infectious diseases of public health importance. These systems include laboratory-based surveillance; population-based active surveillance; sentinel physician networks; hospital-based networks for surveillance of infections; analyses of national databases; and serosurveys and studies of special populations and settings. The Center also collaborates with international organizations and agencies in the global surveillance of selected pathogens.

Partnerships

NCID provides epidemiological, microbiologic, and consultative services to federal agencies, state and local health departments, medical and biomedical science institutions, schools of public health, health care providers, and the World Health Organization (WHO) and other international agencies.

D.1.4 Appropriate Analogies/Examples in the National Responses to the Threat of Infectious Disease

The similarities that the threats to the U.S. Information Technology (IT) infrastructure bear to the emerging infectious disease threat to public health suggest that the CDC experience can provide lessons in responding to national IT security threats. Below are elements of the CDC approach to the threat to U.S. public health which appear to apply to any formulation of a response to IT threats.

Formulating a National Strategic Response Plan

CDC's NCID strategic plan of 1994 has identified need to:

- improve public health infrastructure at local, state and national level
- recognize the global nature of the problem
- institute global surveillance.

The Plan's goals are:

> Goal I—*Surveillance*: Detect, promptly investigate, and monitor emerging pathogens? the diseases they cause, and the factors influencing their emergence.

Goal II—*Applied Research*: Integrate laboratory science and epidemiology to optimize public health practice.

Goal III—*Prevention and Control*: Enhance communication of public health information about emerging diseases and ensure prompt implementation of prevention strategies.

Goal IV—*Infrastructure*: Strengthen local, state, and federal public health infrastructures to support surveillance and implement prevention and control programs.

Similarly, the Federal Government must have a strategic plan to respond to the increasing IT threat, a plan to:

- improve IT infrastructure security at the national level,
- recognize the ubiquitous nature of the problem and
- institute national (and even global) surveillance.

The goals of such a plan could be expected to closely parallel those of CDC's NCID strategic plan:

Goal I—*Surveillance*: Detect, promptly investigate, and monitor Information Technology Infrastructure threats, and the factors influencing their occurrence a national consortium of IT providers and users to promote rapid interchange of event occurrence information a near real time monitoring and assessment function

Goal II—*Applied Research*: Integrate private industry, standards body and government research and development to optimize public and private security practice. Support R&D in IT security. Establish effectiveness studies and disseminate results

Goal III—*Prevention and Control*: Enhance communication of industry and government information about emerging security threats and ensure prompt implementation of prevention and control strategies. Disseminate information. Support security implementation guidelines/standards.

Goal IV—*Infrastructure*: Strengthen national and international infrastructures to support surveillance and implement prevention and control programs. Promote establishment of procedures and policies with supporting legislation and industry, government and intergovernmental agreements. Promote establishment of IT security centers (analogous to Carnegie Mellon's role in S/W process improvement) for research, standards development and training

Establishing an Information Exchange Infrastructure

The Information Network for Public Health Officials (INPHO) was initiated by the Centers for Disease Control and Prevention (CDC) in 1992 as part of its strategy to strengthen the infrastructure of public health in the

United States. The ultimate goal of INPHO is to improve the health of Americans through more effective public health practice. CDC's role in the INPHO initiative is to provide policy and technical assistance states can use to develop INPHO projects for their own public health needs.

The INPHO initiative addresses the serious national problem that public health professionals have lacked ready access to much of the authoritative, technical information they need to identify health dangers, implement prevention and health promotion strategies, and evaluate health program effectiveness. INPHO utilizes state-of-the-art telecommunications and computer networks to give state and community public health practitioners new command over information resources.

As the U.S. health care system shifts towards a managed care model, the role of public health agencies increasingly will center on the provision and use of information. Public health will be responsible for key functions that health care providers themselves cannot perform: 1) systematic surveillance and assessment of health trends, 2) assurance that those in need receive health services, that health care is not excessively costly, and that community health goals are met, and 3) clarifying policy options and implications for public and private decision makers. INPHO helps states build strategic information partnerships between people and organizations that are critical to achieving these goals.

There are three essential components of the INPHO vision: linkage, information access, and data exchange. INPHO computer networks and software link local clinics, state and federal health agencies, hospitals, managed care organizations and other providers, eliminating geographic and bureaucratic barriers to communication and information exchange. Public health practitioners have unprecedented electronic access to health publications, reports, databases, directories, and other information. High speed communications capacity enables them to communicate and exchange data locally and across the nation on the full universe of public health issues. (The INPHO is described further in Attachment 1.)

Similarly, the Federal Government might promote or sponsor systematic information and data exchange among national, state and local IT users and providers to respond to the increasing IT threat.

Convening an Inter-Agency Working Group to Recommend U.S. Government Actions

A U.S. Government interagency working group was convened on December 14, 1994, to consider the global threat of emerging and re-emerging infectious diseases. The working group was established under the aegis of the Committee on International Science, Engineering, and Technology Policy (CISET) of President Clinton's National Science and Technology Council. Dr. David Satcher, the Director of the Centers for Disease Control and

Prevention (CDC), chaired the CISET working group, which included five sub-groups with co-chairs from CDC, the Food and Drug Administration (FDA), the National Institutes of Health (NIH), the U.S. Agency for International Development (USAID), the Department of Defense (DoD), and the State Department. The working group's membership, which included representatives from more than 17 different Government agencies and departments, reviewed the U.S. role in detection, reporting, and response to outbreaks of new and re-emerging infectious diseases and made a number of recommendations which are described in Global Microbial Threats in the 1990s, published in late 1995 by the President's National Science and Technology Council.

As with the National Science and Technology Council's Government interagency working group on the global microbial threat, a multi-agency Government advisory panel to recommend U.S. Government responses to the IT threat might be appropriate.

Forming Partnerships for Interaction, Cooperation and Coordination

Effective public health policy results from interaction, cooperation, and coordination among a wide range of public and private organizations and individuals. Particularly critical to this process are CDC's partnerships with state and territorial health departments; other federal agencies; professional organizations; academic institutions; private health care providers; health maintenance organizations and health alliances; local community organizations; private industry; and international partners, including the World Health Organization (WHO) and international service organizations and foundations. Each of these partners play an integral role in the cooperative efforts required to safeguard the public's health from emerging infectious disease threats.

CDC partnerships at the federal level have been helpful in confronting infectious diseases of public health importance in the United States. For example, CDC and NIH developed improved diagnostic tests for Lyme disease and various fungal infections. CDC has also worked closely with FDA and USDA in controlling emerging food-borne illnesses. Recent CDC collaborations with EPA have been instrumental in recognizing and controlling water-borne outbreaks of giardiasis and cryptosporidiosis in several states.

In addition, CDC has often joined forces with USDA and DoD to control or prevent vector-borne infectious disease threats. Such cooperative efforts were used successfully to address potential mosquito-borne illness following Hurricane Andrew in Florida and Louisiana in 1992.

Clear, well-established lines of communication and responsibility between appropriate personnel in federal agencies, such as CDC, NIH, EPA, FDA, USDA, DoD, and others, are essential to the development of efficient, cost-effective prevention and control strategies. Such links help eliminate costly duplication of effort and focus limited federal resources on the early recognition and timely control of new infectious disease problems.

Similarly any U.S. Government effort to meet the IT threat would require active, long-term partnerships among Federal agencies and with elements of the IT industry.

Assume International Leadership

The CDC is actively promoting U.S. leadership in the development of an international partnership to address emerging infectious diseases. This leadership role is a natural one for the United States since American business leaders and scientists are in the forefront of the computer communications and biomedical research communities that must provide the technical and scientific underpinning for disease surveillance. The United States maintains more medical facilities and personnel abroad than any other country, in terms of both civilian and military, and public and private sector institutions. Furthermore, American scientists and public health professionals have been among the most important contributors to the international efforts to eradicate smallpox and polio.

Similar arguments would support U.S. leadership in the formulation of a global response to what will surely become a global IT threat.

D.1.5 References

Addressing Emerging Infectious Disease Threats: A Prevention Strategy For The United States, 1994.

Global Microbial Threats in the 1990s, 1995.

ATTACHMENT 1

CDC's Information Network for Public Health Officials (INPHO): A Framework for Integrated Public Health Information and Practice.
Baker EL, Friede A, Moulton AD, Ross DA.

J Public Health Management Practice, 1995; 1(1):43-7.

CDC's Information Network for Public Health Officials (INPHO): A Framework for Integrated Public Health Information and Practice

Contents

Summary

To strengthen the public health infrastructure, the Centers for Disease Control and Prevention (CDC) initiated the Information Network for Public Health Officials (INPHO). CDC INPHO has three goals: (1) to make communication among public health practitioners throughout the United States easy, (2) to make information accessible, and (3) to make secure data exchange as swift and smooth as contemporary technology will allow. Based on a systems approach to supporting the core functions of public health, CDC INPHO achieves its goals by creating a flexible and user-responsive infrastructure of open communications and information exchange.

> "Where is the wisdom we have lost in knowledge? Where is the knowledge we have lost in information?" T.S. Eliot, *The Rock*

Vision and Goals

The Centers for Disease Control and Prevention (CDC) initiated the Information Network for Public Health Officials (INPHO) in 1992 as part of its strategy to strengthen the infrastructure of public health in the United

States.[1] The vision driving CDC INPHO is that of a new, integrated public health information system based on a state-of-the-art telecommunications network linking the public health community and providing seamless exchange of information (see the box titled, "INPHO: The Vision, the Need, the Basic Concepts"). When fully deployed, CDC INPHO will become the common pathway for public health practitioners throughout the United States—at the community, state, and national levels alike—to exchange information with each other, with CDC, and with colleagues globally. As a result, every public health worker in the United States should be linked to every other public health worker through telecommunications technology.

CDC INPHO has three goals: (1) to make communication easy, (2) to make information accessible. and (3) to make secure data exchange as swift and smooth as contemporary technology will allow. Achieving those goals will involve a variety of activities in the states, depending on the status of their public health information strategy, telecommunications networks, end-user priorities, and other factors. Similarly, the CDC role will vary from state to state to serve the needs of their public health agencies. All INPHO activities, however, will focus on building a common public health information network linking all public health practitioners across the nation.

Why We Need Better Communication

A particularly insightful way to conceptualize the value of improved public health information comes from Harlan Cleveland, author of The Knowledge Executive: Leadership in an Information Society.[2] Cleveland makes the distinction between data, information, and knowledge. Data are undigested observations and unvarnished facts -basically the raw material of public health. Information is organized data. In public health, however, information typically is assembled not by the practitioners who are the end users but by others who are often in remote, centralized agencies. Knowledge, in turn, is the product of information the end user organizes, internalizes, and integrates with everything else she or he knows from experience, study, or intuition. Knowledge, ultimately, is the best guide to our practice of public health. What public health professionals are interested in is creating access to information that will expand our knowledge base and guide our work.

In thinking about developing an information network for public health officials, CDC focused on four critical needs (see the box):

- Connecting a fragmented system. Everyone familiar with the Institute of Medicine report on the future of public health recognizes its diagnosis that the public health system is in disarray.[3] This clearly indicates the need to take action that will [re] connect the elements of the fragmented system. One way of doing this is through telecommunications technology.
- Linking public health professionals. Many public health professionals operate in significant isolation. One way to break down isolation is by connecting public health professionals through telecommunications

technology. Two examples are CDC's WONDER/PC electronic mail and forums and the national telecommunications network CDC has created as part of the Public Health Leadership Institute.

- Leading and responding to health reform. Clearly, the public health community is in the information business and specifically in the business of providing information to the communities that public health serves.
- Activating public health for the health reform environment. As health reform advances-whether legislated in Washington and the states or propelled by market forces—public health needs to ensure that its core functions continue to be performed.

INPHO: The Vision, the Need, the Basic Concepts

The Vision

- An integrated telecommunications network linking the public health community and providing exchange of data and information

The Need

- Connecting a fragmented system
- Linking public health professionals
- Empowering communities with information
- Leading and responding to health reform

The Basic Concepts

- Linkage
- Information access
- Data exchange

Three Key Concepts

CDC INPHO embodies three concepts key to generating the data, information, and knowledge to address the needs outlined above (see the box). Linkage is the first key concept. Here CDC is active on several fronts. CDC is working with state and local. health agencies to build local and wide-area networks—actual physical construction of networks, supported in some cases through outside resources. Second, CDC is expanding "virtual networks" through the use of CDC WONDER PC, a software system that allows public health professionals to communicate across the globe through electronic mail and that also provides unprecedented access to data and information maintained in CDC's large public health databases.[4, 5] Third, CDC is emphasizing the strategy of connecting to the Internet. CDC encourages each state to identify ways to connect with the Internet and have access to the information superhighway.

In partnership with the Georgia Division of Public Health, CDC is implementing an INPHO project to electronically link all parts of the public health system—the state health agency, district health departments, and county health departments. CDC is providing those offices access to the CDC information bases and other sources of information that the state public health agency and its project partners deem valuable. CDC will work with additional states in a similar manner beginning in late 1994, emphasizing development of network capabilities and applications defined by the states themselves. CDC also is linking its information system initiatives with its Distance Learning Program. A clear linkage exists between the INPHO concept of an information network and the notion of a public health training and distance learning network for public health professionals.

The second key concept is information access. CDC generates a large body of information that is published in various forms, but not always in the form most accessible to end users. In this respect, the CDC INPHO is focused on improving practitioners' access to existing and future CDC information bases. The principal approach is to expand the number of information bases accessible through the CDC WONDER PC system. Areas that warrant particular mention are (1.) The prevention guidelines database, (2.) The training resource directory that will enable public health professionals to identify upcoming training offered by CDC and other organizations, and (3.) On-line access to the Morbidity and Mortality Weekly Report, complete with tables and graphs.

CDC is not attempting to expand access to information exclusively through the CDC WONDER PC system. Public health professionals currently access information in many other ways and from many other sources that have great value. It is CDC's hope that its own efforts will help public health professionals maximize their use of multiple access routes so they can achieve access to the information they want as rapidly as possible.

Exchange of data and information is the third key INPHO concept. Many different types of data are involved, among them health status data, health risk information, and particularly data on health care services. As the era of health care reform advances, it will be vital for public health to have rapid, electronic access to health care services information from personal care providers. One important issue is that of automating data entry. Many health departments do not have access to automated data entry systems. Protecting personal privacy and ensuring confidentiality may be one of the most important issues of all. The structure of the data exchange system also is important. Currently, public health has many disparate data systems in place and needs to look to a more integrated approach.

As the era of health care reform advances, it will be vital for public health to have rapid, electronic access to health care services information from personal care providers.

Finally, as health care reform becomes reality, related information systems are being created. It is essential that the public health community understand the implications of those systems and ensure that they generate information to support and enhance the ongoing core functions of population-based health assessment and assurance.

David Satcher, CDC Director, has identified the obstacles public health faces in fulfilling the concept of data and information exchange:

> First, public health agencies at the local, state, and federal levels have a fragmented set of public information systems that threaten to overwhelm the capacity of state and local health departments to respond to the information needs they face.
>
> Second, there is variable access to technology. Some health departments do not have or cannot make ready use of the telecommunications technologies that the INPHO project envisions.
>
> Third, the issue of confidentiality is significant not only as a complex policy issue but also for its symbolic, perceptual importance. The American public is legitimately concerned about issues of confidentiality. The public health community must address this concern squarely and responsively.
>
> Fourth, public health does not have a wealth of existing integrated systems on which to model its own integrated information initiative. The lack of precedents clearly presents an obstacle but, at the same time, a professional challenge to "reinvent" public health using a "bootstraps" approach that draws on the creativity and energy characteristic of the public health profession.6

CDC Strategies

How is CDC confronting these obstacles? To address the problem of fragmented information systems, Martha Katz, CDC's Associate Director for Policy, Planning, and Evaluation, formed a collaborative committee in 1993 that drafted the Report on Public Health Information and Surveillance Systems.[7] The report contains a set of recommendations for action toward integrated health surveillance and information systems that was issued for review and reaction by state and local public health agencies in the spring of 1994. Initial responses were gathered during the March 1994 first annual CDC INPHO conference held in Atlanta, Georgia, and attended by public health representatives from across the nation.

CDC is also working with states to support network development and address the obstacle of variable access to contemporary technology. The Georgia INPHO project is an invaluable prototype for the nation. CDC is mobilizing funding and other resources to help other states initiate similar projects that speak to their specific needs. CDC will support "knowledge transfer" from Georgia and the succeeding INPHO states.

In 1994, CDC organized a confidentiality work group and charged it to assess the legal and technological dimensions of the issue and to develop recommendations and guidelines for protection of confidentiality in the context of integrated information and health surveillance systems.

CDC's approach to dealing with the lack of precedents has two parts. The first is to proceed with the state INPHO projects and to learn from their experience. Second, and of equal importance, is to learn from the complementary projects that a number of state and local public health agencies have underway. These projects focus directly on integrated information systems, data exchange across categorical program lines, data exchange with hospitals and managed care providers, and other issues integral to the INPHO vision. A key role that CDC can play is to disseminate to the national public health community the innovations, successes, and lessons learned by innovative local and state projects.

The INPHO Project and the Systems Approach

A central tenet of systems thinking, as represented, for example, in the work of Peter Senge, is that today's solutions create the issues of tomorrow.[8]

This insight is germane to the CDC INPHO initiative. If cautions that the goal of INPHO should not be to increase the sheer volume of data and information available to public health professionals. Instead, it is to increase their ability to generate and access the information and knowledge they need to guard the health of the public.

Information overload, already a reality in the lives of many public health professionals, threatens to become the leading occupational disease in the 21st century. Unfocused electronic information systems are a threat, not a boon, to public health. The rainfall of electronic mail that seemingly descends on users' computers overnight is a telling symptom. Surgeon General Joycelyn Elders recently remarked that a symptom of information overload is that the quantity of information in her professional life sometimes prevents her from enjoying the work that she knows in her heart she truly values.

Confronted with the challenges of the 1990s and the 21st century, the public health community ultimately needs wisdom on which to base its decisions and choices of action. Harlan Cleveland defines wisdom as "Integrated knowledge, information made super useful by theory which relates bits and fields of knowledge to each other, which in turn enables us to use the knowledge to do something."[2 (p.23)] Only the human mind can synthesize wisdom from data and information. The vision of CDC INPHO necessarily is more modest.

The key to building successful, integrated public health information systems is to focus on a vision consistent with the core mission and core functions of the profession. CDC INPHO is based on a systems approach to supporting the core functions of public health. It does that by creating a rich, flexible, and user-responsive infrastructure of open communications and information exchange. The CDC INPHO team is developing specific, valuable software and computer/telecommunications networks. The heart of the initiative, however, is the conceptual framework it provides for truly integrated health assessment and assurance both within the public health community and in conjunction with the evolving health care sector.

REFERENCES

[1] Roper, W.L. "Strengthening the Public Health Infrastructure." Speech to Association of State and Territorial Health Officials. Atlanta, Ga.: Centers for Disease Control, May 1990.

[2] Cleveland, H. *The Knowledge Executive: Leadership in an Information Society*. New York, NY: Truman Talley Books, E.P. Dutton, 1985.

[3] Institute of Medicine. *The Future of Public Health*. Washington, D.C.: National Academy Press, 1988.

[4] Friede, A., Reid, J.A., Ory, H.W. CDC WONDER: A Comprehensive Online Public Health System of the Centers for Disease Control and Prevention. *American Journal of Public Health*, 1993; 83: 1 ,289-94.

[5] Friede, A., Rosen D.R., Reid, J.A. CDC WONDER/PC: Cooperative Processing for Public Health Informatics. *Journal of the American Medical Informatics Association*, 1994; 1 :303312.

[6] Address to the "First INPHO Workshop: Creating the Public Health Information Highway," Atlanta, Georgia, March 29,1994.

[7] Centers for Disease Control and Prevention. A Report on Public Health Information and Surveillance Systems. Atlanta. Ga.: CDC, 1994.

[8] Senge, P.M. *The Fifth Discipline: The Art and Practice of the Learning Organization*. New York, N.Y.: Doubleday Currency, 1990.

Footnotes

This article is adapted from an address given by Edward L. Baker, M.D., M.P.H., Director, Public Health Practice Program Office, Centers for Disease Control and Prevention, at the "First INPHO Workshop: Creating the Public Health Information Highway," in Atlanta, Georgia, March 29, 1994.

The authors wish to acknowledge the contributions made by a number of parties at the Centers for Disease Control and Prevention (CDC), universities, and public health agencies. The concepts, mission, and vision underlying the CDC Information Network for Public Health Officials (INPHO) have been shaped by members of the CDC INPHO lead team. They are Mr. James Seligman, Drs. Patrick O'Carroll, and Howard Ory, Information Resources Management Office; Ms. Barbara R. Holloway, Drs. Edwin Kilbourne.

Donna Stroup. and Demetri Vacalis, Epidemiology Program Office; and Mr. Thomas Lacher and Mr. Wallace Wilhoite, Public Health Practice Program Office. The following members of the Georgia INPHO Project Steering Committee have also contributed in shaping our approach: Dr. Karen Chapman, Georgia Division of Public Health; Dr. Kathy Minor, Ms. Melissa Alprin, and Ms. Gail Horlick, Emory University School of Public Health; Drs. Dan Ward and Hartmut Gross, Medical College of Georgia; and Mr. Richard K. Snelling and Mr. Keith Bernhardt, Georgia Center for Advanced Telecommunications Technology. The Robert W. Woodruff Foundation has given generous support to advance the Georgia INPHO project and the national CDC INPHO initiative.

About the Authors

Edward L. Baker, M.D., M.P.H., serves as Director of the CDC Public Health Practice Program Office. The mission of this office is to strengthen the public health system through information systems development, distance learning, leadership development, community planning, and systems research. Prior to taking this position, Dr. Baker served as Deputy Director and Assistant Director of the National Institute for Occupational Safety and Health (NIOSH), a CDC component, from 1985 to 1990. In that capacity, he provided leadership in occupational health surveillance and in development of the OSHA standard for prevention of blood-borne disease in the workplace.

Andrew Friede, MD., M.P.H., is the Chief of the Public Health Information Systems Branch, Information Resources Management Office, Centers for Disease Control and Prevention (CDC). He joined CDC's Information Resources Management Office in 1987 where he has led a large group in development of CDC WONDER/PC, an integrated information and communications public health information system that provides access to some 40 databases for 3,000 users as well as specialized features used by many CDC surveillance programs. Dr. Friede is also a principal participant in the CDC INPHO project.

Anthony D. Moulton, Ph.D., Robert W. Woodruff Health Sciences Center, Emory University, is an assistant to the Information Network for Public Health Officials (INPHO) initiatives of the Centers for Disease Control and Prevention and the Georgia Division of Public Health.

David A. Ross, Sc.D., is Assistant Director for Information and Communication Services in the Public Health Practice Program Office, CDC. Dr. Ross is directing the CDC Information Network for Public Health Officials (CDC INPHO) program.

D.2 FEDERAL EMERGENCY MANAGEMENT AGENCY FEDERAL RESPONSE PLAN ORGANIZATIONAL MODEL

FEMA Experience:

Applicability To The National Center For Information Systems Security Assurance

D.2.1 Background

FEMA is an independent federal agency with more than 2,600 full time employees: at FEMA headquarters in Washington D.C., at regional and area offices across the country, at the Mount Weather Emergency Assistance Center, and at the FEMA training center in Emmitsburg, Maryland. FEMA also has nearly 4,000 standby disaster assistance employees who are available to help out after disasters. Often FEMA works in partnership with other organizations that are part of the nation's emergency management system. These partners include state and local emergency management agencies, 27 federal agencies and American Red Cross.

FEMA's Mission is to provide leadership and support to reduce the loss of life and property and protect our nation's institutions from all types of hazards through a comprehensive, risk-based, all-hazards emergency management program of mitigation, preparedness, response and recovery.

FEMA accomplishes its mission through a very broad range of activities, including:

- helping equip local and state emergency preparedness . . .
- coordinating the federal response to a disaster . . .
- making disaster assistance available to states, communities, businesses and individuals . . .
- advising on building codes and flood plain management . . .
- teaching people how to get through a disaster . . .
- training emergency managers . . . supporting the nation's fire service . . .
- administering the national flood and crime insurance programs . . .

In particular, FEMA fully or partially funds emergency management programs and staff in all 56 states and territories, and helps design and equip emergency operations in thousands of localities. An important objective of this assistance is effective preparedness through planning. Emergency Operations Plans are updated periodically and submitted to FEMA for review.

D.2.2 Concept of Operations

The Federal Emergency Management Agency's Federal Response Plan (for Public Law 93-288, as amended) describes FEMA's Concept of Operations to address the consequences of any disaster or emergency situation in which there is a need for Federal response assistance under the authorities of the Stafford Act. It is applicable to natural disasters; technological emergencies involving radiological or hazardous material releases; and other incidents requiring Federal assistance under the Act

The Response Plan describes the basic mechanisms and structures by which the Federal government will mobilize resources and conduct activities to augment State and local response efforts. To facilitate the provision of Federal assistance, the Plan uses a functional approach to group the types of Federal assistance which a State is most likely to need under twelve Emergency Support Functions (ESFs). Each ESF is headed by a primary agency, which has been selected based on its authorities, resources and capabilities in the particular functional area. Other agencies have been designated as support agencies for one or more ESF based on their resources and capabilities to support the functional area. The twelve ESFs serve as the primary mechanism through which Federal response assistance will be provided to assist the State in meeting response requirements in an affected area. Federal assistance will be provided to the affected State under the overall coordination of the Federal Coordinating Officer (FCO) appointed by the Director of FEMA on behalf of the President.

Federal assistance provided under P.L. 93-288, as amended, is to supplement State and local government response efforts. ESFs will coordinate with the FCO and the affected State to identify specific response requirements and will provide Federal response assistance based on State-identified priorities.

Each ESF will provide resources using its primary and support agency authorities and capabilities, in coordination with other ESFs, to support its missions. ESFs will allocate available resources to each declared State based on priorities identified in conjunction with the State and in coordination with the FCO. If resources are not available within the declared State, the ESF will seek to provide them from a primary or support agency area or region. If the resource is unavailable from an area or region, the requirement will be forwarded to the appropriate ESF headquarters office for further action.

One or more disasters may affect a number of States and regions concurrently. In those instances, the Federal government will conduct multi-State response operations; for each declared State, an FCO will be appointed to coordinate the specific requirements for Federal response and recovery within that State. Under multiple State declarations, ESF departments and agencies will be required to coordinate the provision of resources to support the operations of all of the declared States.

D.2.3 Legislative History/Authorities

In 1988, Public Law 93-288 was amended by Public Law 100-707 and retitled as the Robert T. Stafford Disaster Relief and Emergency Assistance Act (Public Law 93-288, as amended). The Stafford Act provides the authority for the Federal government to respond to disasters and emergencies in order to provide assistance to save lives and protect public health, safety, and property.

In providing response assistance under the Federal Response Plan, Federal departments and agencies are covered under the authorities of P.L. 93-288, as amended. Under P.L. 93-288, the President may direct any Federal agency to utilize its authorities and resources in support of State and local assistance efforts. This authority has been further delegated to the Director, FEMA. the Associate Director, State and Local Programs and Support (SLPS), and to the FEMA Regional Directors in carrying out the provisions of the Stafford Act.

Response by departments and agencies to lifesaving and life protecting requirements under the Plan has precedence over other Federal response activities, except where national security implications are determined to be of a higher priority. Support from departments and agencies will be provided to the extent that it does not conflict with other emergency.

D.2.4 Relationships with Other Government Agencies

General Information

Numerous federal agencies and departments are partners in the nation's emergency management system. In planning, they participate in training exercises and conduct a variety of activities to help the nation prepare for disasters. For example, the Federal Communications Commission and the Commerce Department's National Weather Service provide on-going warning and disaster tracking services. In a catastrophic disaster, FEMA coordinates the federal response, working with 27 federal partners and the American Red Cross to provide emergency food and water, medical supplies and services, search and rescue operations, transportation assistance, environmental assessment, and more. The National Disaster Medical System is a partnership set up to provide emergency medical services in a disaster, involving FEMA, the Department of Health and Human Services, the Department of Defense, the Veterans Administration, as well as public and private hospitals across the country.

- National emergency management organizations. Emergency preparedness and response requires the efforts of many people. FEMA works in partnership with national organizations dedicated to assisting the public in preparation for and response to a disaster. FEMA supports the efforts of the National Emergency Management Association (NEMA), whose membership includes state emergency managers, and the National Coordinating Council on Emergency Management (NCCEM), whose membership includes local emergency managers.

- State emergency management departments. When a disaster overwhelms local resources, the task of coordinating response moves to the next level—the state. States take a leading role in response to any large-scale disaster, even those so major that federal assistance is requested. FEMA supports the state emergency management in many ways, from funding state planning to working directly with state agencies to managing a large-scale response.
- Local emergency management agencies. Local emergency management programs are the heart of the nation's emergency management system. FEMA supports them with funding for emergency planning and equipment, by offering training courses for emergency managers and firefighters, by conducting exercises for localities to practice their response, and by promoting ways to minimize disasters' effects. FEMA also builds partnerships with mayors, county boards and other elected and appointed officials who share responsibility for emergency management.
- Partnerships with the private sector. Disaster requires the full resources of a community to help people respond and recover. FEMA encourages all sectors of society—from business and industry to volunteer organizations—to work together in disaster preparation, response and recovery. FEMA assists in coordinating activities of a variety of players, including private contractors, hospitals, volunteer organizations and area businesses. It is through these partnerships of people working together that communities are able to put the pieces back together.

Relationships with Other U.S. Government Agencies

The Federal Emergency Management Agency's Federal Response Plan provides standing mission assignments to the designated departments and agencies with primary and support responsibilities to carry out Emergency Support Functions (ESFs). Federal departments and agencies designated as primary agencies serve as Federal executive agents under the FCO in accomplishing the ESF response missions. Upon activation of an ESF, a primary agency is authorized, in coordination with the Federal Coordinating Officer (FCO) and the State, to initiate and continue actions to carry out the ESF missions described in the ESF Annexes to the Plan, including tasking of designated support agencies to carry out assigned ESF missions.

At the national level, primary agencies are responsible to plan and coordinate with their support agencies for the delivery of ESF-related assistance.

Primary agencies are responsible for preparing and maintaining the ESF annexes and appendices to the Plan to reflect the policies, procedures regarding assistance to be provided, and associated responsibilities of the designated primary and support agencies.

Support agencies will assist the primary agencies in preparing and maintaining ESF annexes and appendices, developing national and regional operating procedures, and providing support for ESF operations.

EMERGENCY SUPPORT FUNCTION #1: TRANSPORTATION

The purpose of this Emergency Support Function (ESF) is to provide for the coordination of Federal transportation support to State and local governmental entities, voluntary organizations, and Federal agencies requiring transportation capacity to perform disaster assistance missions following a catastrophic earthquake, significant natural disaster, or other event requiring Federal response.

PRIMARY AGENCY: Department of Transportation

SUPPORT AGENCIES:

- Department of Agriculture
- Department of Defense
- Department of Energy
- Department of State
- General Services Administration
- Interstate Commerce Commission
- Tennessee Valley Authority
- Postal Service

EMERGENCY SUPPORT FUNCTION #2: COMMUNICATIONS

The purpose of this Emergency Support Function (ESF) is to assure the provision of Federal telecommunications support to Federal, State, and local response efforts following a Presidentially declared emergency, major disaster, extraordinary situation and other emergencies under the Federal Response Plan. This ESF supplements the provisions of the National Plan for Telecommunications Support in Non-Wartime Emergencies, hereafter referred to as the National Telecommunications Support Plan (NTSP).

PRIMARY AGENCY: National Communications System

SUPPORT AGENCIES:

- Department of Agriculture
- Department of Commerce

- Department of Defense
- Department of the Interior
- Department of Transportation
- Federal Communications
- Commission
- Federal Emergency Management Agency
- General Services Administration

EMERGENCY SUPPORT FUNCTION #3: PUBLIC WORKS AND ENGINEERING

The purpose of this Emergency Support Function (ESF) is to provide Public Works and Engineering support to assist the State(s) in needs related to lifesaving or life protecting following a major or catastrophic disaster.

PRIMARY AGENCY: Department of Defense; U.S. Army Corps of Engineers

SUPPORT AGENCIES:

- Department of Agriculture
- Department of Commerce
- Department of Energy
- Department of Health and Human Services
- Department of the Interior
- Department of Labor
- Department of Transportation
- Department of Veterans Affairs
- Environmental Protection Agency
- General Services Administration
- Tennessee Valley Authority

EMERGENCY SUPPORT FUNCTION #4: FIREFIGHTING

The purpose of this Emergency Support Function (ESF) is to detect and suppress wildland, rural, and urban fires resulting from, or occurring coincidentally with, a catastrophic earthquake, significant natural disaster or other event requiring Federal response assistance.

PRIMARY AGENCY: Department of Agriculture; Forest Service

SUPPORT AGENCIES:

- Department of Commerce

- Department of Defense
- Department of the Interior
- Environmental Protection Agency
- Federal Emergency Management Agency

EMERGENCY SUPPORT FUNCTION #5: INFORMATION AND PLANNING

Information and Planning: collect, process and disseminate information about a potential or actual disaster or emergency to facilitate the overall activities of the Federal government in providing response assistance to an affected State.

PRIMARY AGENCY: Federal Emergency Management Agency

SUPPORT AGENCIES:

- Department of Agriculture
- Department of Commerce
- Department of Defense
- Department of Education
- Department of Energy
- Department of Health and Human Services
- Department of the Interior
- Department of Justice
- Department of Transportation
- Department of the Treasury
- American Red Cross
- Environmental Protection Agency
- General Services Administration
- National Aeronautics and Space Administration
- National Communications System
- Nuclear Regulatory Commission

EMERGENCY SUPPORT FUNCTION #6: MASS CARE

The purpose of this Emergency Support Function (ESF) is to coordinate efforts to provide sheltering, feeding, and emergency first aid following a catastrophic earthquake, significant natural disaster or other event requiring Federal response assistance; to operate a Disaster Welfare Information (DWI) System to collect, receive, and report information about the status of victims and assist with family reunification supplies to disaster victims following a disaster.

PRIMARY AGENCY: American Red Cross

SUPPORT AGENCIES:

- Department of Agriculture
- Department of Commerce
- Department of Defense
- Department of Health and Human Services
- Department of Housing and Urban Development
- Department of Transportation
- Department of Veterans Affairs
- Federal Emergency Management Agency
- General Services Administration
- Postal Service

EMERGENCY SUPPORT FUNCTION #7: RESOURCE SUPPORT

The purpose of this Emergency Support Function (ESF) is to provide logistical/resource support following a catastrophic earthquake, other significant natural disaster or other event requiring Federal response.

PRIMARY AGENCY: General Services Administration

SUPPORT AGENCIES:

- Department of Agriculture
- Department of Commerce
- Department of Defense
- Department of Energy
- Department of Health and Human Services
- Department of Labor
- Department of Transportation
- Department of Veterans Affairs
- Federal Emergency Management Agency
- National Communications System
- Office of Personnel Management

EMERGENCY SUPPORT FUNCTION #8: HEALTH AND MEDICAL SERVICES

The purpose of this Emergency Support Function (ESF) is to provide United States Government coordinated assistance to supplement State and local resources in response to public health and medical care needs following a significant natural disaster or man-made event. Assistance provided under ESF #8—Health and Medical Services, is directed by the Department of Health and Human Services (HHS) through its Executive Agent, the Assistant Secretary for Health (ASH), who heads the United States Public Health Service (PHS). Resources will be furnished when State and local resources are overwhelmed and medical and/or public health assistance is requested from the Federal Government.

PRIMARY AGENCY: Department of Health and Human Services; U.S. Public Health Service

SUPPORT AGENCIES:

- Department of Agriculture
- Department of Defense
- Department of Justice
- Department of Transportation
- Department of Veterans Affairs
- Agency for International Development
- American Red Cross
- Environmental Protection Agency
- Federal Emergency Management Agency
- General Services Administration
- National Communications System
- Postal Service

EMERGENCY SUPPORT FUNCTION #9: URBAN SEARCH AND RESCUE

The purpose of this Emergency Support Function (ESF) is to describe the use of Federal Urban Search and Rescue (US&R) assets following an event requiring a Federal response, including locating, extricating and providing for the immediate medical treatment of victims trapped in collapsed structures.

PRIMARY AGENCY: Department of Defense

SUPPORT AGENCIES:

- Department of Agriculture
- Department of Health and Human Services

- Department of Labor
- Department of Transportation
- Agency for International Development
- Environmental Protection Agency
- Federal Emergency Management Agency
- General Services Administration

EMERGENCY SUPPORT FUNCTION #10: HAZARDOUS MATERIALS

The purpose of this Emergency Support Function (ESF) is to provide Federal support to State and local governments in response to an actual or potential discharge and/or release of hazardous materials following a catastrophic earthquake or other catastrophic disaster.

PRIMARY AGENCY: Environmental Protection Agency

SUPPORT AGENCIES:

- Department of Agriculture
- Department of Commerce
- Department of Defense
- Department of Energy
- Department of Health and Human Services
- Department of the Interior
- Department of Justice
- Department of Labor
- Department of State
- Department of Transportation
- Federal Emergency Management Agency
- General Services Administration
- Nuclear Regulatory Commission

EMERGENCY SUPPORT FUNCTION #11: FOOD

The purpose of this Emergency Support Function (ESF) is to identify, secure, and arrange for the transportation of food assistance to affected areas following a major disaster or emergency or other event requiring Federal response.

PRIMARY AGENCY: Department of Agriculture

SUPPORT AGENCIES:

- Department of Defense
- Department of Health and Human Services
- Department of Transportation
- American Red Cross
- Environmental Protection Agency
- Federal Emergency Management Agency

EMERGENCY SUPPORT FUNCTION #12: ENERGY

The purpose of this Emergency Support Function (ESF) is to facilitate restoration of the Nation's energy systems following a catastrophic earthquake, natural disaster, or other significant event requiring Federal response assistance. Power and fuel are critical to save lives and protect health, safety, and property, as well as carry out other emergency response functions.

PRIMARY AGENCY: Department of Energy

SUPPORT AGENCIES:

- Department of Agriculture
- Department of Defense
- Department of State
- Department of Transportation
- General Services Administration

Information Warfare—Defense Appendix D

- National Communications System
- Nuclear Regulatory Commission
- Tennessee Valley Authority

ATTACHMENT 1
COMPENDIUM OF EMERGENCY AUTHORITIES AND DIRECTIVES

PUBLIC LAW 78-410, "PUBLIC HEALTH SERVICE ACT," SECTION 216, 42 U.S.C. 217—

This provision authorizes the President, in time of war or upon Presidential declaration of an emergency, to utilize the Public Health Service to the extent and in the manner that in his judgment will promote the public interest.

PUBLIC LAW 78-410, "PUBLIC HEALTH SERVICE ACT," SECTION 311 U.S.C. 243—

This provision authorizes the Secretary of Health and Human Services to develop (and may take such action as may be necessary to implement) a plan under which personnel, equipment, medical services, and other resources of the Public Health Service and other agencies under the jurisdiction of the Secretary may be effectively used to control epidemics of any disease or condition, as specified, and to meet other health emergencies or problems involving or resulting from disasters or any such disease.

PUBLIC LAW 78-410, "DEFENSE HEALTH SERVICE ACT," SECTION 319—

This provision authorizes the Secretary of Health and Human Services to take appropriate action to respond to a "public health emergency" resulting from disease, disorder, or other cause. The Secretary must consult with the Director of the National Institute of Health, Administrator of the Alcohol, Drug Abuse, and Mental Health Administration, Commissioner of the Food and Drug Administration, or the Director of the Center, for Disease Control before determining that an emergency exists, and he must act through that official in responding to the emergency.

PUBLIC LAW 81-774, "DEFENSE PRODUCTION ACT OF 1950, AS AMENDED," 50 U.S.C. 2061, TITLE I, SECTION 101(a) AND 101(b)—

This provision authorizes the President to establish performance priorities and to allocate materials and facilities to promote the national defense.

PUBLIC LAW 93-288, AS AMENDED BY PUBLIC LAW 100-707, "ROBERT T. STAFFORD DISASTER RELIEF AND EMERGENCY ASSISTANCE ACT," NOVEMBER 23, 1988—

The Robert T. Stafford Disaster Relief and Emergency Assistance Act, P.L. 93-288 as amended, provides an orderly and continuing means of assistance by the Federal Government to State and local governments in carrying out their responsibilities to alleviate the suffering and damage which result from disasters. The President, in response to a State Governor's request. may declare an "emergency" or "major disaster," in order to provide Federal assistance under the Act.

The President, in Executive Order 12148, delegated all functions, except those in Section 301, 401, and 409, to the Director, Federal Emergency Management Agency (FEMA). The Act provides for the appointment of a Federal Coordinating Officer who will operate in the designated area with a State Coordinating Officer for the purpose of coordinating state and local disaster assistance efforts with those of the Federal Government

PUBLIC LAW 95-124, "EARTHQUAKE HAZARDS REDUCTION ACT OF 1977," 42 U.S.C. 7701 AND 7704—

The Earthquake Hazards Reduction Act of 1977, as amended by P.L. 96472 and P.L. 99-105, provides for the establishment of the National Earthquake Hazards Reduction Program (NEHRP) to reduce the risk to life and property from future earthquakes in the United States. FEMA is designated as the agency with primary responsibilities to plan and coordinate the NEHRP, which has five major elements: Hazard Delineation and Assessment; Earthquake Prediction Research; Seismic Design and Engineering Research; Preparedness Planning and Hazard Awareness; and, Fundamental Seismological Studies. Planning for the Federal response to a catastrophic earthquake is a major aspect of Preparedness Planning and Hazard Awareness under the NEHRP.

PUBLIC LAW 95-313, "COOPERATIVE FORESTRY ASSISTANCE ACT OF 1978"—

This Act authorizes the Secretary of Agriculture to assist in the prevention and control of rural fires through coordination among Federal, State, and local agencies; and to provide prompt and adequate assistance whenever a rural fire emergency overwhelms, or threatens to overwhelm, the firefighting capability of the affected State or rural area.

PUBLIC LAW 96-510, "COMPREHENSIVE ENVIRONMENTAL RESPONSE, COMPENSATION, AND LIABILITY ACT OF 1980," SECTION 104(i), 42 U.S.C. 9604(i)—

More popularly known as "Superfund", CERCLA was passed to provide the needed general authority for Federal and State governments to respond directly to hazardous substances incidents.

PUBLIC LAW 101-640, "WATER RESOURCES DEVELOPMENT ACT OF 1990," TITLE III, SECTION 302, 5(A)(1), NOVEMBER 28,1990—

This Act amends 33 U.S.C. 701n)a)(1) by replacing the term "flood emergency preparation" to include "preparation for emergency response to any disaster" and includes a provision that "The emergency fund may be expended for emergency dredging for restoration of authorized projects for Federal navigable channels and waterways made necessary by flood, drought, earthquake, or other natural disasters."

UNITED STATES CONGRESS ACT OF JANUARY 5 1905 AS AMENDED. 36 U.S.C.—

The American National Red Cross Congressional Charter assigning the authority and responsibility for the American Red Cross to undertake activities for the relief of individuals suffering from a disaster.

COMMUNICATIONS ACT OF 1934, AS AMENDED—

This Act gives the Federal Communications Commission emergency authority to grant Special Temporary Authority on an expedited basis to operate radio frequency devices.

OLDER AMERICANS ACT OF 1965, AS AMENDED, SECTION 310, 42 U.S.C. 3030—

This provision authorizes the Commissioner of the Administration on Aging to reimburse States for social services provided to older Americans following a Presidentially—declared disaster.

FOOD STAMP ACT OF 1977, SECTION 5(h)(1), IMPLEMENTED BY PROPOSED FINAL RULEMAKING AT 46 CFR 8922 AND 46 CFR 8923—

Authorizes the Department of Agriculture to make food stamps available to low income households in any disaster situation in which normal channels of retail food distribution have been restored and the existing Food Stamp Program cannot handle applications from affected households. Food stamp assistance must be requested by a State.

INTERSTATE COMMERCE ACT, EMERGENCY RATES, 49 U.S.C. 10724 AND 11121 TO 11128—

These authorities allow the Interstate Commerce Commission (ICC) to authorize a common carrier to give reduced rates for service and transportation in an emergency. Further, these authorities permit the ICC to suspend any car service rule or practice, take action during emergencies to promote car service in the interest of the public and commerce; to require joint or common use of facilities when that action will best meet the emergency; to direct preferences or priorities in transportation, embargoes, or movement of traffic under permits; and to reroute traffic.

"ROBERT T. STAFFORD DISASTER RELIEF AND EMERGENCY ASSISTANCE ACT (P.L. 93-288, AS AMENDED)," IMPLEMENTED BY FOOD DISTRIBUTION REGULATIONS, PARTS 250.1(b) AND 250.8(e)—

These provisions allow any person/household temporarily displaced by a disaster to obtain USDA foods in congregate feeding provided by volunteer organizations such as the American Red Cross and the Salvation Army; no formal approval is required from USDA. Additionally, low income families can receive household distributions of food in situations where a Food Stamp Program is not available (e.g., commercial channels of trade are disrupted); formal USDA approval is required.

EXECUTIVE ORDER 10480, AS AMENDED, "FURTHER PROVIDING FOR THE ADMINISTRATION OF THE DEFENSE MOBILIZATION PROGRAM," AUGUST 14, 1953—

Part II of the Order delegates to the Director, FEMA, with authority to redelegate, the priorities and allocation functions conferred on the President by Title I of the Defense Production Act of 1950, as amended

EXECUTIVE ORDER 12148, "FEDERAL EMERGENCY MANAGEMENT," JULY 20, 1979—

Executive Order 12148 transferred functions and responsibilities associated with Federal emergency management to the Director, FEMA. Assigns the Director, FEMA, the responsibility to establish Federal policies for and to coordinate all civil defense and civil emergency planning, management, mitigation, and assistance functions of Executive Agencies.

EXECUTIVE ORDER 12472, "ASSIGNMENT OF NATIONAL SECURITY AND EMERGENCY PREPAREDNESS TELECOMMUNICATIONS FUNCTIONS," APRIL 3, 1984—

Executive Order 12472 establishes the National Communications System (NCS). The NCS consists of the telecommunications assets of the entities represented on the NCS Committee of Principals and an administrative structure consisting of the Executive Agent, the NCS Committee of Principals, and the Manager. The NCS Committee of Principals consists of representatives from those Federal departments, agencies, or entities, designated by the President, which lease or own telecommunications facilities or services of significance to national security or emergency preparedness.

EXECUTIVE ORDER 12656, "ASSIGNMENT OF EMERGENCY PREPAREDNESS RESPONSIBILITIES," NOVEMBER 18,1988—

Assigns emergency preparedness responsibilities to Federal departments and agencies.

EXECUTIVE ORDER 12657, "FEMA ASSISTANCE IN EMERGENCY PREPAREDNESS PLANNING AT COMMERCIAL NUCLEAR POWER PLANTS," NOVEMBER 18, 1988—

Assigns FEMA and other Federal agencies certain emergency planning responsibilities related to commercial nuclear power plants.

EXECUTIVE ORDER 12777, "IMPLEMENTATION OF SECTION 311 OF THE FEDERAL WATER POLLUTION ACT OF OCTOBER 18,

1972,AS AMENDED, AND THE OIL POLLUTION ACT OF 1990," OCTOBER 18, 1991—

Refers to certain activities of the National Response Team and the Regional Response Team under the National Contingency Plan.

7 CFR, PART 250.1(B)(10)&(11)—

Refers to Section 409 and 410 b of P.L. 93-288, as amended, Robert T. Stafford Disaster Relief and Emergency Assistance Act, which reads, "The Secretary of Agriculture shall utilize funds appropriated under Section 32 of the Act of August 1935 (7 USC 612 c) to purchase food commodities necessary to provide adequate supplies for use in any area of the United States in the event of a major disaster or emergency in such area."

28 CFR, PART 65, "EMERGENCY FEDERAL LAW ENFORCEMENT ASSISTANCE"; FINAL RULE—

These Department of Justice regulations implement the Emergency Federal Law Enforcement Assistance functions vested in the Attorney General by the Justice Assistance Act of 1984 (Public Law 98-473). Those functions were established to assist State and/or local units of government in responding to a law enforcement emergency. The Act defines the term "law enforcement emergency" as an uncommon situation which requires law enforcement, which is or threatens to become of serious or epidemic proportions, and with respect to which State and local resources are inadequate to protect the lives and property of citizens, or to enforce the criminal law. Emergencies which are not of an ongoing or chronic nature, such as the Mount Saint Helens volcanic eruption, are eligible for Federal law enforcement assistance. Such assistance is defined as funds, equipment, training, intelligence information, and personnel. Requests for assistance must be submitted in writing to the Attorney General by the chief executive officer of a State. The Plan does not cover the provision of law enforcement assistance. Such assistance will be provided in accordance with the regulations referred to in this paragraph [28 CFR Part 65, implementing the Justice Assistance Act of 1984] or pursuant to any other applicable authority if the Department of Justice.

40 CFR PART 300, "NATIONAL OIL AND HAZARDOUS SUBSTANCES POLLUTION CONTINGENCY PLAN" (NCP)—

The purpose of the NCP is to effectuate the powers and responsibilities for responding to non-radiological oil and hazardous substances discharges, releases, or substantial threats of releases as specified in the Comprehensive Environmental Response, Compensation and Liability Act, as amended, (CERCLA) and the authorities established by Section 311 of the Clean Water Act, as amended. The plan is required by section 105 of

CERCLA, 42 U.S.C. 9605, and by section 31 1(c)(2) of the Clean Water Act, as amended, 33 U.S.C. 1321(c)(2).

44 CFR PART 322, AS AMENDED, "DEFENSE PRODUCTION: PRIORITIES AND ALLOCATION AUTHORITY (DMA-3)"—

The Order delegates the functions of the Director, FEMA, under Title I of the Defense Production Act, as amended, to those offices and agencies named in Section 201 of Executive Order 10480 with respect to the areas of responsibility designated and to the Secretary of Transportation with respect to priorities and allocations for civil transportation services.

FEDERAL COMMUNICATIONS COMMISSION REPORT AND ORDER OF AUGUST 4, 1981—

This order modified parts 2, 90, and 99 of the Commission Rules and Regulations to establish a disaster radio response capability for local government and State radio services.

"FEDERAL RADIOLOGICAL EMERGENCY RESPONSE PLAN"—

This document is to be used by Federal agencies in peacetime radiological emergencies. It primarily concerns the off-site Federal response in support of State and local governments with jurisdiction for the emergency. The Federal Radiological Emergency Response Plan (FRERP) provides the Federal government's concept of operations based on specific authorities for responding to radiological emergencies, outlines Federal policies and planning assumptions that underlie this concept of operations and on which Federal agency response plans were based, and specifies authorities and responsibilities of each Federal agency that may have a significant role in such emergencies.

"NATIONAL PLAN FOR TELECOMMUNICATIONS SUPPORT IN NON-WARTIME EMERGENCES," JANUARY 1992—

This plan provides guidance in planning for and providing telecommunications support for Federal agencies involved in emergencies, major disasters, and other exigencies, excluding war.

DEPARTMENT OF DEFENSE DIRECTIVE 3025.1, "MILITARY SUPPORT TO CIVIL AUTHORITIES (MSCA)," 1992—

This directive outlines Department of Defense (DOD) policy on assistance to the civilian sector during disasters and other emergencies. Use of DOD military resources in civil emergency relief operations will be limited to those resources not immediately required for the execution of the primary defense mission. Normally, DOD military resources will be committed as a supplement to non-DOD resources which are required to cope with the humanitarian and property protection requirement caused by the emergency. In any emergency, commanders are authorized to employ DOD resources to save lives, prevent human suffering, or mitigate great property loss. Upon declaration of a major disaster under the provisions of P.L. 93-288, as amended, the Secretary of the Army is the DOD Executive Agent, and the Director of Military Support is the action agent for civil emergency relief operations. Military personnel will be under command of and directly responsible to their military superiors and will not be used to enforce or execute civil law in violation of 18 U.S.C. 1385 except as otherwise authorized by law. Military resources shall not be procured, stockpiled, or developed solely to provide assistance to civil authorities during emergencies.

FEDERAL PREPAREDNESS CIRCULAR 8, "PUBLIC AFFAIRS IN EMERGENCIES"—

This Circular establishes the Interagency Committee on Public Affairs in Emergencies (ICPAE) to coordinate public information planning and operations for management of emergency information. The Circular was reviewed in draft by the ICPAE and will receive formal department and agency review.

AMERICAN RED CROSS DISASTER SERVICES REGULATIONS AND PROCEDURES, ARC 3003, JANUARY 1984—

This document details the delegation of disaster services program responsibilities to officials and units of the American Red Cross. Also defined are Red Cross administrative regulations and procedures for disaster planning, preparedness, and response.

AMERICAN NATIONAL RED CROSS MASS CARE PREPAREDNESS AND OPERATION PROCEDURES AND REGULATIONS, ARC 3031"—

This document details the Red Cross mass care preparedness and operating regulations and procedures.

AMERICAN NATIONAL RED CROSS NATIONAL BOARD OF GOVERNORS DISASTER SERVICES POLICY STATEMENT, JULY 1, 1977"—

This document outlines the basic policies of the American Red Cross disaster services program, and the disaster relief services to be provided by units of the American Red Cross on a uniform and nationwide basis.

STATEMENT OF UNDERSTANDING BETWEEN THE FEDERAL EMERGENCY MANAGEMENT AGENCY AND THE AMERICAN NATIONAL RED CROSS, JANUARY 22. 1982—

The statement of understanding between FEMA and the American National Red Cross describes major responsibilities in disaster preparedness planning and operations in the event of a war-caused national emergency or a peacetime disaster, outlines areas of mutual support and cooperation, and provides a frame of reference for similar cooperative agreements between State and local governments and the operations headquarters and chapters of the ARC.

D.3 NATIONAL DRUG INTELLIGENCE CENTER

A Quick Look at the National Drug Intelligence Center (NDIC) for Lessons Applicable to the Formation of a National Defensive Information Warfare Center

D.3.1 Background and Legislative History

During the cocaine epidemic of the late 1980s, U.S. public opinion demanded greater Federal Government efforts to combat a nationwide drug problem. Members of Congress and the Executive Branch both reacted with pronouncements and policy moves. In 1988, the Office of National Drug control policy (ONDCP) was created and the Defense Department was given increased responsibility for counter-narcotics support actions. As policy makers attempted to cope with the increased public interest, the dimensions and dynamics of the situation were not fully understood, partially because of the lack of strategic intelligence regarding narcotics organizations. The National Drug Control Strategy of 1989 noted:

> A comprehensive thrust against drug trafficking enterprises and organizations requires a different kind of intelligence. . . . Greater emphasis needs to be devoted to automating this information for law enforcement purposes and analyzing it [and other data] to produce a better understanding of the structure and infrastructure of trafficking organizations and their allied enterprises.

In 1989 and early 1990, the ONDCP negotiated a constituency supporting the case for establishing a National Drug Intelligence Center (NDIC). In January 1990, the ONDCP publicly endorsed the NDIC and, in June 1990, the Administration introduced legislation to establish the organization. The NDIC outlined by ONDCP emphasized modernization of law enforcement intelligence rather than making narcotics intelligence the purview of the intelligence community as some critics had feared. The proposal envisioned the Center as a focal point for consolidating and coordinating relevant intelligence gathered by law enforcement agencies and analyzing it to develop a full understanding of the drug trafficking organizations. The processed intelligence would be distributed to Federal, State and local officials for use. NDIC would maintain computer databases, coordinate collection and tasking and assess interagency efforts. The NDIC was seen to be an interagency organization to include Treasury, State, Justice, and Defense assets. Supervision of the NDIC would be the responsibility of the Attorney General. The Intelligence Community's supporting role included foreign collection and methodological and technical assistance. The NDIC was envisioned as being a small, efficient organization in Washington, DC.

With the formal Administration proposal to create the NDIC, the field of action for forming it shifted to the Congress. The House passed the measure, but differences arose in the Senate concerning the need for and location of the Center. In the end, after significant Congressional negotiations and compromise, the NDIC was authorized. The compromise placed the NDIC in Johnstown, Pennsylvania, made the DOD the executive authority for the project, and restricted the Justice Department role in the Center itself to participation. A summary of relevant key dates and legislation is provided in Table 1.[1]

D.3.2 Concept of Operations

The multi-agency National Drug Intelligence Center is located in Johnstown, Pennsylvania. It is organized with a Director and three Deputy Directors. The Director is a Department of Justice position. The Deputy Director for Operations is a DEA position; the Deputy Director for Administration is an FBI position; and the Deputy Director for Technology is a DOD position currently filled by DIA. The staff of approximately 300 is composed of intelligence analysts (from Federal law enforcement agencies [LEAs]), special agents (from DOJ), technical experts (from DOD), administrative support, liaison staff from other agencies, and specialized contractor support. The Center also has a small liaison office in the Washington DC area to facilitate coordination.

Generally, the Federal LEAs have stand-alone terminals at the Center which can be used to receive data released to the Center and send material to the owning agency, but cannot directly access agency network systems or databases. However, the Center has made some progress in negotiating direct access in some cases. PCs in a designated Operational Research Center allow analysts access to open source material such as Reuters, AP, and Nexis/Lexis. Desktop PCs throughout the NDIC allow analysts to exchange information among themselves via a LAN, but they are not connected outside the facility. Analysts generally focus on specific organizations as targets. They correlate and fuse information on crop production and facilities, financial practices, chemical sources, transportation and distribution assets, communications and other topics to produce strategic organizational drug intelligence (SODI) pertaining to the infrastructure of a drug trafficking organization.

The Center both responds to specific requests for intelligence products and strives to develop and maintain a strategic organizational drug intelligence database, library and index system. The Center also has a deployable docu-

1 This paragraph abstracted from *Executive-Legislative Relations in the Creation of the National Drug Intelligence Center*, Donald J. Carey, LT., U.S. Navy, September 1991.

ment exploitation team that can assist LEAs with reviewing, cataloging, analyzing and exploiting various documents which are seized in drug raids.

Senior personnel at the Center acknowledge that rivalry among the LEAs—largely as a result of a "scoring system" that keys future funding to arrest and prosecution statistics adversely affects the degree of information sharing and coordination that is achieved today. However, they indicate a belief in a positive trend as the mutual confidence builds from personal interaction by representatives from the different agencies.

Table 1. A Summary of Milestones in Establishing the NDIC

1986	PL 99-570$ 1.7 Million approved for anti-drug measures.
1988	PL 11-463: Defense Appropriations Bill includes $ 300 Million for narcotics interdiction.
	PL 100-690: $ 2.8 Billion approved for anti-drug measures; creation of Office of National Drug Control Policy with Cabinet-level "Drug Czar" position; required national drug control strategy be submitted to Congress within 180 days of confirmation; provided death penalty for traffickers.
1989	PL 101-164: Authorized $ 3.18 Billion in new anti-drug funding
	PL 101-231: Authorized drug fighting assistance for Columbia, Bolivia and Peru
	September 1989: 1989 Drug Control Strategy released
	December 1989: Panama invaded, Gen. Noriega arrested on drug charges
1990	January 1990: 1990 Drug Control Strategy released
	June 1990: Legislation to establish NDIC sent to Congress
	PL 101-511: FY 1991 Defense Appropriations Act provided $ 10 Million for NDIC in Johnstown
	PL 101-515: Department of Justice prevented from expending funds on NDIC.
1991	February 1991: 1991 National Drug Control Strategy released
	FY 1992 Defense Appropriations Bill Provided $ 40 Million for NDIC
	October 1991: NDIC opened in Johnstown, PA.

D.3.3 Relationships Between NDIC and Other Government Agencies

The NDIC has the responsibility for developing technical and organizational protocols (Memoranda of Agreement) required for access to information provided by other organizations. Technical protocols specify the hardware and software interfaces to allow NDIC access to the Agencies' information. Organizational protocols, documented in memoranda of agreement, specify restrictive procedures for accessing data and assure the protection by NDIC of both data source and success as specified by the originator of the information. The other Government agencies NDIC is working to establish protocols with to preclude duplication of effort and redundancy include: Treasury, U.S. Coast Guard, Immigration and naturalization Service, Customs Service, CIA, NSA, FBI, DEA and selected DOD organizations.

D.3.4 Relationships Between NDIC and International Agencies

Currently, NDIC has no direct relationships with international agencies such as Interpol or with law enforcement agencies of other nations, although they are deemed desirable. At this time, such relationships are the closely guarded province of other Federal agencies. This situation exists regarding State and local authorities as well such relationships are the province of the Federal law enforcement agencies.

D.3.5 Observations on Potential Lessons Learned and Pitfalls

- It is essential to develop a constituency in both the Congress and the Administration in order to establish a IW-D Center.
- A high level advocate who can articulate the need for the Center is essential.
- There are likely to be concerns regarding the integration of the intelligence community or its use in support of a IW-D Center.
- The preliminary operations concept of the Center needs to allow for Congressional compromises regarding physical location.
- Interagency sensitivities regarding information use and sharing may be nearly as strong as those of civilian organizations that may be involved in the Center.
- Funding for the Center should be as stable as possible through the formative period for establishing a capability.
- High quality "human capital" is a must.

APPENDIX E
THINK PIECES

The following discussions were a part of the Task Force deliberations and judged worthy of inclusion in the Task Force Report for reference only.

E.1 INFORMATION INFRASTRUCTURE ASSURANCE PRINCIPLES

Information assurance is a term which can be used to describe the needed IW-D capabilities (and associated protection) of an information infrastructure. Some basic definitions are needed to understand the principles:

- Availability of Service—An assured level of service, capacity, quality, timeliness, and reliability.
- Denial of Service—The opposite of availability of service.
- Information Integrity—Complete, sound, unaltered, and unimpaired information.
- Corruption of Information—The opposite of information integrity.
- Information Assurance—The availability of services and information integrity.
- Disruption—Denial of service or corruption of information resulting from a single event, cause, or source; whether direct or indirect; whether accidental, intentional, rare or common.
- Stress Level—Military situations under which the infrastructure is expected to operate. These include:
  - Peacetime (natural disasters, sabotage, equipment and service failures, unintentional acts)
  - Crisis/mobilization (terrorism, low intensity conflict, conventional war)
  - Simultaneous two-theater engagements
  - Limited nuclear war (nuclear terrorism, uncoordinated/accidental, theater nuclear) Expanded nuclear (coordinated attack)
  - Post-attack (recovery and reconstitution).

In the traditional systems engineering context, availability is a function of the reliability and maintainability of the system while integrity of data is a function of the quality (or grade of service) of the system transporting the data. In addition, these measures of system performance are traditionally based on design assumptions that disruptions are random in nature (e.g., component failures, human errors, and acts of nature).

Information assurance is not just a function of the reliability, maintainability, and quality of the network or infrastructure. Information assurance addresses the capability of an infrastructure to endure a variety of disruptions ranging from natural disasters to accidents to intentional disruptions by the enemies or by insiders. For example:

- A lightning strike on a critical node in the network can cause node failure; or, an earthquake or hurricane cannot only physically disrupt the network but can also cause network congestion, another source of disruption.
- Inadvertently erasing a data base containing terrain data critically needed for a cruise missile strike can compromise a key part of an offensive strike.
- Corruption of key network management data by a network manager can cause many networks to fail.
- An enemy agent located in a safe haven can introduce viruses that can cause a network to become overloaded and ineffective or cause the entire network to break down at a critical juncture.

This perspective on disruptions poses challenges for the intelligence, operations, and training communities in defining the threat, which is essential for a reasonable articulation of information assurance principles.

There are substantial differences between designing a typical information system and designing a resilient information infrastructure capable of enduring in the face of intentional disruptions. A typical information system design assumes that all of the system components will normally operate properly, with the common failure mode being failure of individual components. A resilient information infrastructure design must be based on the assumption that only some of the components will operate properly at any point in time. A typical information system design will incorporate central control mechanisms, synchronized clocks, and other techniques to use resources efficiently. A resilient information infrastructure design must be based on some decentralization of control and independent operation of portions of the infrastructure. Information system design is typically based on efficiency while a resilient information infrastructure design must be based on effectiveness. For example, the entire field of fault tolerant computing is based on the introduction of redundancy into otherwise efficient systems in order to make them more effective, particularly against random disruptions. Similarly, the design of a resilient infrastructure will assure diversity of hardware and software so that a common failure mode will not result in an infrastructure failure.

In the context of information assurance, network operation, management, and maintenance should be viewed from a war fighting perspective. Personnel performing these functions (and users in some cases), should be able to detect, differentiate among, warn of, respond to, and recover from disruptions. Recovery from disruptions resulting from failures or attacks might involve repair, reconstitution, or the employment of reserve assets.

In some cases, network managers may have to isolate portions of the network to preclude the spread of disruption. Given the speed with which disruptions can propagate through networks, these capabilities may need to be available in automated form within the network itself. Finally, there must be some means to manage and control these capabilities.

The underlying philosophy in information assurance and in satisfying the IW-D need must be that of risk management and not of risk avoidance. There are not enough resources to armor plate the infrastructure. Risk management suggests that the threat be defined, that measures be undertaken to reduce the realization of the threat, that countermeasures to threat occurrence be based on realistic application of resources and that response to and recovery from threat occurrences be part of the infrastructure. Finally, it will be necessary to assume some degree of risk while maintaining some minimum infrastructure operating capability.

Based on a review of existing documentation, a list of information assurance principles has been developed and is presented below. Because the infrastructure and the concept of information assurance are still under development, the list is not exhaustive.

The following operational information is required from CJCS and the Commanders-in-Chief (CINCs) of the Unified and Specified (U&S) Commands to quantify some of the principles:

- Information Transfer Priorities—Priorities for the transfer of voice, data, imagery, and video information based on a process developed by the JCS and based on the existing process used to establish priorities for voice and messages.
- Minimum Operating Capability—The minimum set of fixed and deployed capabilities required for each stress level, based on operations tempo and forces supported.
- Normal Operating Capability—A specified set of fixed and deployed capabilities required for peacetime and crisis/mobilization stress levels, based on operations tempo and forces supported. (In coordination with CJCS and the CINCs, DISA will, in its role as the central manager of the DII, specify this set.)
- Expected Disruptions—The expected level of disruptions to be sustained over time at each stress level. (This is normally based on intelligence estimates of enemy capabilities, insider threats, natural disasters, and other anticipated causes.)
- Minimum Assured Resiliency—The capability to sustain a specified number of simultaneous, worst-case disruptions at each stress level while still maintaining the Minimum Operating Capability.

- Desired Resiliency—The capability to sustain Expected Disruptions while maintaining a Normal Operating Capability. (In coordination with CJCS and the CINCs, DISA will, in its role as the central manager of the DII, specify this set.)

Information Assurance Principles:

- The infrastructure shall be considered a potential battlefield.
- The infrastructure shall provide Minimum Resiliency.
- The infrastructure shall detect substantial disruption, differentiate accidental disruption from intentional disruption, provide ample warning of disruption, respond to and recover from disruption, and be repairable at a rate sufficient to sustain Minimum Operating Capability under Expected Disruptions.
- The infrastructure shall detect large classes of event sequences that are likely or anticipated to lead to disruption and provide mechanisms so that disruptions from these events are:
  - Prevented when possible within cost constraints
  - Limited in the extent of their effect when prevention is not feasible
  - Responded to prior to actual disruption when detected in time
  - Traced to their source whenever possible within cost constraints.
- The infrastructure network and system control functions shall be designed to operate without dependence on the normal operation of the network or processes being controlled.
- The infrastructure responses to disruption shall be prioritized and shall take into account factors such as time, value, criticality, and locality as related to the information being transported.
- Changes to the infrastructure shall be analyzed and simulated prior to implementation to ensure that the infrastructure maintains assurance attributes during and after these changes.
- The infrastructure operations, management, and maintenance personnel and information assurance capabilities shall be regularly tested under realistic conditions to ensure that they perform and operate properly. Prior to testing, proposed tests must be simulated to assess expected behavior and ensure that the tests do not unduly degrade the infrastructure. After testing, expected and actual behavior must be reconciled and addressed.
- The infrastructure shall be designed to be flexible with respect to information assurance attributes so that as requirements, technologies, and processes are altered over time, the infrastructure will retain the Desired Resiliency specified by DISA.

- The infrastructure shall be capable of retaining the Desired Resiliency during infrastructure expansion, contraction, modification, and connection to combined forces infrastructures.
- New infrastructure components shall be designed such that:
  - If they are disrupted, they do not react so as to disrupt neighboring components
  - Disrupted neighboring components do not disrupt the new component regardless of the neighboring component's behavior
  - Disrupted components are quarantined until they return to normal operating behavior
  - Network and system management services are notified of disruptions and quarantines.
- Techniques for limiting the spread of disruptions (e.g., firewalls) shall be used where applicable, particularly in the design of network protocols and in gateways between networks.
- The infrastructure training and readiness programs shall be designed to ensure that personnel tasked with operating, managing, and maintaining the infrastructure are prepared for operations under stress, and that ample personnel and resources are available to operate and sustain the infrastructure at the Minimum Operating Capability during Expected Disruptions.
- Sufficient inventory of and/or manufacturing capability for parts, equipment, tools, supplies, and support systems shall be maintained to enable operation, repair, and reconstitution of the infrastructure under all stress levels.
- The infrastructure users shall be licensed to operate on the information highway. Licensing procedures shall include knowledge of the network, rules of the road, information assurance, and incident response processes and capabilities.

The goal in postulating these information assurance principles is to eventually outline a set of specifications (on the order of A-Level specifications) that will shape the design and integration of the infrastructure or that can be used as a part of the specifications for the acquisition of services from the local and long-distance carriers and from information processing vendors. In order to bridge the gap between the information assurance principles and a set of specifications, it will be necessary to develop strategies for providing the attributes. Some elements that might be considered in developing those strategies include:

- Capacity
- Diversity

- Co-location of network components at hardened subscriber sites
- Provision of uninterruptable power to selected sites
- Selected redundancy in network components
- Use of diverse transmission media
- Redundant network access links for key subscribers
- Precedence (priority) mechanisms
- Congestion control mechanisms
- Transportable reserve assets for reconstitution of damaged portions of the network
- Infrastructure restoration and reconstitution
- Multiple inter-network gateways
- Personal reliability program for network managers
- End-to-end network control (that does not depend on the network to operate)
- Scalable infrastructure components
- Repairability.

Successful implementation of information assurance will require a multi-disciplinary team capable of formulating a comprehensive set of requirements, knowledgeable of current and emerging technologies, capable of overseeing the design of the infrastructure from an information assurance perspective, and capable of managing the implementation of information assurance in the infrastructure.

E.2 "Raise the Bar" Exercise

The goal is to maximally improve DoD's information assurance as quickly as possible but "do it on the cheap" without involving unnecessarily complex technology, and without awaiting the outcome of R&D efforts now underway or that could be imagined.

It can be played two ways:

1. Assume that a given pot of money is available, take as a goal maximizing the protection of DoD information assets and internal systems soonest (*i.e.*, little or no R&D), and decide how and on what to spend it.

2. As above in item [1] except first compile a reasonable list of actions to be taken, and then estimate the cost to do them.

Below are some options from which to select, but not a comprehensive or complete list by any means. The sequence in the list is happenstance.

1. Provide users of the most sensitive systems commercially available tokens of some sort to improve the user identification/ authentication act of logging on; e.g., SecurID cards.

2. The same as item [1] except do it for all users in an operational entity; e.g., the command-control chain, tactical logistics, forward air bases.

3. Increase the level of effort in the USAF program (briefed to us) by a factor of 3 to get it done sooner. Alternately, pick a different factor of speedup.

4. Examine the other military services to ascertain whether corresponding programs would be effective for them, or whether variations on the USAF approach would be more sensible.

5. Implement [4] with a projected time-to-complete of X years.

6. Industrial organizations who have had serious intrusions into their systems and who appreciate the importance of protecting against them have mounted massive internal programs to make every employee aware of the issue, of individual responsibility, and of the actions being taken by the organization. Notable among such examples is Citibank.

Mount an intensive all-hands awareness program of information assurance in some/all/each of the military services. Alternately, confine the program to those organizational entities that are "closest" to the information assets and in best position to take appropriate steps if informed.

7. Survey all installed info-systems in the military structure that are based on COTS software and/or hardware. Compile a corresponding list of the known security flaws and fixes for each of them, and institute an aggressive effort to make sure that all such fixes are properly installed, tested, and made operational in (say) 18 months, and that the relevant operational staffs are also well informed and trained.

8. Make the recently published NIST Handbook of computer security required reading for all personnel associated with the operations, maintenance, installation, design, procurement and upgrade of both hardware and software in key [or: all] information systems [Alternate: do this initially for all information systems based on COTS; but later, add the embedded systems as well].

Make this handbook also required reading for every training or educational course given to military personnel.

9. Survey all acquisitions of information systems and computer-containing weapon systems now underway and take such steps as necessary to

guarantee that up-front design consideration has been given to information assurance, netsec, infosec and opsec.

10. Compile an inventory of all weapon systems that contain embedded computers and for each, define and characterize the line of responsibility, organization(s) and physical locations which support the deployed system. Hence, identify vulnerabilities and weak spots that might be exploited by an opponent; create plans to remedy these risks on a quick response basis.

11. Survey all deployed weapon systems that are computer-based with especial attention to all phases of maintenance and upgrades of software and hardware and to daily operations. The object is to identify places and means by which subversive actions could be taken to degrade or perturb weapon performance. The level of effort might be such that candidates for this examination will need to be ranked in order of importance and operational vulnerability.

12. As in item [11] but do for all support systems, whether CONUS or field deployed, that are not COTS-based but use specialized software and/or hardware.

13. As in [12] but for COTS-based systems.

14. Reconsider any/all of the prior suggestions from the point of view of likely geographic, cultural and infrastructure circumstances in which U.S. military forces might have to operate in the next (say) decade; e.g., SWA, Adriatic theater, mid-East, Korea. Object: to judge whether a different prioritization of effort would be suggested or warranted.

15. Begin an assessment of the civilian-infrastructure aspect of the issue; e.g., identify the military bases essential for an OCONUS deployment and do so for several different durations of engagement (e.g., weeks, months, years). Identify for each the present arrangements for provision of electrical power, of other energy sources, of communications—especially telephone and PSN-based, and of off-base medical, personnel, or commissary requirements.

16. As in [14], but for long-term overseas bases; e.g., Europe, Japan/Korea/Okinawa.

17. Any/all of the above for the intelligence systems (sensors, ground stations, antenna farms. electronic establishments) rather than for the operational forces and the support structure.

APPENDIX F

TECHNOLOGY ISSUES

This appendix provides background information on issues and recommendations developed by the Technology Panel of the Task Force. It is provided for reference and further discussion of the issues and recommendations contained in the basic report. Technology recommendations made by the full Task Force are included in the basic report.

TECHNOLOGY ISSUES FOR THE DSB TASK FORCE ON INFORMATION WARFARE DEFENSE

- System, Architecture, and Technology for Security and Survivability
  - 1-Security & Survivability of New and Emerging Technology
  - 2-Robust Survivable System Architecture Development and Demonstration
  - 3-COTS Information System Technology Evaluation Capability
  - 4-Maturity Models
- Modeling, Simulation, Training, and Exercises
  - 5-Training of System and Network Administrators
  - 6-Modeling and Simulation
  - 7-red Teaming
  - 8-Coordination Among Offensive IW, Defensive IW, and Intelligence
- Warning, Monitoring & Surveillance, and Damage Assessment
  - 9-National Capability for IW Indications and Warning
  - 10-Monitoring and Surveillance
  - 11-Damage Assessment
- 12-Minimum Essential Information Infrastructure (MEII)
- 13-Comprehensive Research Effort

Issues developed by the Technology Panel are presented in a set of key technology areas for Information Warfare Defense, which are grouped as shown.

SECURITY & SURVIVABILITY OF NEW AND EMERGING TECHNOLOGY

- Issue:
  - Systems Based on Current Technology are Vulnerable Due to Lack of Attention to Security and Survivability During Design and Development
- Recommendations:
  - Incorporate Information Security Early on in New Information Systems Technology Development
  - Develop and Mandate Use of Widely Accepted Robustness Standards (Cooperatively Developed by Government & Commercial Interests)
    - Components
    - Interface Standards
    - Policies, Procedures & Processes
      - Compliance Assurance
      - Configuration Management
      - Administrative Oversight
    - Operational Training
  - Require Vulnerability & Countermeasure Analysis During R&D and Systems Development

SECURITY AND SURVIVABILITY OF NEW AND EMERGING TECHNOLOGY

Current system vulnerabilities are due in part to lack of attention to security and survivability issues during design and development of computing and communicating technologies. Now that the collective vulnerability due to dependence on these technologies is recognized, it is equally important to recognize the need to address security and survivability concerns in the development of new technologies. Security and survivability must be treated as critical requirements in the conceptualization and development of new and emerging technologies. While new technology is in its earliest conceptual stages, there are unique opportunities to influence developments so as to minimize vulnerabilities and strengthen security.

Information security needs to be incorporated early on in new information systems technology development. It is essential that the government and commercial developers of products for information systems cooperate in the evolution of common standards for robust products and practices. Information security and survivability should be incorporated early on in the development of new information systems. It is recommended that significant attention be given to stimulating and encouraging this process. Areas where commonality of robustness standards and practices should be pursued include: component hardware and software products; security interfaces; system management policies, procedures, and processes addressing such issues as compliance assurance, configuration management, administrative oversight, and robust systems operational training programs.

Since a significant level of research in the information technology area is funded by the DoD, security and survivability should become required aspects of funded programs. In addition, a DoD funded research activity should be directed at vulnerability—and countermeasures-oriented analyses of new ideas and emerging technologies, and making the results widely available to the research community.

ROBUST SURVIVABLE SYSTEM ARCHITECTURE DEVELOPMENT AND DEMONSTRATION

- Issue:
  - Lack of Comprehensive, Principled, Demonstrably-effective Approach for Architecture, Design, and Analysis of Secure, Survivable Information Systems.
- Recommendations:
  - Build Upon Existing/Emerging Information and Software Engineering Principles
    - Fault Tolerant Systems
    - Trusted Systems
    - Architecture for Secure, Distributed Systems Which Can Operate While Utilizing Insecure Subnets and Subsystems
    - Consistency Mechanisms For Distributed Systems
  - Develop New Theory For Robust Systems
    - Models for Robust Systems: Attacks and Survival Techniques
    - Formal Methods for Distributed, Heterogeneous Systems
    - Analysis Techniques for Red/Blue Conflicts
  - Develop Security/Survivability Architecture Principles
    - Adaptive System-of-systems Architecture With Allocation of High-priority Task to Surviving Subsystems
    - Integration of Security Management and Systems Management
    - Accommodation of Legacy & COTS Subsystems (e.g. Via Wrappers)
  - Demonstrate Effectiveness of Survivable Architecture Principles, Theory, and Design
    - Show Experimentally in Existing and Emerging Testbeds and Networks

COTS INFORMATION SYSTEM TECHNOLOGY EVALUATION CAPABILITY

- Issues:
  - There Is Growing Risk Of Vulnerability Due To Increased Reliance On COTS Information System Products; Robustness & Security Features Not Generally a Priority For Vendors
- Recommendations:
  - Establish For DoD a COTS Information System Technology Evaluation Capability To:
    - Identify Vulnerabilities, Find Workarounds, and Disseminate Results
    - Help DoD Be an Informed Buyer
    - Understand Risks and How to Operate in Face of Risk
    - Screen for Viruses
    - Conduct Vulnerability Analyses
    - Develop Mitigation Techniques For Existing Problems
    - Evaluate Integrated Security Architectures
    - Provide Risk Assessment/Advisory Services to Users/System Developers
    - Provide Informal Rankings of COTS Information Technology Products to Create a Market Incentive For Vendors to Improve Their Products
  - Develop Long Range Plan to Migrate to a National Capability
- Note—This is an open-ended problem because the number of COTS products is growing rapidly. Funding is identified to develop the basic capability—applicator of it would be distributed.

COTS INFORMATION SYSTEM TECHNOLOGY EVALUATION CAPABILITY

Economic pressures are driving the DoD toward use of COTS information systems technology, rather than custom mil-spec systems. Unfortunately manufacturers are not motivated to develop defensive IW features in their products, since commercial customers generally are not demanding them, and such features typically impact performance. Thus the DoD must take special measures to insure that the COTS approach provides adequate DIW protection for DoD applications. It is recommended that a COTS information system technology evaluation capability be established within the DoD, in order to characterize vulnerabilities in COTS products, and to develop means for dealing with their deficiencies. Basic DIW performance/certification criteria should be developed, focusing initially on DoD needs but conforming to best commercial practices insofar as possible. a major long term goal is to foster collaboration with the commercial marketplace, and plans should be developed to migrate toward a national joint DoD/commercial technology evaluation capability, rather than unilaterally setting rigid DoD requirements that ultimately will be resisted or ignored by industry. This organization or set of organizations should identify product vulnerabilities, discover workarounds, and disseminate the results. The idea is to understand the risks and learn to operate in the face of them. Currently many DoD organizations would have to analyze these products themselves; a central facility would leverage scarce expertise and save money. Such a center could serve a role like a "Consumer's Union," and informal rankings of products could be provided, which could act as a spur to vendors to improve their products.

R&D is needed, preferably with joint government/industry support and working with both the offensive and defensive IW communities, to develop means for identifying product vulnerabilities to both established and emerging threats, disseminating information on such weaknesses, and developing corrective measures. Such a technology evaluation center should also provide risk assessment/advisory services to system developers and users, perhaps based on the current Internet model of cooperation.

Note—Implementation of this recommendation is not trivial. The intent is to develop the capability, which will undoubtedly need to be tailored for different products. The tailoring/application of the capability should be distributed. One possibility is to require that before a COTS product is incorporated into a DoD system,, the capability must have been applied. Program offices could take results from previous application of the capability or pay for its application. The expectation is that this will create a market incentive for vendors to submit (and even pay for) their system for test. The funding proposed is only for development of the capability.

MATURITY MODELS

- Issue:
  - Lack of Capture and Promotion of Best Security Practices to Guide Robustness Improvement in System Acquisition, Engineering, and Management
- Recommendations:
  - Develop Maturity Models For Robustness and Security (Build on Software & System Engineering Maturity Models)
  - Extend Acquisition Maturity Model to Include Practices For Improving Robustness of Acquired Systems
  - Develop Maturity Model For System Management Practices
  - Recognize System Management As a Readiness Issue
  - Develop Assessment Methods to Support Each Model
  - Include "Red-teaming" of The Maturity Models
  - Develop Toolkits to Aid Implementation of Practices Defined by The Models
  - Apply Models to Assess The Maturity to The Critical National Infrastructure (E.g., Telecommunications, Energy Distribution, Transportation, Etc.)

MATURITY MODELS

The trend toward increased use of commercial off-the-shelf software, open systems and wide area networks, is placing the information assets of many organizations at risk. These organizations may not be aware of the risks associated with these new environments, and may not be aware of the key engineering and network management practices that can be used to mitigate the risks. Acquisition and engineering managers lack comprehensive models and analytic techniques to evaluate the impact of architectural and other design choices on system robustness before major implementation investments have been made. Once networked systems are placed into operation, network operators often depend on practices and tools that were developed to assure the integrity of proprietary networks that had limited external connectivity and that were based on custom-designed software. Integrity assurance techniques developed for these restricted environments are not adequate for open, wide area networks or for an environment characterized by rapidly changing technologies and threats, and are typically focused on classical security issues.

Organizations that acquire and operate networked systems are in need of models, guidelines and tools that are effective at helping them acquire and operate systems that are highly resistant to attack, that are able to limit the damage from successful attacks, and that are capable of rapid recovery from attack. As missions, technology and threats evolve, these organizations also need system robustness assessment methods that allow them to adapt to the changing environment.

Models, methods, and tools should be developed and refined concurrently to insure that management practices are aligned with the technology that supports them. The areas indicated below should he addressed

Robustness Engineering Models

It is recommended that existing Software and Systems Engineering Capability Maturity Models be extended to describe the key engineering practices and technologies needed to allow organizations to specify and analyze non-functional attributes, such as security, as the system architecture and implementation design develops. The models should provide guidance on the process of analyzing robustness attributes, and making engineering tradeoffs. The models should be validated through empirical tests that demonstrate their ability to reflect the desired robustness attributes of the delivered systems.

Robust Systems Acquisition Maturity Model

It is recommended that the existing acquisition maturity model be extended to provide descriptions of the key practices acquisition organizations should follow to improve the robustness (including security) of acquired systems. This model should insure that the acquisition process specifies and monitors the use of robust system engineering practices for custom software, and specifies the use of robustness evaluation for off-the-shelf software.

Survivable Network Management Model

While it is unrealistic to expect that robust survivable systems can be achieved only through good management practices, it is widely recognized that without good management practices it will not be possible to secure critical systems against information warfare attacks. There is a need to develop and adhere to standardized policies, procedures and practices for management of information systems. Management and validation of information systems must be given the same priority as that given to achieving and maintaining readiness of weapons systems. It is recommended that a model be developed that provides a description of the key practices an organization should have in place to improve the robustness and survivability of its installed, networked information systems. The model should be structured as a set of key practices where each practice definition contains: (1) a description of the key practice; (2) a rationale for implementing the process; (3) guidance on how to implement the practices that helps organizations balance the value of the assets being protected with the costs of various forms of implementing the practice, the effectiveness of each form or practice, and the magnitude of the threat

But more than a Network Management Model is needed. Automated tools are needed which will allow integrated management of large, complex, heterogeneous networks, with automated enforcement of an organization's survivability and security management models.

Robustness Assessment Methods

It is recommended that robustness assessment methods be designed to allow an organization, with minimal outside expertise, to analyze its practices against each of the system robustness models for the purpose of identifying its current state and developing robustness/survivability improvement strategies and plans. The assessment methods must:

- be suitable for self-assessment;
- yield detailed results that tell an organization where it is, where it should be, and how to get there;
- take advantage of a knowledge base that tracks threats and vulnerabilities; and
- be self-tailoring to the organization being assessed.

Robustness Improvement Toolkits

It is recommended that robustness improvement toolkits be developed that provide the tools needed to support the assessment methods and the key practices defined by the models. Tools must be structured to encapsulate knowledge of system robustness practices to leverage scarce human resources in order to help people understand which tools to use for what purposes, and promote commercialization of the tools and a community of vendors to extend and maintain them over time.

As these models and practices evolve, it is recommended that they be applied to and evaluated for effectiveness against critical elements of the national, information-dependent infrastructure, such as energy distribution, telecommunications, and transportation systems.

TRAINING OF SYSTEM AND NETWORK ADMINISTRATORS

- Issues:
  - IW Vulnerability is Usually The Result of Human Error, Insufficient Training and Lack of Knowledge; Technology Alone is Insufficient to Close The Gap
- Recommendations:
  - Give High Priority to Training of System and Network Administrators to Form Skilled Cadre
  - Establish Recognized Career Path
    - Criteria For Selection and Certification
    - NEC/MOS/AFSC
  - Develop Infrastructure For Security Training
    - Techniques, Curriculum, Tools Test Beds
    - Exploit Simulation Technology

MODELING AND SIMULATION

- Issue:
  - Current Modeling and Simulation Efforts Do Not Include DIW
- Recommendations
  - Develop Modeling and Simulation Environments to Support DIW Training, Tactics Development, Technology & Tools Evaluation
  - Task The Defense Modelling and Simulation Office (DMSO) and J8 As Actionable Organizations to Manage This Effort
  - Specifically Include Models & Simulations of C4I System and Its Critical Components & Resources, and Modes of Failure Under IW Attack; These C4I Models Are Lacking in Current Simulation Systems
  - Integrate Whit Planning and C4i Functions to Facilitate "What If" Exercises For Planning, Operations Analysis, Mission Rehearsal and Training, and IW Gaming
  - Pursue Development of Simulations That Minimize Built-in Assumptions on Human Tactical Behavior by Including Dynamic Human Interplay
  - Leverage Off Current Distributed Interactive Simulation Efforts (Coordinate with DMSO)

RED TEAMING

- Issue:
  - DIW Red Teams Are Not Used Routinely in Operations & Exercises
- Recommendations:
  - Establish Red Teaming & Anti-red Teaming As Key Components of DIW Technology and Strategy Development Process
    - Conduct Red Team Exercises Under Properly-constituted Rules of Engagement to Avoid Unnecessary Damage/disruption
    - Include Vulnerability Analysis &Robustness Engineering As Well As in Attacks
    - Provide Vulnerability Analysis to The Anti-red Team
    - Spectrum of Attack Should Include:
      - Deception, Destruction, Corruption, As Well As Exploitation
      - Software and Database Attacks As Well As Comm/Jamming Attacks
  - Develop New Attack Methodologies in Addition to Applying Known Techniques
  - Develop Anti-red Team Tactics/SOPS
  - Must Distinguish Red Team Penetrations From Real Penetrations

RED TEAMING

Red Teaming is an essential component of the DIW strategy and technology development process, but it is recommended that the concept be extended to include vulnerability analyses as well as carefully planned attacks during experimental activities in controlled testbeds and during training/planning exercises. The Red Team exercises should be conducted under proper rules of engagement to avoid unnecessary damage or disruption to information systems. The vulnerability analyses should be utilized by an Anti-Red Team to perform robustness engineering and to plan for fighting the Information War during the exercises as well as during operations.

Emphasis should be given to developing new attack methodologies in addition to reuse and application of current attacker techniques. For example, attacks should be designed which exploit the system's survivability features, which must be assumed known to a sophisticated attacker. In formulating these attack strategies, models should first be developed for system vulnerability and its likely defenses, and these models should be exploited in the attack strategies. Vulnerability analyses and Red Team attacks should be conducted at the application and system level, as well as at the subsystem level, with the goal of uncovering how operations can be perturbed (e.g., the planning and execution of an air tasking order or the deployment of sensors and communication assets), and how supporting communication links, or specific computers and network nodes can be compromised.

In addition to Red Teams, it is recommended that Anti-Red Teams (DIW Teams) be formed and tasked to prepare for and fight Red Team attacks. These activities will provide the basis for developing strategies and tools for use during operations to detect and respond to Information Warfare attacks. The Anti-Red Team should also be charged with providing inputs to the system designers and builders to assure the incorporation of robustness features. Network managers should be included as part of the DIW teams to assure that damage containment and service restoral techniques are effectively exercised as part of the counter-IW operation.

COORDINATION AMONG OFFENSIVE IW, DEFENSIVE IW, AND INTELLIGENCE

- Issue:
  - DIW Vulnerability Assessments Could Be Significantly Improved Through Dynamic Interplay With The IW Offense Side
  - Lack of Information Sharing Among Offense, Defense, and Intelligence Introduces Unnecessary IW Vulnerabilities
  - IW Offense, DIW, and Intelligence Could All Benefit From Increased Coordination
- Recommendations:
  - Require and Motivate Offense and Intelligence Units to Play a Role in Evaluating Vulnerability Assessments
  - Promote Dynamic Interplay Among Offensive & Defensive Sides and Intelligence, To:
    - Evaluate Emerging DIW Technologies & Practices;
    - Help The Offense Hone Its Tactics and Techniques and Anticipate Potential Counter-IW Measures
  - Establish an Independent Orange Team, Including Offense, Defense, and Intelligence, to Facilitate Coordination
  - Secure Channels Will Be Needed to Protect Highly Sensitive Information

COORDINATION AMONG OFFENSIVE IW, DEFENSIVE IW, AND INTELLIGENCE

In any adversarial situation, the offensive side enjoys an inherent advantage over the defensive side, since they own the initiative, and need discover and exploit only one flaw in a defensive perimeter to achieve their ends. In contrast, the defensive side must attempt to foresee and protect against every conceivable form of assault. The situation prevails in the IW context, and it is clear that the defensive side would greatly benefit in developing risk management strategies and protective mechanisms from an awareness of the offensive side's view of the vulnerability profile and preferred set of attack techniques and approaches. The friendly offensive IW team would also benefit by interplay with its DIW allies, which would enable the offense to hone its tactics and techniques and anticipate potential counter-IW tactics. In addition, intelligence units should work together with both offense and defense to increase overall effectiveness of IW and to avoid unnecessary vulnerabilities.

It is recommended that a mechanism and a forum be developed to support a continuing, dynamic dialogue and interplay among the offensive, defensive, and intelligence communities. This activity should include information exchange on new vulnerability discoveries and attack techniques on the one hand, and red team vulnerability assessments of emerging defensive techniques and technologies on the other hand.

To provide an objective mechanism for facilitating and coordinating this dynamic interplay, an independent "ORANGE" team could be formed, devoid of vested interests on either the offensive, defensive, or intelligence sides. The orange team can also play the role of umpire and objective score keeper in red-team/vulnerability assessment war gaming activities. In general this type of 3-way interaction will lead to a better understanding of the fundamental exploitable flaws typically occurring in system and communication software, distributed system architecture, communications infrastructure, and system management policies and procedures. This will also lead to new tools to address these particular areas of weakness, such as a tool for scanning developmental software to uncover design and/or implementation flaws, and leading ultimately to more reliable, robust end products.

NATIONAL CAPABILITY FOR IW INDICATIONS AND WARNING

- Issue:
  - No National Capability Exists to Detect and Assess Possible Attacks in Progress Or Preparation
- Recommendations:
  - Establish National Capability to Perform IW Indications and Warning
    - Must Be Broad-based, With Industry/Private/Government Cooperation (No Dominant Interests)
    - Scalable, With Interfaces to Outside World
    - Create a Dod Capability as Part of a Broader System to Include Private Sector, Other Government Agencies, and International Agencies
  - Consider Interconnection of Existing Certs as Initial Pilot Program
  - Conduct Research and Development in Underlying Theory and Techniques For Detection and Analysis

NATIONAL CAPABILITY FOR IW INDICATIONS AND WARNING

At present, no integrated, national capability exists to detect and assess possible IW attacks in progress or preparation. Several civilian and military computer emergency response team centers have evolved, however, to provide expert diagnosis and recovery assistance for computer systems and networks that have been attacked and seriously affected by hostile actions. There is a need to detect and counter such attacks before costly information corruption and network damage occurs.

It is recommended that a National Capability Center for Indications and Warning be formed, capable of continuously gathering and analyzing monitoring data derived from government as well as commercial infrastructure systems. The center should be charged with searching for and detecting early signs and precursors of a wide scale, coordinated attack and providing warnings to U.S. government and private sector organizations. Towards that end, a phased approach is recommended, beginning with a DoD-specific organization which is scalable and extensible, and evolving towards a pan government and private sector organization. Roles of the organization should include gathering and analysis of voluntarily contributed data, dissemination of findings, and acting as a clearing house to coordinate feedback and responses from the community. The center should also act as focal point for conducting R&D on techniques and tools for attack detection and analysis. As an initial, limited scale pilot program, an interconnection of existing DoD emergency response centers should be considered.

MONITORING AND SURVEILLANCE

- Issue:
  - Technology to Measure and Monitor Penetrations and Wide Scale Attacks on The NII is Inadequate
- Recommendations
  - Develop Automated, Distributed, Collaborative Monitoring and Surveillance Strategy
  - Establish Broad-based R&D Effort To:
    - Create Tools to Filter Network Audit Data
    - Create Tools to Discriminate Between Normal and Abnormal Behavior, Easily Extendible For Changing Threats
    - Develop Automated, Distributed, Cooperative Techniques For Correlating and Exploiting Data Across Multiple Sites
  - Develop, Evaluate and Transition Intrusion Detection Technology to Critical Infrastructure Systems
  - Develop, Evaluate and Transition Intrusion Detection Technology to Critical Infrastructure Systems
    - Develop Techniques For and Investigate Use of Cooperating Intrusion Detection Systems in Large Heterogeneous Networks
    - Develop Analysis and Evaluation Techniques For Intrusion Detection Systems
    - Develop Models to Characterize IW Attacks
    - Develop Techniques For Automated Response to IW Attacks

MONITORING AND SURVEILLANCE

Current technology to detect, monitor and characterize local penetrations and wide scale attacks on the National Information Infrastructure (NII) is inadequate. a wide scale, coordinated, multifaceted IW attack on the national information-dependent infrastructure represents a major distributed measurement and analysis challenge. In order to detect attacks of such scale and likely degree of subtlety, it will be necessary to extract and correlate data across many sites, since measurements at any single site may not be sufficient to reveal the emerging overall pattern. The types of attack mounted may involve techniques and degrees of sophistication beyond simple, standard intrusion detection tactics.

It is recommended that an investment be made in developing a distributed monitoring and surveillance strategy for large scale networks, along with an associated set of supporting network architectural and instrumentation principles. Further, it is recommended that a broad based research and development effort be established to develop: 1) flexible, field modifiable, trainable tools to leverage human network and security administrators in filtering network audit data, discriminating between normal and abnormal behavior, and recognizing network attacks; 2) applied pattern recognition techniques (e.g., statistical model based, or neural net) capable of adaptation, learning and coping with temporal pattern sequences; and 3) techniques and strategies for automated, collaborative, distributed pattern recognition and problem solving, supporting the correlation and exploitation of data gathered across multiple sites in a large scale network.

There is a critical need to develop, evaluate and transition intrusion detection technology and methodology to critical infrastructure systems, in particular telecommunication systems. To meet this need it is recommended that significant R&D efforts be focused in 1.) development and investigation of techniques for cooperative intrusion detection in large scale heterogeneous networks characterized by different transmission speeds, different networking technologies and various security policies 2) development of analysis and evaluation techniques and standard metrics for intrusion detection systems 3.) development of models to characterize IW attacks 4.) development of techniques for automatic response to incoming attacks. Such techniques should include strategies for degraded modes of operation, determining attack origin and restoral of services.

Specific emphasis should be given to demonstrating and transitioning results of the development efforts to the telecommunications and other critical infrastructure industries. This will involve adapting technologies for specific environments and evaluating and demonstrating performance in realistic scenarios and testbeds.

DAMAGE ASSESSMENT

- Issue:
  - Current Damage Assessment Techniques Are Inadequate and Lead to Over Or Under Reaction to Information Warfare Attacks
- Recommendations:
  - Develop Comprehensive Damage Assessment Techniques/Measures That Assess Lost Information and Services, and Corrupted Information of Software
  - Develop Tools to Support Assessment Techniques
    - Develop Multiple Levels of Damage Assessment Tools: Application, Data, Middleware, Networks
    - Develop Secure Logging Tools
    - Develop Tools That Identify:
      - Scope & Time Frames If Intrusions
      - Fault/intrusion Locations
      - Scope, Time Frame and Impact of Service Losses
      - Corrupted Data and Software
      - Data, Software, and Systems Which Are Intact

DAMAGE ASSESSMENT

In order to determine the appropriate response for a detected attack, it is important to correctly assess the associated damage. Failure to correctly assess damage could lead to costly over reaction (e.g., removing operational systems from service and/or unnecessarily rebuilding software and data bases), or dangerous under reaction (e.g., attempts to continue operations with corrupted data and software). Currently, there are no proven methods for reliably assessing the extent and nature of damage associated with information warfare attacks.

It is recommended that research and associated tool development be pursued with the objective of producing acceptable measures and techniques for damage assessment of both technological and business assets. These tools need to be able to assess damage at multiple levels, from application to networks, and to coalesce the results of the assessments at these levels. It is also recommended that secure logging tools and standard instrumentation packages for damage assessment be developed, which can be provided to all DoD sites where they are needed. Attention will have to be paid to adequately protecting such logs from tampering by an intruder. In addition, to coherently understand and deal with existing and potential future damage, it should be part of damage assessment to locate fault/intrusion sites for containment and purging purposes.

An important sub-problem in damage assessment is to identify information system components which remain undamaged and operational. Those components must be used to continue operations, as well as to help in the damage assessment process. Reliable damage assessment methods are needed for the information warfare communities and for other government and business interests, for a wide range of threats.

MINIMUM ESSENTIAL INFORMATION INFRASTRUCTURE (MEII)

- Issue:
  - The Current Information Infrastructure is Vulnerable to IW Attacks; a Means For Restoring and Establishing MEII is Needed
- Recommendations:
  - Define a Concept For a MEII:
    - Define Minimum Essential Services
    - Develop Secure Gateways to Hardcore Network
    - Develop a Network Management Architecture
  - Develop a Restoration Strategy For The MEII Based on The Use of a Hardcore Network (e.g., Milstar) to Exchange Critical Information in Order to Restore Connectivity
  - Conduct Prototype MEII Exercises

MINIMUM ESSENTIAL INFORMATION INFRASTRUCTURE

The current information infrastructure which supports telecommunications, power, transportation, etc., is susceptible to IW attacks, and in particular to wide scale coordinated attacks aimed at disabling or disrupting government as well as commercial systems. It is recommended that a strategy and overall architecture concept be developed for a minimum essential information infrastructure which can serve as a means for restoring services and adapting to wide scale outages. The technical feasibility of using Milstar assets as a means for determining available connectivity and providing modest but critical packet data service for exchange of routing, node status, and other essential network management information, should be investigated. In this role, Milstar would be supplemented with available commercial resources as possible and as needed.

The concept should consider the applications and deployment of secure gateways connected to Milstar ground station equipment and re-allocated Milstar assets as a hardcore network for use in restoring critical connectivity. The authentication of commercial wireline and wireless network access through the gateway to the hardcore network is a critical issue which must be addressed.

In addition to an overall MEII architectural concept, minimum essential services, an operational concept, and a management structure must be developed. a strategy must be developed for transitioning from peacetime or normal operational activities to the minimum essential information infrastructure. It will be important to execute the transition strategy in the context of exercises.

This activity spans government, industry, and private interests, therefore it is recommended that an organization like NSTAC be commissioned to develop and define the concept of an MEII in cooperation with the DoD.

COMPREHENSIVE RESEARCH EFFORT

- Issues:
  - A Comprehensive, Unified R&D Effort is Needed in Architecture, Analysis, and Synthesis of Survivable Information Systems, Similar to the Earlier Investment Which Established U.S. Preeminence in Cryptography
  - Previous R&D Efforts Have Focussed Separately on Specific Areas (e.g. Computer Security, Encryption, Operating Systems)
- Recommendations:
  - Establish a Comprehensive Research Effort to Significantly Advance the State-of-the-art in Theory, Analysis, & Science for High Assurance Systems
    - Develop Rigorous Mathematical Approaches for Analyzing and Synthesizing Complex Information Systems
    - Develop Advanced Modelling & Analysis Techniques Building Upon, But Extending Beyond, Prior Research in Formal Methods; Include a Focus on Formal Methods Which Can Cross Layers of Abstraction in a Large-scale System Design
    - Develop Techniques for System Synthesis, and for Predicting and Evaluating Performance; Include Formal Approaches to Design of Appropriate System Tests
  - Establish a Brad-based R&D Effort Focussed on the Design, Monitoring, and Management of Large Scale Distributed Systems, Including:
    - Architectures, Design Tools, & Methodologies for Robust Survivable Distributed Systems
    - Techniques & Tools for Monitoring & Managing Large-scale Distributed/Networked Systems
    - Techniques for Detecting Local Or Large-scale Attacks, and for Adaptation to Support Graceful Degradation
    - Testbeds and Simulation-based Mechanisms for Evaluating Emerging DIW Technology and Tactics
  - Incentivize Industry and Academia to Participate in Broad-based R&d Efforts
  - Establish a Cross-government Effort to Coordinate DIW Research and Deployment Efforts

COMPREHENSIVE RESEARCH EFFORT

The development of robust survivable distributed systems resistant to information warfare attack, as well as other types of failure, requires major advances in theory, modeling and technology, and the combined efforts of a vigorous research community embracing academia, industry and government. Prior R&D efforts have focused on specific areas, such as computer and network security, encryption technology, operating system environments with multi-level security features, and coping with benign network outages caused by single node failures, etc. Little attention has been paid to the *ab initio* design and implementation of systems capable of surviving willful malicious attack, or detecting and tolerating corrupted software. Even less attention has been paid to the non-*ab-initio* case, where the system must incorporate legacy subsystems which are not under the designer's control. a comprehensive research effort is required, similar to the earlier investment in cryptographic theory, higher mathematics and associated technology, which led to U.S. preeminence in cryptography. The area of robust survivable systems offers an opportunity for a unifying theme to constitute a broad-based research effort covering the full range of 6.1, 6.2, 6.3 research, to stimulate fresh and/or revolutionary ideas and comprehensive problem solutions.

A fundamental and essential underpinning of any proposed technology base for designing and implementing large scale, robust, survivable distributed systems is a science and associated suite of design technologies for high-confidence/high assurance systems. Ideally such a set of tools would afford designers and implementers a means for describing, constructing and verifying the anticipated behavior of a complex system at all levels of abstraction. These design technologies must be capable of capturing behavioral descriptions, system properties and design descriptions in ways which enable the timely creation and performance validation of a given system implementation. Such a capability is needed because it is impossible to either anticipate or exercise all possible interactions among the large number of constituent elements and subsystems that typically comprise any system of meaningful, real world complexity. These technologies should be bound together by a unifying, fundamental mathematical logic which would allow an integrated treatment of all hierarchies of a complex system design, from logic gates to networks of computer networks.

APPENDIX G

LIST OF ACRONYMS

ABIS	Advanced Battlefield Information System
ACTD	Advanced Concepts Technology Demonstration
Active X	See Appendix H, Glossary
Arch	Architecture
ASD(C3I)	Assistant Secretary of Defense for Command, Control, Communications and Intelligence
ATD	Advanced Technology Demonstration
C2	Command and Control
C3	Command, Control and Communications
C3I	Command, Control, Communications and Intelligence
C4I	Command and Control, Communications, Computers and Intelligence
C4ISR	Command and Control, Communications, Computer Intelligence, Surveillance and Reconnaissance
CDR	Commander (USN designation of rank)
CIA	Central Intelligence Agency
CINC	Commander in Chief
CIO	Chief Information Officer
CIP	Critical Infrastructure Protection
CJCS	Chairman, Joint Chiefs of Staff
Conv.	Conventional
CONUS	Continental United States
Coord.	Coordination
CSA	CINCs/Service/Agencies
CSAAS	Combat Support Agency Assessment System
CSPAR	CINCs Preparedness Assessment Report
Ctr	Center

DASD	Deputy Assistant Secretary of Defense
DCI	Director of Central Intelligence
DEPSECDEF	Deputy Secretary of Defense
Des	Design
DIA	Defense Intelligence Agency
DII	Defense Information Infrastructure
DIS	Defense Investigative Service
DISA	Defense Information Systems Agency
DoC	Department of Commerce
DoD	Department of Defense
DoDD	Department of Defense Directive
DoE	Department of Energy
DoJ	Department of Justice
DoT	Department of Transportation
EEI	Essential Elements of Information
FBI	Federal Bureau of Investigation
FEMA	Federal Emergency Management Agency
GAO	Government Accounting Office
GII	Global Information Infrastructure
HUMINT	Human Intelligence
I&W/TA	Indication and Warning/Threat Assessment
IC	Intelligence Community
Info.	Information
Intel	Intelligence
IT	Information Technology

IW	Information Warfare
IW-D	Information Warfare-Defense
JAVA	See Appendix H, Glossary
JWCA	Joint Warfare Analysis Center
MEII	Minimum Essential Information Infrastructure
Mil Deps	Military Departments
NCS	National Communications System
NEC	National Economic Council
NII	National Information Infrastructure
NRC	National Research Council
NSA	National Security Agency
NSC	National Security Council
NSIE	Network Security Information Exchange
NSTAC	National Security Telecommunications Advisory Board
Nuc.	Nuclear
OCONUS	Outside of CONUS
Off	Office
OMB	Office of Management and Budget
Ops	Operations
OSTP	Office of Science and Technology Policy
OUSD (A&T)	Office of the USD (A&T)
OUSD(P)	Office of the USD (P)
Plan	Planning
PSA	Principle Staff Assistant
PSN	Public Switched Network

Ret.	Retired
SECDEF	Secretary of Defense
SORTS	Status of Resources and Training System
TOR	Terms of Reference
Treas	Department of the Treasury
U.S.	United States
USAF	United States Air Force
USD (A&T)	Under Secretary of Defense for Acquisition and Technology
USD (C)	Under Secretary of Defense, Comptroller
USD (P&R)	Under Secretary of Defense for Personnel and Readiness
USD (P)	Under Secretary of Defense for Policy
USN	United States Navy
VADM	Vice Admiral
WARM	War-time Mode

APPENDIX H

GLOSSARY

Source: Joint Pub 1-02, Department of Defense Dictionary of Military and Associated Terms, 23 March 1994. The DOD Dictionary of Military and Associated Terms," is promulgated for mandatory use by the Office of the Secretary of Defense, Military Departments, Joint Staff, combatant commands, and Defense agencies. Those terms approved for both DOD and NATO use are marked with an asterisk within parentheses, i.e.. (*). Other sources are indicated by brackets, e.g., [CJCSI 3210.01, 1996].

acoustic warfare(*)—Action involving the use of underwater acoustic energy to determine, exploit, reduce or prevent hostile use of the underwater acoustic spectrum and actions which retain friendly use of the underwater acoustic spectrum. There are three divisions within acoustic warfare: 1. acoustic warfare support measures. That aspect of acoustic warfare involving actions to search for, intercept, locate, record and analyze radiated acoustic energy in water for purpose of exploiting such radiations. The use of acoustic warfare support measures involves no intentional underwater acoustic emission and is generally not detectable by the enemy. 2. acoustic warfare countermeasures. That aspect of acoustic warfare involving actions taken to prevent or reduce an enemy's effective use of the underwater acoustic spectrum. Acoustic warfare countermeasures involve intentional underwater acoustic emissions for deception and jamming. 3. acoustic warfare counter-countermeasures. That aspect of acoustic warfare involving actions taken to ensure friendly effective use of the underwater acoustic spectrum despite the enemy's use of underwater acoustic warfare. Acoustic warfare counter-countermeasures involve anti-acoustic warfare support measures and anti-acoustic warfare countermeasures, and ma not involve underwater acoustic emissions.

acoustic warfare counter-countermeasures—See acoustic warfare Part 3.

acoustic warfare countermeasures—See acoustic warfare Part 2.

acoustic warfare support measures—See acoustic warfare Part 1.

active air defense(*)—Direct defensive action taken to nullify or reduce the effectiveness of hostile air action. It includes such measures as the use of aircraft, air defense weapons, weapons not used primarily in an air defense role, and electronic warfare. See also air defense.

Active X—A name for a version of Distributed Object Linking and Embedding (OLE) that enables sharing of data, links, and controls over a network (primarily for interoperability among Windows-based software).

anti-air warfare—A U.S. Navy/U.S. Marine Corps term used to indicate that action required to destroy or reduce to an acceptable level the enemy air and missile threat. It includes such measures as the use of interceptors, bombers, antiaircraft guns, surface-to-air and air-to-air missiles, electronic attack, and destruction of the air or missile threat both before and after it is launched. Other measures which are taken to minimize the effects of hostile air action are cover, concealment, dispersion, deception (including electronic), and mobility. See also counter air.

antisubmarine operation—Operation contributing to the conduct of antisubmarine warfare.

antisubmarine warfare(*)—Operations conducted with the intention of denying the enemy the effective use of submarines.

attack assessment—An evaluation of information to determine the potential or actual nature and objectives of an attack for the purpose of providing information for timely decisions. See also damage estimation.

biological operation(*)—Employment of biological agents to produce casualties in personnel or animals and damage to plants or materiel; or defense against such employment

biological operation(*)—Employment of biological agents to produce casualties in personnel or animals and damage to plants or materiel; or defense against such employment. biological warfare—See biological operation.

C2-protection—See command and control warfare.

chemical warfare—All aspects of military operations involving the employment of lethal and incapacitating munitions/agents and the warning and protective measures associated with such offensive operations. Since riot control agents and herbicides are not considered to be chemical warfare agents, those two items will be referred to separately or under the broader term "chemical," which will be used to include all types of chemical munitions/agents collectively. The term "chemical warfare weapons" may be used when it is desired to reflect both lethal and incapacitating munitions/agents of either chemical or biological origin. Also called CW. See also chemical operations, herbicide, riot control agent.

combined warfare—Warfare conducted by forces of two or more allied nations in coordinated action toward common objectives.

command and control warfare—The integrated use of operations security (OPSEC), military deception, psychological operations (PSYOP), electronic warfare (EW), and physical destruction, mutually supported by intelligence, to deny information to, influence, degrade, or destroy adversary command and control capabilities, while protecting friendly command and control capabilities against such actions. Command and control warfare applies across the operational continuum and all levels of conflict. Also called C2W. C2W is both offensive and defensive: a. counter-C2—To prevent effective C2 of adversary forces by denying information to, influencing, degrading, or destroying the adversary C2 system. b. C2-protection—To maintain effective command and control of own forces by turning to friendly advantage or negating adversary efforts to deny information to, influence, degrade, or destroy the friendly C2 system. See also command and control; electronic warfare; intelligence; military deception; operations security; psychological operations.

counterguerrilla warfare(*)—Operations and activities conducted by armed forces, paramilitary forces, or nonmilitary agencies against guerrillas.

damage estimation—A preliminary appraisal of the potential effects of an attack. See also attack assessment.

directed-energy protective measures—That division of directed-energy warfare involving actions taken to protect friendly equipment, facilities, and personnel to ensure friendly effective uses of the electromagnetic spectrum that are threatened by hostile directed-energy weapons and devices.

directed-energy warfare—Military action involving the use of directed-energy weapons, devices, and countermeasures to either cause direct damage or destruction of enemy equipment, facilities, and personnel, or to determine, exploit, reduce, or prevent hostile use of the electromagnetic spectrum through damage, destruction, and disruption. It also includes actions taken to protect friendly equipment, facilities, and personnel and retain friendly use of the electromagnetic spectrum. Also called DEW. See also directed energy; directed-energy device; directed-energy weapon; electromagnetic spectrum; electronic warfare.

directed-energy weapon—A system using directed energy primarily as a direct means to damage or destroy enemy equipment, facilities, and personnel. See also directed energy; directed energy device. economic warfare—Aggressive use of economic means to achieve national objectives.

electromagnetic intrusion—The intentional insertion of electromagnetic energy into transmission paths in any manner, with the objective of deceiving operators or of causing confusion. See also electronic warfare.

electronic warfare—Any military action involving the use of electromagnetic and directed energy to control the electromagnetic spectrum or to attack the enemy. Also called EW. The three major subdivisions within electronic warfare are: electronic attack, electronic protection, and electronic warfare support. a. electronic attack—That division of electronic warfare involving the use of electromagnetic or directed energy to attack personnel, facilities, or equipment with the intent of degrading, neutralizing, or destroying enemy combat capability. Also called EA. EA includes: 1) actions taken to prevent or reduce an enemy's effective use of the electromagnetic spectrum, such as jamming and electromagnetic deception, and

2) employment of weapons that use either electromagnetic or directed energy as their primary destructive mechanism (lasers, radio frequency weapons, particle beams). b. electronic protection—That division of electronic warfare involving actions taken to protect personnel, facilities, and equipment from any effects of friendly or enemy employment of electronic warfare that degrade, neutralize, or destroy friendly combat capability. Also called EP. c. electronic warfare support—That division of electronic warfare involving actions tasked by, or under direct control of, an operational commander to search for, intercept, identify, and locate sources of intentional and unintentional radiated electromagnetic energy for the purpose of immediate threat recognition. Thus, electronic warfare support provides information required for immediate decisions involving electronic warfare operations and other tactical actions such as threat avoidance, targeting, and homing. Also called ES. Electronic warfare support data can be used to produce signals intelligence (SIGINT), both communications intelligence (COMINT), and electronics intelligence (ELINT). See also command and control warfare; communications intelligence; directed energy; directed-energy device; directed-energy warfare; directed-energy weapon; electromagnetic compatibility; electromagnetic deception; electromagnetic hardening; electromagnetic jamming; electromagnetic spectrum; electronics intelligence; frequency deconfliction; signals intelligence; spectrum management; suppression of enemy air defenses.

guerrilla warfare(*)—Military and paramilitary operations conducted in enemy-held or hostile territory by irregular, predominantly indigenous forces. See also unconventional warfare.

indications and warning—Those intelligence activities intended to detect and report time-sensitive intelligence information on foreign developments that could involve a threat to the United States or allied military, political, or economic interests or to U.S. citizens abroad. It includes forewarning of enemy actions or intentions; the imminence of hostilities; insurgency; nuclear/non-nuclear attack on the United States, its overseas forces, or allied nations; hostile reactions to United States reconnaissance activities; terrorists' attacks; and other similar events.

information warfare—Actions taken to achieve information superiority by affecting adversary information, information-based processes, information systems, and computer-based networks while defending one's own information, information-based processes, information systems, and computer-based networks. [CJCSI 3210.01, 1996]

integrated warfare—The conduct of military operations in any combat environment wherein opposing forces employ non-conventional weapons in combination with conventional weapons.

JAVA—An object-oriented, platform-independent programming language, often used to create small cross-program executable software applications called applets that are downloaded from remote sites and that execute automatically.

mine warfare—The strategic, operational, and tactical use of mines and mine countermeasures. Mine warfare is divided into two basic subdivisions: the laying of mines to degrade the enemy's capabilities to wage land, air, and maritime warfare; and the countering of enemy-laid mines to permit friendly maneuver or use of selected land or sea areas.

naval coastal warfare—Coastal sea control, harbor defense, and port security, executed both in coastal areas outside the United States in support of national policy and in the United States as part of this Nation's defense. Also called NCW.

naval special warfare—A specific term describing a designated naval warfare specialty and covering operations generally accepted as being unconventional in nature and, in many cases, covert or clandestine in character. These operations include using specially trained forces assigned to conduct unconventional warfare, psychological operations, beach and coastal reconnaissance, operational deception operations, counterinsurgency operations, coastal and river interdiction, and certain special tactical intelligence collection operations that are in addition to those intelligence functions normally required for planning and conducting special operations in a hostile environment. Also called NSW.

nuclear warfare(*)—Warfare involving the employment of nuclear weapons. See also post-attack period; trans-attack period.

operations security—A process of identifying critical information and subsequently analyzing friendly actions attendant to military operations and other activities to: a. Identify those actions that can be observed by adversary intelligence systems. b. Determine indicators hostile intelligence systems might obtain that could be interpreted or pieced together to derive critical information in time to be useful to adversaries. c. Select and execute measures that eliminate or reduce to an acceptable level the vulnerabilities of friendly actions to adversary exploitation. Also called OPSEC. See also command and control warfare; operations security indicators; operations security measures; operations security planning guidance; operations security vulnerability.

perception management—Actions to convey and/or deny selected information and indicators to foreign audiences to influence their emotions, motives, and objective reasoning; and to intelligence systems and leaders at all levels to influence official estimates, ultimately resulting in foreign behaviors and official actions favorable to the originator's objectives. In various ways, perception management combines truth projection, operations security, cover and deception, and psychological operations. See also psychological operations. political warfare—Aggressive use of political means to achieve national objectives.

psychological operations—Planned operations to convey selected information and indicators to foreign audiences to influence their emotions, motives, objective reasoning, and ultimately the behavior of foreign governments, organizations, groups, and individuals. The purpose of psychological operations is to induce or reinforce foreign attitudes and behavior favorable to the originator's objectives. Also called PSYOP. See also perception management.

psychological warfare—The planned use of propaganda and other psychological actions having the primary purpose of influencing the opinions, emotions, attitudes, and behavior of hostile foreign groups in such a way as to support the achievement of national objectives. Also called PSYWAR.

strategic air warfare—Air combat and supporting operations designed to effect, through the systematic application of force to a selected series of vital targets, the progressive destruction and disintegration of the enemy's war-making capacity to a point where the enemy no longer retains the ability or the will to wage war. Vital targets may include key manufacturing systems, sources of raw material, critical material, stockpiles, power systems, transportation systems, communication facilities, concentration of uncommitted elements of enemy armed forces, key agricultural areas, and other such target systems.

tactical warning—I. A warning after initiation of a threatening or hostile act based on an evaluation of information from all available sources. 2. In satellite and missile surveillance, a notification to operational command centers that a specific threat event is occurring. The component elements that describe threat events are: Country of origin—country or countries initiating hostilities. Event type and size—identification of the type of event and determination of the size or number of weapons. Country under attack—determined by observing trajectory of an object and predicting its impact point. Event time—time the hostile event occurred. Also called integrated tactical warning. See also attack assessment; strategic warning.

tactical warning and assessment—A composite term. See separate definitions for tactical warning and for attack assessment.

unconventional warfare—A broad spectrum of military and paramilitary operations, normally of long duration, predominantly conducted by indigenous or surrogate forces who are organized, trained, equipped, supported, and directed in varying degrees by an external source. It includes guerrilla warfare and other direct offensive, low visibility, covert, or clandestine operations, as well as the indirect activities of subversion, sabotage, intelligence activities, and evasion and escape. Also called UW.

John Gibbon's Testitimony on Technology Policy

The White House
Office of Science and Technology Policy

FOR IMMEDIATE RELEASE **April 27, 1993**

Statement of John H. Gibbons
Director, Office of Science and Technology Policy
before the
Committee on Science, Space, and Technology
U.S. House of Representatives

April 27, 1993

INFORMATION INFRASTRUCTURE AND HR1757, THE "HIGH PERFORMANCE COMPUTING AND HIGH SPEED NETWORKING APPLICATIONS ACT OF 1993"

Mr. Chairman, Members of the Committee, thank you for this opportunity to testify on the Administration's High Performance Computing and Communications Program and its important role in assisting the development of the National Information Infrastructure.

This information infrastructure consists of computers, computer data banks, fax machines, telephones, and video displays linked by high-speed telecommunication links capable of transmitting billions of bits of information in a second—an entire Encyclopedia Britannica in a few seconds. The computing and networking technology that makes this possible is improving at an unprecedented rate, expanding both our imaginations for its use and its effectiveness. Using these technologies, a doctor who needs a second opinion could transmit a patient's entire medical record—x-rays and ultrasound scans included—to a colleague thousands of miles away, in less time that it takes to send a fax today. A school child in a small town could come home and through a personal computer, reach into an electronic Library of Congress—thousands of books, records, videos, and photographs, all stored electronically. At home, viewers could choose whenever they wanted from thousands of different television programs or movies.

As you know, the Administration is committed to accelerating the development of the National Information Infrastructure (NII) this Nation needs for the 21st Century. This infrastructure will provide Americans the information they need, when they need it, whether in the form of text, images, video, or sound. These "information superhighways" will revolutionize the way we work, learn, shop, and live. They promise to have an even greater impact than the interstate freeways or the telephone system. This infrastructure will be as ubiquitous as the telephone system, but will be able to carry information at least 1,000 times faster. It will be able to transmit not only voice and fax, but will also provide hundreds of channels of interactive high-definition TV programming, teleconferencing, and access to huge volumes of information.

This technology is already in use in many of our research laboratories where it is transforming the way research is done. It allows scientists and engineers to access information from computer databases scattered throughout the country and enables them to use supercomputers and research equipment thousands of miles away. Perhaps most importantly, it enables researchers to collaborate with colleagues around the country and around the world almost as easily as if they were in the same building.

This same telecommunications and computing technology could someday be available to all Americans, provided there is adequate public and private investment and forward-looking government policies that promote its deployment and use.

The Clinton Administration believes that the Federal government has several important roles to play in assisting the development of this infrastructure, which will be built and run primarily by the private sector. In many ways, the High Performance Computing and Communications (HPCC) Program provides the technological foundation upon which the Administration's strategy for the NII rests. On February 22, the President and the Vice President unveiled a Technology Initiative which outlined the five parts of the Administration's strategy for building the National Information Infrastructure:

> 1) Implement the High-Performance Computing and Communications Program, which is helping develop the basic technology needed for the NII. 2) Through the Information Infrastructure Technology and Applications program, work with industry, universities, and Federal labs to develop technologies needed to effectively utilize the NII for a wide range of applications. 3) Provide funding for networking pilot projects through the National Telecommunications and Information Administration (NTIA) of the Department of Commerce. NTIA will provide matching grants to assist states, local governments, universities and school systems, hospitals and other health care providers, and other non-profit entities in purchasing equipment and in undertaking planning related to telecommunications infrastructure development. These pilot projects will demonstrate and explore the benefits of networking in the educational and library communities. 4) Promote dissemination of Federal information. Every year, the Federal government spends billions of dollars collecting and processing information (e.g. economic data, environmental data, and technical information). Unfortunately, while much of this information is very valuable, many potential users either do not

know that it exists or do not know how to access it. The Administration is committed to using new computer and networking technology to make this information more available to the taxpayers who paid for it. This will require consistent Federal information policies designed to ensure that Federal information is made available at a fair price to as many users as possible while encouraging the growth of the information industry. 5) Reform telecommunications policies. Government telecommunication policy has not kept pace with new developments in telecommunications and computer technology. As a result, government regulations have tended to inhibit competition and delay deployment of new technology and services. For instance, without a consistent, stable regulatory environment, the private sector will hesitate to make the investments necessary to build the high- speed national telecommunications network that this country needs to compete successfully in the 21st Century. To address this problem and others, the Administration is creating a high-level, interagency Information Infrastructure Task Force at the White House that will work with Congress, the private sector, and state and local governments to find consensus on and implement policy changes needed to accelerate deployment of the NII.

As you can see, the HPCC Program is a critical part of the Administration's effort to build the NII. It is a key part of a comprehensive strategy that will not only develop and demonstrate new information technology, but also ensure that we have intelligent, forward-looking policies that encourage the private sector to deploy it and the public to use it. Over the next four years, the Administration is proposing to spend over $5 billion on this Program.

Scope

For more than 10 years, the Science, Space, and Technology Committee has been a leader in the area of information technology and information policy. Legislation sponsored by this Committee, the "Supercomputer Networking Study Act of 1986," mandated a report by the White House Office of Science and Technology Policy (OSTP) that helped lay the foundation of the High Performance Computing and Communications Program. The High-Performance Computing Act of 1991, first introduced by then-Senator Gore and championed in the House by Representatives Brown, Boucher, Valentine, and others, authorized and defined that program. When then-Senator Gore proposed the idea of a High-Performance Computing Initiative more than five years ago, most people believed it would primarily benefit the research and higher education communities. And its first users have been scientists, engineers, and university educators. The supercomputer technology developed under this program has helped users to improve our understanding of global warming, develop new drugs, design safer and more fuel-efficient cars and aircraft, and access huge "digital libraries" of information. The high-speed networking technology developed and demonstrated by the HPCC Program and industry has accelerated the growth of the Internet computer network and enabled millions of users not just to exchange electronic mail, but to access computers, digital libraries, and research equipment around the world. This technology, which allows network users to conduct video conferences at their desk, is enabling researchers around the country to collaborate more effectively.

The technology now used by the research and development community could provide huge benefits in other sectors of our economy. Unfortunately, much of this technology is "leading- edge" technology that is still experimental and difficult to use. That is why, in the Technology Initiative of February 22, the Administration announced creation of a program to "assist industry in the development of the hardware and software needed to fully apply advanced computing and networking technology in manufacturing, in health care, in life-long learning, and libraries." The Administration requested $47 million for FY93 and $96 million for FY94 for this program. The legislation we are considering today, as well as provisions in S. 4 and S. 473 being considered by the Senate, support these efforts.

This new program will be part of the High-Performance Computing and Communications Program, which is coordinated by the High Performance Computing and Communications and Information Technology (HPCCIT) Subcommittee of the Federal Coordinating Council for Science, Engineering and Technology (FCCSET), which I chair. The HPCCIT has incorporated this new program into the HPCC Program by adding a fifth component to the program for FY94 and by putting more emphasis on applications throughout the program. This new component, Information Infrastructure Technology and Applications (IITA), will develop and apply high performance computing and communications technologies to improve information systems needed to address what we call "National Challenges"—major societal needs that computing and communications technology can help us address—and include design and manufacturing, health care, education, digital libraries, environmental monitoring, energy demand management, public safety, and national security. These National Challenges are analogous to the "Grand Challenge" research problems which have been the primary focus of the HPCC Program to date. In addition to addressing these problems, this new component will support the development, with industry, of the NII and the development of the computer, network, and database technology needed to provide appropriate privacy and security protection for users.

Components of the HPCC Program

The President's FY 1994 budget requests $1 billion for the HPCC Program plus $96 million for the new IITA component. As I mentioned, the Program consists of five integrated components. Let me outline the goals and strategic priorities of each.

1. HPCS (High Performance Computing Systems)—Its goal is to extend U.S. technological leadership in high performance computing through the development of scalable computing systems, with associated software, capable of sustaining at least one trillion operations per second (teraops) performance. Scalable parallel and distributed computing systems will be able to support workstation users through the largest-scale highest-performance systems. Workstations will extend into portable wireless interfaces as technology advances.

2. NREN (National Research and Education Network)—Its goal is to extend U.S. technology leadership in computer communications by a program of research and development that advances the leading edge of networking technology and services. NREN will widen the research and education community's access to high performance computing and research centers and to electronic information resources and libraries. This will accelerate the development and deployment of networking technologies by the telecommunications industry. This includes nationwide prototypes for terrestrial, satellite, wireless and wireline communications systems, including fiber optics, with common protoc.htmlol support and application interfaces.

3. ASTA (Advance Software Technology and Algorithms)—Its purpose is to demonstrate prototype solutions to Grand Challenge problems through the development of advanced algorithms and software and the use of HPCC resources. Grand Challenge problems are computationally intensive problems such as forecasting weather, understanding climate changes, improving environmental quality, building more energy-efficient cars and airplanes, designing better drugs, and conducting basic scientific research.

4. BRHR (Basic Research and Human Resources)—This element supports research, training, and education in computer science, computer engineering and the computational sciences and enhance the infrastructure through the addition of HPCC resources. Initiation of pilot projects for K-12 and lifelong learning will support expansion of the NII.

5. IITA (Information Infrastructure Technology and Applications)—Its purpose is to demonstrate prototype solutions to National Challenge problems using HPCC enabling technologies. This will support integrated systems technology demonstration projects for critical National Challenge applications through development of intelligent systems interfaces. These will include systems development environments with support for virtual reality, image understanding, language

and speech understanding, and data and object bases for electronic libraries and commerce.

Close cooperation between the federal government and industry is essential if technology developed by the HPCC Initiative is to be effectively used to build an advanced NII. Both individually and as members of the HPCC Initiative, the participating agencies collaborate with industrial partners, fund research and development in the private sector, and work together to plan the HPCC Program. In addition, my office is working on a High Performance Computing Advisory Committee consisting of representatives from the private sector and academia and we note with approval your proposed legislation's broadening of the membership of the advisory group to include representation from the K-12 education community and from consumer and public interest groups.

Agency Roles in the HPCC Program

The HPCC Program has been a model of how Federal agencies with different missions can work together effectively toward a common goal. The participating agencies have built a coherent, coordinated program that is far greater than just the sum of its parts. They have eliminated wasteful duplication of effort and research dollars and found and exploited opportunities for joint projects. This kind of synergy ensures the best use of Federal research dollars.

Allow me next to briefly outline for you the roles, the accomplishments to date, the implementation plans, and the FY 1994 proposed activities for the agencies that take part in the HPCC Program. No single agency has expertise in all areas of HPCC technology; each plays an essential role. Agencies participate in the HPCC Program in support of their individual missions, overall Program goals, or both. The agencies and their roles (as outlined in a recent OSTP summary of the FCCSET programs) are:

- The Advanced Research Projects Agency (ARPA) coordinates the advanced computing and networking technologies needed by the Program. ARPA supports projects throughout academia and industry to accelerate innovation and the transition of advanced concepts into new technologies for use within the Program and the defense and national technology base. The projects are developing the full range of technologies needed for a scalable technology base of interoperating workstations, networks, and parallel computing systems with mass storage, systems software and development tools. This technology will enable a rapid transition from concepts to computational and integrated system solutions in an open heterogeneous computing environment. This will enable solution of the Grand Challenges and other National Challenges while providing the foundation for a NII.

- The National Science Foundation (NSF) supports advanced fundamental research in HPCC technologies and their application to science and engineering problems. While coordinating the NREN component, NSF is upgrading NSFNET backbone services, deploying networking information services, increasing network connections, and expanding gigabit research and development. NSF supercomputer centers are collaborating towards a "metacenter." NSF enables coordinated approaches to Grand Challenge problems, and addresses algorithm and software technology issues and providing computational research and educational programs from the secondary schools through the postdoctoral level. Expanded IITA research includes distributed databases and digital libraries, multimedia computing and visualization, and image recognition.
- The Department of Energy (DOE) funds HPCC research on parallel systems, software, and gigabit networks technology. It funds Grand Challenge research in future energy sources, fusion energy, combustion, environmental remediation, ground water flow, petroleum reservoir modeling, atmospheric and ocean modeling, and structural biology. DOE supports high performance computing centers and the Energy Sciences network, ESNET. Collaborations between DOE investigators and industry include a joint effort by DOE, NASA, six major industrial firms to establish the National Storage Laboratory, which addresses the pressing mass data storage problems. The DOE has education and training programs in computational sciences and other HPCC technology areas for secondary schools through the doctoral level. IITA research includes areas such as energy demand management and telecommuting.
- The National Aeronautics and Space Administration's (NASA) high performance computing centers address Grand Challenge problems such as improving advanced aerospace vehicles (including high speed civil transport); simulating an entire vehicle through the full flight envelope; advanced robotics for space exploration; modeling the interactions among the atmosphere, oceans, and land masses; deploying NASA's high performance NREN; and managing huge volumes of space data. NASA coordinates the ASTA component's software sharing activity, and participates in gigabit network research. NASA conducts educational pilot programs with elementary and secondary schools and supports university research. IITA efforts include increasing accessibility of remotely sensed data and developing technologies to manipulate these large volumes of data.
- The National Institutes of Health (NIH), through its National Library of Medicine, National Center for Research Resources, Division of Computer Research and Technology, and the Biomedical Supercomputer Center of the National Cancer Institute, develops algorithms and software in molecular biology (including comparison of genetic and protein

sequence data) and biomedical imaging for high performance systems; develops prototype biomedical digital image libraries; provides NREN access to researchers and medical centers; supports training at all levels in high performance computing for medical applications and provides fellowships in medical informatics. NIH IITA efforts will expand technology development for telemedicine, medical record management, and medical imagery.

- The National Security Agency (NSA), Department of Defense, conducts research in all aspects of highly heterogeneous computing environments, including specialized high speed hardware. NSA focuses on interoperability, increased performance, network and computer security, mass storage, and gigabit networks. NSA promotes research in high performance computing including superconductivity and ultra high-speed-switching at its own facilities, in industry, and at universities.
- The National Oceanic and Atmospheric Administration (NOAA), Department of Commerce, conducts Grand Challenge research in climate prediction and weather forecasting, and archives and disseminates environmental data, including climatic data for the Global Change Research Program. By exploiting the computing power of scalable parallel systems, global ocean and atmosphere models will accurately represent weather fronts and ocean eddies, and distortions due to clouds can be eliminated. In support of this research, NOAA is acquiring scalable systems and enhancing NREN connectivity. Within the IITA component, NOAA is investigating environmental monitoring, prediction and assessment applications, and expanding efforts to make its environmental data more accessible.
- The Environmental Protection Agency (EPA) conducts Grand Challenge research in air and water pollution management and in ecological assessment in a distributed, heterogeneous high performance computing environment. The research focuses on improving environmental decision-making and policy support tools, improving NREN connectivity, and developing and implementing training programs, particularly for state and environmental groups. EPA is integrating user-friendly advanced assessment tools into a high performance computing environment, which will include a scalable parallel system to enable more complex multipollutant and multimedia assessments.
- The National Institute of Standards and Technology (NIST), Department of Commerce, develops instrumentation and performance measurement methods for high performance computing and networking systems; develops security policies and technologies for the NREN; facilitates the development of appropriate voluntary standards; and designs and implements methods for organizing, documenting, and disseminating software. As coordinating agency for manufacturing ap-

plications for the IITA component, NIST will establish an advanced manufacturing systems and networking testbed. NIST will work closely with DOC's National Telecommunications and Information Administration (NTIA). NTIA has a key role in developing Federal telecommunications policy and funding networking pilot projects at schools, libraries, hospitals and health care providers, and other non-profit institutions.

- The Department of Education (ED) sponsors program initiatives and activities through its regional laboratories and research centers. The Department will provide information to educators and students in K-12 and lifelong learning about high performance computing and networking application resources. It promotes initiatives in training, curriculum development, library connectivity and research and development projects that support the emerging information infrastructure.

HR1757—"High Performance Computing and High Speed Networking Applications Act of 1993"

HR1757 is important, forward-looking legislation and is largely consistent with the Administration's proposal for developing the nation's information infrastructure. I commend the Chairman and the cosponsors of this bill for their efforts and their vision. This legislation can help accelerate the development of a National Information Infrastructure.

I believe that the legislation being considered today is generally consistent with the Administration's goals for the HPCC Program and the development of the National Information Infrastructure. It supports the expansion of the HPCC Program and additional emphasis on development of applications technology for manufacturing, health care, education, libraries, and other sectors of the economy. It also authorizes additional Federal funding for connecting schools and non-profit organizations to high-speed networks. The legislation includes very important provisions for improving the dissemination of Federal information.

We share the committee's desire that scientific and technical information be made more available to the taxpayers who paid for it. Scientific and technical information (STI) is a multi-billion dollar annual enterprise which provides the critical raw material of research and development as well as a tangible output of our R&D investment. It is estimated that between 2 percent and 4 percent of each Federal research dollar is spent on the management of STI. When effectively mobilized, STI programs can provide input to the solution of major technical problems. From large earth observing experiments to genome databases to computational science research to using scientific visualization for new scientific insights, rapid advances in information technologies combined with an increased volume of data and information have brought renewed recognition of the importance of information management in the science and technology program life cycle.

HR1757 was just introduced on April 21st. My office and all of the other interested agencies have already begun to carefully review the details of the Bill. The Administration will provide the Committee with its views on the Bill as soon as the review has been completed.

Conclusion

Thank you again for this opportunity to appear here today. The Administration is committed to the rapid development of the National Information Infrastructure. This commitment is reflected in both the President's Technology Initiative and Administration's FY94 budget request. Your proposed legislation will help provide greater access to the Internet and accelerate development of applications of high-performance computing and communications technology in the areas of primary and secondary education, health care, libraries, and access to government information. We look forward to working with the Committee on this legislation and on the development of the National Information Infrastructure. There are few initiatives that offer as many potential benefits to all Americans.

CHAPTER 17

REMARKS BY STEVAN D. MITCHELL COMMISSIONER, PRESIDENT'S COMMISSION ON CRITICAL INFRASTRUCTURE PROTECTION

DoD Worldwide Antiterrorism Conference

August 21, 1997

It is a pleasure and an honor to be here with you this evening. The fact that you would invite a Department of Justice Attorney here to speak to you about infrastructure assurance is indicative, I think, of just how far the federal government has come on this issue in the past year or so. A year ago, someone from DOJ might not have had a whole lot of interesting things to say to many of you—who, after all—have been working these issues for years within DoD. I hope I can do a little better than that!

The creation of the President's Commission on Critical Infrastructure Protection signaled a change in the way our government is preparing to address infrastructure assurance. We are hoping that our Commission report will signal a change in the way our *nation* should prepare to address infrastructure assurance.

I'm going to try to present a near up-to-the-minute update on Commission activities—which are, as of late—heating up quite a bit. As some of you may know, we are currently scheduled to provide our report to the President in mid-October. The Commission is an even busier place than usual these days.

I'll talk a little about the formation of the Commission, our mission and focus, and about the people who make up the Commission. I also hope to touch on some points bearing on Commission methodology, and describe for you what the process looks like from now until we present our report. Finally, I'll give you an overview of some of the major issues that have been driving our Commission deliberations, and that are most likely to end up being the subject matter of some of the Commission's recommendations.

Now here's the disclaimer (after all, what kind of lawyer would I be without a lengthy disclaimer). Keep in mind that even though our October 13 date for reporting to the President is fast approaching, we are still working hard in deliberations to arrive at and refine our final recommendations. So take what I have to say here tonight as the observations of an enthusiastic participant in the process—but not as necessarily indicative of the Commission's final recommendations.

The Commission

With the signing of Executive Order 13010, the President brought together representatives across government and the private sector to address a set of issues that were, quite frankly, new to a number of us. What the President did when he signed the Order was, in retrospect, fairly extraordinary. He extracted the societal component of what was previously known as defensive information warfare—something that had rarely been spoken of outside the Pentagon—something that most private sector folks don't even want to think about much less talk about—and opened it up to very public policy scrutiny.

The bringing together of representatives of 10 key agencies on the Commission was a loud and clear initiation of a cross-governmental response. The joining of this group with representatives of key portions of the private sector was a clear initiation of a cross-societal response.

The Advisory Committee

One of the ways the private sector has had direct, "top-down" input into the process is through the Commission's Advisory Committee. We are very fortunate to have an Advisory Committee of private sector CEOs to assist in our effort and oversee our progress.

Co-chairs of the Advisory Committee are former Senator Sam Nunn and former Deputy Attorney General Jamie Gorelick. They were powerful proponents of infrastructure assurance within government, and we are quite pleased that they have agreed to help us with our most significant challenge: achieving private sector buy-in of our recommendations. As you all know, most of the infrastructures are privately owned and operated. We recognized from the outset that any solution that does not include their input and support is not a viable solution. Our goal is a public-private partnership that works together to protect our future.

The Commissioners

The lion's share of the Commission's work has been conducted by Commissioners representing the public and private sectors. Half are executives from the involved departments and agencies in Washington; half are executives from infrastructure companies and organizations who bring valued industry experience, expertise, and perspective.

Truth be told, we are quite a bunch. We range in age from the mid-thirties to the upper seventies, and our experiences run the gamut. We have engineers, IT specialists, physicists, lawyers, and even a physicist-lawyer! We bring with us to the table not only our personal experiences, but our own agencies and companies' interests, equities and methodologies. However difficult our differences may make things at this stage, I think we all recognize its for the best that we bring all of these perspectives to bear on our difficult challenges.

Commission Focus

Allow me just a word or two on the task that we were given. I should have known something was going on early in our process—during our first few outreach meetings with representatives of major industry trade associations. We would arrive politely enough, introduce ourselves, and begin to describe our mission. Often we would hand our hosts a copy of the executive order that set forth our mission. *That's when the laughter started.* We would hear questions like, "You have to do what?" "You have how much time"? Or my favorite, "Exactly how many of you are there?"

You see, we had been charged with studying, in some degree of detail, eight infrastructures that had been deemed by the President as "critical"—along with any others we happened to discover along the way. The President identified these as critical because their incapacity or destruction would have a debilitating impact on our defense and/or economic security. So what's the first thing you do when asked to look at eight critical infrastructures? Well, we're not crazy—we condensed them down to five. They are:

- Energy—including electric power; natural gas and petroleum production, distribution and storage;
- Physical distribution—including railroads; highways; air traffic; maritime transportation; and pipelines;
- Banking and finance—including banks; financial services; markets and payment systems such as SWIFT and CHIPS
- Information and communications—including telecommunications; computers; software, the Internet; satellite communications; and
- Vital human services, where we combined Water; emergency services such as fire, police, and emergency medical services, and government services–such as social security and welfare.

Commission Methodology

The challenges set forth in the Executive Order required that we employ an eclectic approach. Even several eclectic approaches all at once.

Conduct outreach

We were instructed to "identify and consult with: (i) elements of the public and private sectors that conduct, support, or contribute to infrastructure assurance; (ii) owners and operators of the critical infrastructures; and (iii) other elements of the public and private sectors, including the Congress, that have an interest in critical infrastructure assurance issues." We did that. To date we have met with more than 5,500 individuals, corporations, associations, and government agencies. We held public meetings in Los Angeles, Atlanta, Houston, Boston, and St. Louis. In each of these cities, we talked with hundreds of people from industry, labor, state and local government, academia, as well as private citizens.

Assess threats and vulnerabilities

We were charged with examining the threats to and vulnerabilities of each of the eight enumerated critical infrastructures. We divided ourselves into infrastructure sector teams—and did just that. Much of this research was conducted through our extensive outreach program consulting with stakeholders around the country. As a result of these efforts, we expect that our report to the President will contain infrastructure-specific as well as cross-infrastructural findings and recommendations.

Identify legal issues

We were asked to determine what legal and policy issues are raised by efforts to protect the critical infrastructures; and to propose statutory and regulatory changes necessary to effect our recommendations.

Well it did not take long to realize that all of the issues raised across all of the infrastructures could not easily be captured in a "report." Instead, we built a legal issues database—a database of pertinent federal laws, regulations, executive orders, and even international agreements. We included in the database a sampling of relevant state legislation, and even included information on what appeared to be relevant private sector policies and practices. The result is a rich collection of legal authorities that one day may call to be carefully reexamined in light of the concerns raised by the Commission.

The database was originally intended to serve as an implementation tool for the recommendations of the Commission. In keeping with the concept of "infrastructure assurance" as a series of gradual cultural changes, however, we believe that the database might serve an equally important long-term role. We are exploring the possibility of making the database broadly available, to help raise awareness across the legal community and facilitate implementation of future assurance measures.

Issue assurance policy and implementation strategy

We were tasked to recommend a comprehensive policy and implementation strategy for protecting the critical infrastructures from both physical and cyber threats.

Here's where things got creative. We have met with experts from science, technology, the military, business, and government. We have conducted strategic simulations with Sandia National Labs, the Prosperity Institute, and Booz-Allen & Hamilton. We have compiled databases that are the most robust sources of information on these critical infrastructures in existence. We endeavored to leave no stone unturned as we identified dependencies, threats, and impediments to solutions.

These research efforts concluded earlier this summer, and we have since commenced the most important phase of our work: (1) identifying those fundamental issues that the Commission must address; (2) deliberating the options for how to answer those questions; and (3) writing our report and making our recommendations to the President.

Some Personal "Lessons Learned"

Before sharing with you some of the issues that the Commission is now considering, I hope you don't mind if I take a few minutes to share with you some personal reflections—my own personal "lessons learned" about this thing that we at the Commission now quite casually refer to as "infrastructure assurance."

Commissioners and commission staff have become fairly comfortable talking about whether some action or policy x or y enhances infrastructure assurance, or whether some law, for example, may operate in a way antithetical to infrastructure assurance. It has become a convenient shorthand for us. And we all know pretty much what others are talking about because we have been wrestling with these issues for over a year. I suspect, however, that it is just that, a shorthand for something that we have not characterized as completely as we should.

Sure, we have our "dictionary definition" of infrastructure assurance, and we hope some day to be sharing that with you in our report. Let me offer a different type of definition of "infrastructure assurance," one grounded in experience. One grounded in what we must do to enhance infrastructure assurance—what we must begin to do today to be adequately prepared for tomorrow.

What I think of when I think about "infrastructure assurance," in its broadest sense, is the cultural change that we will have to undergo to preserve our national security, our nation's global competitiveness, and the overall public good in light of newly emerging and intensifying threats and vulnerabilities. This, I believe, is a cultural change, one that will occur gradually but pervasively. It is a change that will bring changes in the way that business does business, government does government, and individuals conduct their lives. It is a change that will see:

- Businesses responding by including emerging threats and vulnerabilities in their risk management methodology;

- Government responding by organizing itself in such a way as to detect and, in appropriate circumstances, respond to emerging threats and vulnerabilities;
- Private citizens, in their homes and offices, modifying the way that they practice computer and information security.

I believe this change will be quite natural, such as the change that took root in the 1960s and 70s, when people learned that they must lock their front doors and no longer leave the keys in the ignition overnight. Some people we have spoken to say this is a cultural change that will occur on its own over the next 20 years. I believe, however, that if we can do certain things to "jump start" this cultural change, and accelerate it by as little as 5 years, we will have done much to preserve this country's global competitiveness and perhaps even save American lives.

A Progress Report on Commission Issues

So where are we exactly in all of this? Well, we are just bringing to a close our deliberation process. We had originally identified some 130+ candidate issues for deliberation. Well eventually we managed to combine, massage, manipulate or just plain drop them until we arrived at a much more manageable number—about 40. We assigned to individual Commissioners responsibility for drafting issues papers, and generating options to feed the deliberation process.

Actually, it has been fairly extraordinary. Every day we see connections between issues—and between potential solution options—that we had not previously seen. The issues are grouping themselves into some broader issues categories, several of which I will touch on now.

I'd like to preview five of our major issue areas, some in greater detail than others: Globalization, Education and Awareness, Research and Development, Legal and Regulatory issues; and what we are calling "National Structures"—how we believe the government should realign itself to more adequately address assurance concerns.

Globalization, Deregulation and Restructuring

One of the things that makes proposing solutions so difficult is that we are aiming at a moving target—several moving targets. Our economy is changing —becoming more global. The infrastructures are changing—they are becoming more deregulated, and companies are restructuring to respond to these and other changes. At the same time, technology is enabling new types of threats, and introducing new vulnerabilities. What are the combined effects of these trends? We are planning to chart some of them out in a high-level risk assessment.

Education and Awareness

This is going to sound trite, but it should be of no surprise that we found education and awareness to once again be a key to meaningful long-term solutions.

I don't need to highlight for any of you that there is a shortage of talented information technology and—most particularly—information security people around today. They're being bought up left and right—so quickly that even the private sector is feeling the pinch. Despite this, very few American colleges and universities include computer and network security within their graduate school computer science and management curricula.

The obvious long-term solution is education. We're taking a broad view of education: We believe benefits can be derived from greater availability of professional degrees, from enhanced focus on security and risk management in engineering and business curricula, and even from enhanced ethics training at the K through 12 level. Is there a public-private sector "partnership" role here? In light of the potential for large economic losses, we suspect there might be.

Research and Development

The Commission is holding out high hopes for research and development. Right now, we simply *do not have the technology to know* when we are under cyber attack.

We have been exploring the appropriate balance between private and public investment. The private sector is seeing to it that there is adequate "D" going into development of product-based solutions. But there is likely not enough "R" going into the discovery of new technologies. Is there a breakthrough on the horizon? Some say probably not not. But our Chairman fondly recalls a time, in the early 50's, when the technology did not exist to detect a hostile missile launch. Experts said there would be no way to differentiate a launch from a volcano. Some well placed R&D changed that, and we believe the same may be true here.

We may never get to find out unless the government steps to the plate to fund some of the long term research that may be a bit too speculative to prompt serious private sector investment. Even if major breakthroughs are not immediately forthcoming, a bit of "invigoration" by the government could help to reduce end user costs for advanced security technology.

Legal and Regulatory Issues

I touched earlier on our responsibilities with respect to legal authorities. You'll recall that one of the things that made going tough was how pertinent legal authorities are located in a number of different places. There is no established body of law for "infrastructure assurance"—and asking lawyers to look it up for you is likely to get them frustrated and angry. Fortunately, I have been graced with a most insightful staff, and have been able to acquire the research services of some very patient people. Together, we have been able to unearth some important legal issues relating to infrastructure assurance—some that support other areas of Commission inquiry, and some that stand on their own.

Legal impediments to information sharing

We recognized early on the importance of a trusted environment for information sharing: Government should alert the private sector about threats, and the private sector should feel comfortable reporting problems to the government so government can better focus its efforts. But what kind of collaborative mechanism could possibly protect classified government intelligence information, confidential law enforcement information, and also protect confidential private sector information affecting reputation, consumer confidence, and liability? There are a host of practical and legal impediments. Interestingly, some of the legal problems may be easier to resolve than the practical ones.

Infrastructure owners and operators tended to raise two areas of predominant concern, one relating to the sharing of information with other private sector people (antitrust concerns), and another relating to the initial acquisition of potentially valuable employee background information (privacy concerns).

With respect to sharing threat and vulnerability information—antitrust concerns are more appearance than reality. Even when shared exclusively between private sector entities, this type of information does not go to the heart of competition (as does, for example, "price fixing"), and so it is simply not a target for antitrust inquiries. Even if it were, there would be workarounds. Just having government around—playing even a modest role in information sharing—does wonders to prevent antitrust exposure. But me saying this doesn't help much. Even having a Presidential Commission saying this might not help much. What might be needed is more definitive guidance, from DOJ or the FTC, for example, to set rough parameters around information-sharing practices and make the private sector feel more at ease.

Privacy is trickier. Our outreach revealed a degree of frustration among infrastructure owners and operators, some of whom wanted to take greater employee security precautions than some laws might otherwise allow. But here, the employer's interest in security bumps up against the employee's interest in privacy. The privacy issues implicated by sharing of employee background information, for example, touch individuals in very personal ways.

The importance of individual privacy has made obtaining background information on a potential or current employee a fertile area for strong state laws. Though grounded in best intentions, this has resulted in unpredictable disparities of approaches across state lines. The bottom line: a critical infrastructure owner and operator may in some instances be discouraged if not prevented from asking many of the types of questions that could help identify and prevent a growing insider threat to our infrastructures. The federal government already conducts background investigations on employees going into sensitive positions. Should employers looking to fill certain key positions within the critical infrastructures be permitted to access some of the records or use some of the tools available to the federal government?

The Commission is currently looking at how best to facilitate inquiries into employee backgrounds for select positions within critical infrastructures. An astounding number of factors—privacy, fair employment, post-conviction rehabilitation, collective bargaining, federalism—come into play when attempting to strike the appropriate balance. It may be necessary to recommend follow on studies to find ways to balance the equities across 51 different legal jurisdictions.

Adequacy of criminal law and procedure

We have also looked at the adequacy of existing criminal laws governing cyber abuse. We have noted, as have others, that some aspects of the criminal law have not kept pace with technology. We have looked at procedural laws to make sure that they do not unduly hinder the conduct of investigations in electronic environments. The Commission's recommendations may highlight some perceived deficiencies. We are also likely to advocate that domestic law enforcement continue to work closely with the international law enforcement community to insure that there will be a strong network of criminal laws addressing unauthorized intrusions across national boundaries, and procedural agreements to insure speedy investigative response.

We are also looking at that body of "traditional" criminal law that protect the critical infrastructures against physical attack. While we are reassured that the federal law is capable of reaching a broad range of terrorist threats, we want to be similarly assured that criminal penalties reflect the true costs associated with these crimes, and the gravity of harm done to all of the victims.

Enhancing deterrence

We have also been asking tough questions about deterrence. There should be little doubt that a deterrent solution beats a reactive solution every time. After all, what is the value of a reactive solution against the likes of someone willing to take their own life in furtherance of a cause? We are taking a look at the deterrent policies of the country and the deterrent effects of criminal law and law enforcement to see whether they might require enhancement.

How do you go about enhancing the deterrent value of law enforcement? For one, you see that the laws themselves, the criminal prohibitions, the "do's and don'ts," keep pace with prevailing technology. You check to see that penalties adequately reflect the seriousness of the harm and cost to the victim in today's terms, taking into account any losses resulting from denial of service or the compromise of sensitive or proprietary information. You also see to it that adequate resources are available to investigate an adequate percentage of criminal incidents. Adequate and timely laws will not deter if enforcement is known to be sporadic.

In this regard, a key question that the Commission has begun to explore is whether a conventional law enforcement response can possibly keep pace with expected increases in the volume and severity of cyber-intrusion incidents. Is "staffing up" sufficient in and of itself? Chances are that the Commission will not be around long enough to take the measurements and evaluate the trends. But it may be the case, that at some point, new and novel approaches will be needed to really and truly *deter* unlawful behavior, rather then merely react to its consequences.

National Structures

Our largest challenge involves redefining the role of government in light of all of the emerging trends we have noted. What is the proper role of government? Since most infrastructures are privately owned and operated, can and should we rely exclusively on market forces to assure delivery of vital services? The private sector must guard against commonplace intrusion, theft and fraud, but what about state-sponsored terrorism or hostile attack? What's the federal government's responsibility vis-à-vis the states and the private sector?

Well, being a public-private Commission, we went round and round on this one (The truth is we're still hammering out details). Some of it just depends on who you talk to. I personally think our disagreements relate more to methodology than substance. Interestingly, our search for a productive methodology led to something of a breakthrough. Here's my own, admittedly somewhat stylized version of what happened:

Infrastructure assurance "functions"

Our first reflex was to talk in terms of "entities"—as in, "What is really needed is an entity for this, and an entity for that." And because it took us about 4 or 5 months to start bringing on board our private sector Commissioners, it was just the government side Commissioners who were participating at this point. So you know what that means: What we were really saying, in our own self-satisfied way, was "WHAT IS REALLY NEEDED IS A *GOVERNMENTAL* ENTITY FOR THIS AND A *GOVERNMENTAL* ENTITY FOR THAT." Well, our Chairman and some of the private sector commissioners—thankfully, in retrospect —pulled us back to earth.

We then entered into what can be called our "functionality" period. We tried consciously to wean ourselves away from the word "entity," since it had become something of a loaded term. Instead we forced ourselves to speak in terms of "follow on functions." That freed us up to play an extended game of "Wouldn't it be nice if." Wouldn't it be nice if there were *some way* for *someone* to perform function *x*, or function *y*. *That* was actually a very productive exercise, because it enabled us to focus on what was needed in light of new threats and vulnerabilities, without the need to jump immediately to the question of "who should be doing what."

It was actually a very productive period. We were able to isolate, identify and describe some 31 separate infrastructure assurance follow-on function, all grouped into five major function categories. The five categories isolated thus far (and some of their component parts) are:

1. **Policy formulation functions, including:**
   - assessing national risk
   - proposing national objectives and developing strategies
   - proposing and promoting legislation
   - assessing and promoting regulations
   - influencing private sector investments
   - preparing, recommending, and advocating budget requests
   - managing and enforcing implementation
   - shaping the international environment
2. **Prevention and mitigation functions, including**
   - promoting education and awareness
   - monitoring assurance standards, certifications, and best practices
   - assessing risk of system components
   - proposing protection and mitigation objectives and strategy
   - developing concepts, promoting research, and funding R&D and acquisition
   - providing funding for protection and mitigation operations
   - transferring government information to private sector for better risk analysis

3. **Operational Warning functions, a necessary precursor to which is better information sharing**
   - need to disseminate strategic and tactical warning information
   - jurisdictional concerns (limitations of NSA / FBI CIA)
4. **Counteraction and Incident Management**
   - plan for and manage counteraction to deter, halt, or minimize an attack
   - plan for and manage an integrated law enforcement, intelligence, and military response
5. **Response and Recovery.**
   - plan for and manage response to consequence of an attack;
   - plan for and manage restoration of disrupted infrastructures

"Functional assignment"

After settling on a representative set of "functions," we entered into (and in some sense, are still in) our "post-functionality" period—the most painful phase of all. We had identified 31 important functions that needed to be performed by *someone*, but we were at an impasse when it came to identifying who should be doing what. We were stuck, until we devised a way that we could continue to focus on each of the functions, but perhaps array them in such away that we could then discern some patterns. If we were to be successful at grouping the functions, then maybe some structure would begin to suggest itself.

I suggested that we begin with a very simple two-by-two matrix, along two dimensions: Public-led versus private-led, and centralized versus decentralized authority. See Appendix A. The result—our "assignment space," so to speak, was an interesting representation of the world today. Consider the four quadrants suggested by the matrix. The upper left hand corner—Public-led on the horizontal axis and centralized on the vertical axis—well, that's the federal government. Upper right hand corner—private-led but centralized, that's the space reserved for trade and industry associations and alliances. The bottom left is public-led decentralized. That's space reserved for responsibilities of the often-overlooked but integrally important state and local governments. And the bottom right, private and decentralized, well that's the space that represents responsibilities of individual companies, entities and market forces.

In plotting functions against the matrix, we had in mind some high-level guidance. We knew (taking welfare reform as just one example) that Congress has been of late inclined to relinquish more and more previously centralized authority to state and local governments. We also knew, from Mr. Magaziner's Framework for Global Electronic Commerce, that the White House might be more amenable to private sector-led solutions in this areas. So in placing the various functions on the matrix, we made a concerted effort to force them down and to the right, away from traditional, centralized, public-led initiatives and toward decentralized, private-led solutions.

We were also prepared for disagreement between the respective infrastructures. We were all ready for someone to argue that some function *x* was more adequately performed in one place for a given infrastructure, say, telecommunications, but that it was more adequately performed elsewhere for another. What happened was mildly shocking. We found ourselves—public and private folks alike—by and large agreeing to where certain functions fit on the matrix. Answers to the question of "Who should have responsibility for what" were taking shape—effortlessly and in relatively short order.

Here are some broad conclusions from this "functional assignment" exercise:

The federal government

The most important role that American citizens look to the federal government to perform is national defense and security—protecting the United States from "attack." But fulfilling these functions is no longer simply a matter of maintaining arsenals and defending borders. For one thing its hard to identify, let alone defend, borders in cyberspace. If borders are defined as potential entry points, well then, with respect to the critical infrastructures, the majority of these borders are clearly not subject to immediate federal government control and supervision. Nor should they be. The point is that the government's provision of a service previously taken for granted has been significantly disrupted.

Other centralized public functions include the assessment of national risk and issuance of national policy, preparation of threat advisories, and the coordination of research and development efforts. Each of these functions may be in danger of breaking down when decentralized or privatized out toward the far reaches of the functional assignment chart.

State and local government

Of the follow-on infrastructure assurance functions, several make the most sense when they are public and decentralized. These functions include what we have termed defensive counter-action, response management, and response planning. The states have in the past, and should continue in the future, to take the role of first responder in event of emergencies that bring serious consequences to people and property. They are, and will continue to be, the first responders—hours or even days before additional support can arrive. But while our public outreach has demonstrated that they are clearly ready to fulfill these responsibilities, state and local responders are all too often understaffed, undertrained, and underfunded. We hope that by highlighting their crucial role, we might encourage delivery of a degree of relief.

Trade and industry associations

Private industry is becoming ever-more adept at forming the partnerships, associations and alliances to respond to arising needs. These partnerships take many forms and perform many roles. Many of their traditional roles—setting standards, certifications or best practices; sharing information; forming alliances to engage in research and development—are necessary elements of a concerted national response to threats posed to, and vulnerabilities of the critical infrastructures.

Internal processes and market forces

Much of the burden of infrastructure assurance will fall on the individual infrastructure owners and operators. This is no different than similar burdens that come from the need to address the potential for damage from natural disasters, accidents, or other previously envisioned sources of threats. As they have done in the past, companies must continue to assess their risks and vulnerabilities, implement best practices, protect their systems, engage in research and development and share information. These are individual responsibilities that are necessary components of an effective national response.

"Borders and crosshairs"

As interesting as the functions that fit cleanly into quadrants are those that did not: Roughly fifty percent of the functions did not fit cleanly into one box or another. It is important to note that this did not result from indecision or lack of agreement. Rather, there appeared to be collective agreement that *certain functions* must be placed on the borders between boxes in order that they make sense.

I'm going to refer to "border" functions as those that are settled to one side or another along one axis, but not the other. Proposing a national strategy or national objectives of an infrastructure assurance program should be done from a centralized perspective, but with roughly equal input from the government and private sector. It therefore resides on a boundary. It is a centralized public-private activity. By contrast, the planning, management and delivery of law enforcement functions is a combination of centralized (federal) and decentralized (state) public-sector activity.

"Crosshair" functions is my term for those follow-on functions that appear squarely in the middle. The only way they can work is through mutual cooperation of the four other quadrants. They also appear to be, perhaps not coincidentally, the functions that are most vital to an effective infrastructure assurance effort. They include information sharing, education and awareness, and funding issues. These functions are the hub of our national efforts, signaling in a dramatic way the need for public-private *and* centralized-decentralized cooperation and partnership that is not often called for in policy initiatives of a lesser scope and complexity.

Organizing to perform new roles

The four quadrants of the matrix represent the way our societal and governmental structures are ordered today. Our laws and conventions divide the world into entities and organizations that are either public or private, and centralized or decentralized. But many of the most critical infrastructure assurance functions must be—of necessity—performed on the borders and crosshairs—in places where existing entities do not reside. It was this dramatic realization that enabled many of us to finally articulate why new relationships, partnerships, and structures may well be needed to respond to emerging concerns. Our conclusion? That government and the private sector should enable structures that reside on the borders and crosshairs to accept and act upon these new and different responsibilities.

Our next step is to flesh out what these structures should look like—and we're still hammering out the details.

Conclusion

In the coming weeks, we will be finalizing and polishing our recommendations for the President. I hope that I have given you a feeling for the process (blood, sweat and tears) that has driven those recommendations and an appreciation of how, when our final report lands on the President's desk, the Commission will have truly gone full circle—beginning with an Executive Order calling for an unprecedented degree of partnership and coordination with the private sector, through some difficult struggles, to a final and well founded conclusion that the real future our of national security, of our critical infrastructures, lies in successful collaboration across all levels of government and the private sector.

Appendix A
Public and Private Roles in Assurance

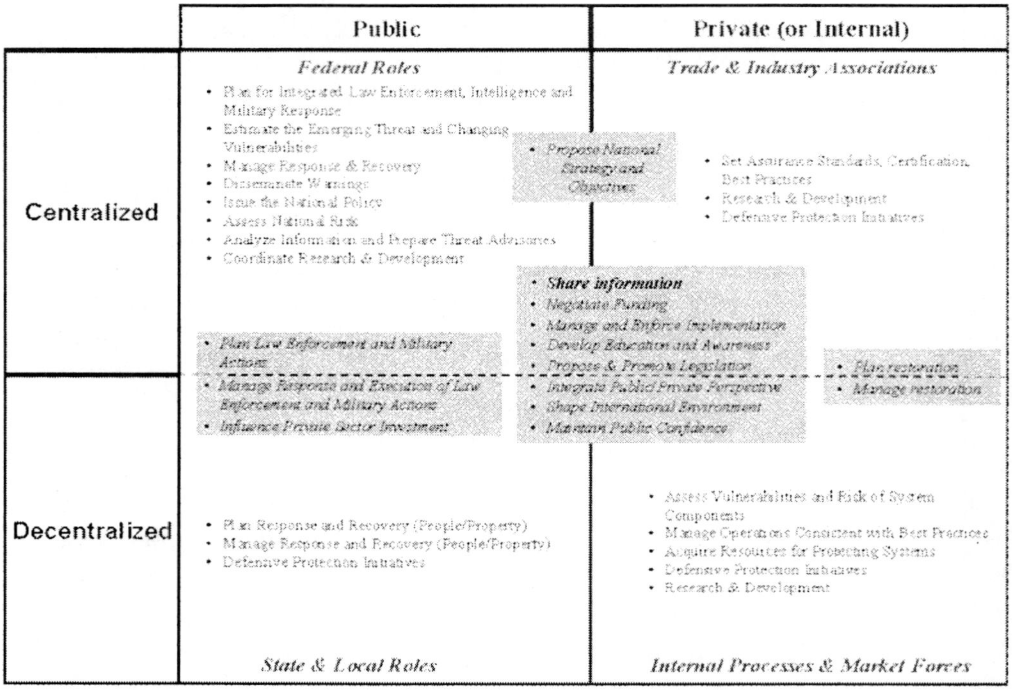

CHAPTER 18

DoD SPEECH TO FORTUNE 500 CHIEF INFORMATION OFFICERS FORUM DEPUTY SECRETARY OF DEFENSE JOHN J. HAMRE

Aspen, Colorado

Tuesday, July 21,1998

Secretary Hamre: Thank you, Peter (Petre, Executive Editor, Fortune Magazine) You're very kind. And I appreciate being invited to come out to be with all of you. More than anything it is great just to get out of Washington, to be blunt. (Laughter.)

Let me apologize at the outset. I did have a speech prepared to give you, but this morning, I had a very interesting experience on the way. I stopped at Cincinnati to talk with folks at General Electric who do engine work to learn about how they're doing six sigma. Frankly, that visit changed a lot of my thinking. On the flight here from Cincinnati I re-wrote my speech and so I'm afraid you're going to get a much rougher product. I'm going to try to bring together a number of different strands, talk both about our war fighting side as well as our business side and how we use information technology. In the process I hope to give you a sense of what we're facing.

First, by way of introduction, and as Peter said, I function as the chief operating officer for the Defense Department. Let me, if I may, give you a sense of the size and complexity of this organization. As Peter mentioned, it's a large organization. We have 1.4 million men and women who serve on active duty in the uniformed services. And about 800,000 civilians that work for the department. Every year, we recruit about 200,000 new people to join the armed forces. And we separate about 220,000 people. So about 30% of our organization is either coming or going every year. We are spread out all over the world, as you know. We have about 250 major installations. We operate 550 public utility systems, everything from gas, water, electricity, and natural gas distribution.

We support one of the largest school systems in the world. We have 126 high schools and elementary schools.

We are the world's largest daycare provider. (Laughter.) Seriously, we have 300,000 kids that go to Department of Defense day schools—daycare centers—because we've become a very different military over the last 25 years. Twenty-five years ago, we were a conscript military composed in large measure of bachelors. Today, it's very much a married military and lots of single parents. And you've got to reassure parents that somebody's going to be taking care of their kids if they get mobilized on short notice. So we spend substantial sums on daycare. We're the largest daycare provider in the world.

We have 28,000 separate computer systems that we're tracking for Year 2000. Twenty-eight hundred of them are mission critical. I used to be the comptroller, so I'm a little more familiar with some of this, but we cut 5 million paychecks a month. We cut about 400,000 bonds every month. We pay about 600,000 travel vouchers a month, 800,000 contract actions a month. Out at Columbus, Ohio, where we do our large contract management administration, we have about 390,000 contracts under administration. We disburse about $43 million an hour. (Laughter.)

We sustain operations in every time zone. There isn't a time zone on the planet that we don't have military personnel operating someplace. Today, there are about 115,000 military personnel who are deployed. And that's in addition to the 200,000 who are permanently stationed overseas. We operate over 400,000 vehicles-everything from sedans and buses, to the street sweepers we use to clean runways, to combat vehicles to tanks to armored bulldozers.

As an organization, one of our real challenges is to manage about 70 years worth of technology at any one point in time. We operate, on a daily basis, aircraft that were designed back in the early '50s and we still have to maintain them, buy spare parts for them and keep them updated. At the same time, we are working on research and development programs for systems that won't be fielded until 2015-2020. Managing that spectrum of technology is a real challenge.

In information technology, we operate some of the most advanced computers.

And yet, just last year, we moved a bunch of Burroughs punch card readers to a new mega center because we were still operating punch cards for some business applications. So, it's an astounding spectrum of technology that we try to manage. It is, I would argue, not only the largest, but probably most complex, organization in the world.

Now, this is an organization that has had its budget cut for 15 consecutive years. In part, this is understandable. The Cold War is over, and I don't think it's inappropriate that it is reduced. But our budget is about 46% smaller than it was in 1986. We've undergone significant reductions. This is, as I mentioned, an organization that is operating at 46% of its budget resources only 12 and 13 years ago, has a third of its personnel coming and going in any one year and is still able, within a month, to send 60,000 people to the Persian Gulf along with 400 combat aircraft and 500 cruise missiles and could carry out war tomorrow if we had to.

So, I say that to frame an observation and a dichotomy: I firmly believe that we are a world class organization in what it is we're supposed to do, which is to fight and win wars. There isn't anybody who is as good as we are. But I've also got to be honest and say we're a second sigma organization when it comes to business practices. And that's the curious dichotomy that I see and that I'd like to talk with you about today.

First, let me briefly discuss the war fighting side of the house. We have been engaged in an unprecedented change in the way we think about warfare. It's been going on for some time and it's ready to enter into a much more sophisticated dimension. It really took off in the late '70s and the early '80s when we were starting to bring microprocessors into weapons systems. That probably wasn't as revolutionary as was the revolution in training technology that we developed in the late '70s and into the early '80s. And that's been, I think, far more important, frankly. Most of the senior commanders in the Persian Gulf have said that we could have switched equipment during Desert Storm and still beaten the Iraqi's, just because of the people and the skills that we developed over that time. For us, training technology, growing our most important asset, people, is by far the most important thing we do, more important than the hardware that's sitting out on a ramp.

So for what we're supposed to do, we're a world class organization. And I've got to tell you, frankly, we're getting better. Dramatically better. We're already without peer. I know this sounds boastful. I don't mean it to be that way, but we're already without peer as a fighting force and ten years from now, we're going to be, I think, significantly stronger. Let me give you an example.

We are on the edge of breaking through in what we call network-centric warfare. I'm sorry, this is a little blunt, a little coarse, but we are in the business of destroying things. In warfare we try to do that in a very focused way without doing lots of damage to things that you don't want to destroy. We've done that before by largely putting very lethal and accurate capabilities in the hands of whoever was doing the shooting at the time. We're now moving into a much more interesting and highly leveraged dimension where the person that launches the missile doesn't have to see the target. We're going to be sharing information across a network and still be able to attack and destroy an opponent. This dramatically improves the survivability of our own forces, of course. It is going to be revolutionary. The situational awareness that will be on our side of the battlefield will be three or four orders of magnitude better than our opponents.

This really comes from the way we've brought information technology into the core of war fighting. For example, we are examining technology that is not with us today, but I don't think it's many years away. We will have a device about the size of a watch that will be worm by our average soldier that will monitor vital signs. When a squad is deployed at night, the sergeant doesn't have to wonder if a private has fallen asleep in a foxhole. He can see it on a monitor. If one of our soldiers got hit, you don't necessarily have to send out a couple of guys to try to get him and find out that he's already dead and then put more loves at risk when you're in a firefight. It's that kind of knowledge and that kind of skill-based control that we're going to bring to the fight that will be without peer in the world of warfare.

We're doing this with a very systematic application of information technology and warfare. It's really unbelievable the way it's happening. While just now a prototype, we're putting computers on-board airplanes along with pilots that serve as a copilot for all practical purposes, giving that pilot extra situational awareness, queuing and advice. We're going to be able to put inside a tank a monitor with a graphic picture of the entire battlefield for that tank commander. You can't imagine how limiting your perspective is when you're inside a tank, buttoned up, bouncing all over, maybe wearing chemical protective gear, trying to have some idea where the enemy is and where your own friendly forces are. Now the tank commander is going to be able to look at a computer screen and see the full battlefield. It's going to be absolutely revolutionary for what we can do.

In the past, the dilemma of warfare was always how to bring mass together for its effect over your opponent without giving your opponent lots of targets to shoot at. It is the classic dilemma. One of the reasons there were so many casualties during the Civil War was because firepower technology had gone so much further than communications technology. We were still massing people close to each other, side by side so the soldiers could hear shouted orders. Firepower technology had advanced, however, so that cannon could mow people down.

That's why there were so many casualties in the Civil War, because communications technology still relied on people standing close enough to hear an order. We're now going to be in a wholly different world where people don't have to see each other and yet, they can operate together as a combined arms team. It's dramatic what we're going to be able to do.

Now, I contrast that with where we are in our business operations. Our employees are working very hard trying to make the system work. But I'd have to be honest and say that we're a second sigma organization for most of our supporting structure. We are still largely dominated by stovepipe organizations. We automated our stovepipes and thereby reinforced their bureaucratic rigidity. Does this sound familiar? (Laughter.)

And of course, what happens? When you automate manual processes, you then have to invent interconnections to get them to work together, right? And of course, those are all basically failure points in a system. So, on the average, between a decision to buy something and getting a check out the door, it takes 105 paper transactions in the Department of Defense right now. We've glued together lots of different, old-fashioned, manual procedures that were designed during the '50s and '60s and '70s that have all been automated. We are an enormously paperbound organization.

When I was the comptroller, I was responsible for all the finance and accounting operations of the department, and that included our disbursing operations. I mentioned our finance center out at Columbus, Ohio. We have about 3, 000 people there cutting checks to all of our vendors. They are administering 390,000 contracts. It takes 15 linear miles of shelf space to hold the contracts. When we sign and issue a contract, we print on the average 17 copies of the contract. One of them goes to Columbus. So we've got miles and miles and miles of shelf space every place, administering what's in essence a paper-bound system.

The first time I went out to Columbus, there were great big sorting wheels just to sort the documents that were coming into the mailroom every day. This is 1903's office technology, but that's kind of where we were.

I find these two phenomena very curious. We are so advanced in some areas it just would dazzle you to see some of the things we can do. I can walk into the Pentagon, go down the hall and see real time video footage of a camera taking pictures over Bosnia. It's startling some of the things we can do. And yet, I go to places where we're still using 1930's sorting wheels for the documents.

These phenomena are related in a very important way. As I said, our budget's been cut now for 15 years. We happen to think we're absolutely at rock bottom. But I don't know that there's a lot of support for dramatically increasing our budget. So if we're going to be able to sustain the kind of war fighting modernization that we need for the future, and we're currently not buying enough things for the long term modernization of the department.

We're going to have to create our own spending power. And we're going to have to do that by shaking loose dollars out of the support structure of this department. Too much of our scarce resource is being consumed by old-fashioned business practices.

When Secretary Cohen came in a year ago, I was the comptroller, and he said, "We've got to do something about this." He put me in charge of an effort to try to come up with ways to bring in modern business practices. There are four basic things that we're trying to do. The first is to streamline our headquarters operations. We cut out about a third of the direct office supporting the Secretary. There were actually 3,000 people who were involved in that, and we've cut out a thousand of them. We've cut back about 30,000 out of our defense agencies, our support agencies. We're trying to launch a wide set of changes in our business practices and I'll describe a few of those in just a moment.

Second, we're trying to compete government jobs against the private sector.

We've done about 2,000 competitions. This is governed by a process called the A-76 process, a circular that OMB maintains. We know that when we compete jobs head to head on the average, the government wins half of the time and the other of the time the private sector wins. When the private sector wins, the savings are usually about 40% and when the government wins the savings are usually about 20%. So we know we can save substantial sums. We intend to compete 200,000 of our jobs over the next four years and try to shake dollars out of the system. We will generate, when we're finished, annual savings of over $2.5 billion.

Third, we need to close bases. We've gone through four rounds of base closures. We're still doing base closures from those 4 rounds. But we still have to go for another couple of rounds of base closures. This has not been gladly received by Congress. We did not get permission to proceed. I don't know what we're going to do exactly, but we have too much physical infrastructure that we're having to support. We're going to have to figure some way to streamline that physical infrastructure because it's taking dollars away from what we really need it for, which is modernization of programs in our war fighting.

Within our bases, we're trying everything we can to shrink infrastructure. We're going to, over the next four years, knock down 8,000 buildings that we consider obsolete. We will break even on the fifth year. We've got a lot of old structure, going back to World War II. The phenomena is if you've got a building and you've got heat going into it, people move into it. (Laughter.) So we are knocking down old buildings.

We would like to try to privatize all of our utility systems over the next five years. We have, as I said, about 550 utility systems. Now, some places, that's very plausible—San Diego, Norfolk. But you get into the middie of the desert, it's probably not very plausible that we're going to be able to do that. But again, we're on a detailed plan to try to privatize our utility system.

The fourth dimension of our plan involves new business practices I will cite only a few examples. For example, we're trying to move to what we call a paper free acquisition environment. I described to you a current system that is enormously paper-bound, 15 linear miles of shelf space devoted to contracts at our finance center. We found some very innovative ways to do that. We're essentially shifting over our acquisition system to an Internet-based system. We had some clever folks who said, "You know, before that contract ever turned into paper, it was electrons on it's way to a printer. If you can intercept those electrons and drop them in a server, you can access them with standard search tools and get enterprise-wide imaging, you know, on the cheap. "You don't have to buy all those scanners if you can find a way just to borrow the electrons at the outset. One of the advantages, of course, is that you don't have to have everybody come on board the system to get an enterprise-wide solution. People can get on board it when they want to and when it's in their purposes. We'll break even just with file clerk costs alone in the first two years. We're making very good progress on this. We're probably posting thousands of contracts a month, never printing the contract.

We've been able to shift over to a paper free process for our technical drawings-about 65% of all of our technical drawings, and we have a lot of them. We have 5 million items in our active stock list that we're buying from industry. Many of them have to have technical specs behind them for competition. Now, 68% of them are only in electronic format and we only compete them in electronic format. So, we're trying very hard to shift over to this sort of a world.

I think some of the greatest promise lies in some of the new, but not exotic, technology. For example, electronic malls. We are now shifting over in a very dramatic way to using electronic malls. It's not just cheaper, but it's a revolutionary way of approaching acquisition. Electronic malls are "democratizing" the acquisition process.

Rather than process low value contracts our acquisition professionals are developing the underlying contractual instrument for the acquisition. Then you turn it over to a first sergeant and let him buy his own batteries or his own spark plugs. We don't have to have an acquisition system that's buying it for him and then a finance system that's paying the bill later on. You know, we can integrate all that together into one instrument. For the first time we now have a department-wide electronic mall. Today they are limited. We have virtually all our food stuffs [on-line], and, when you feed 1.2 million people a day, there's an awful lot of food in the messing system. So it's having dramatic implications already.

Now, where are we with this agenda? I'm very gratified at the progress that we've made, but it is still slow. I think if I were to give us a grade, I'd give us a B+ on effort and a B- on progress. You know, we're doing better than, I think, the average but we have a long ways to go. I think there are a lot of promising, leveraging technologies that we are putting in place. We've got a lot of building blocks for dramatically improved performance. But we're still fighting organizations that know the old way of doing business and don't want to give it up. We're having to find ways to push them over the edge to adopt new practices.

Let me shift now to a new subject. I'm actually going to give you two speeches today. This is the end of the first one. I'll be honest, I'm campaigning here. And I want to talk to you about information security and infrastructure protection.

This country is wide open to attack electronically. A year ago, concerned for this, the department undertook the first systematic exercise to determine the nation's vulnerability and the department's vulnerability to cyber war. and it was startling, frankly. we got about 30, 35 folks who become the attackers, the red team. We gave them enough money to go down to CompUSA or where ever. They only could buy computers off the shelf. They were given no special software. The only software they were allowed to use was stuff they either develop themselves or they downloaded from hacker web sites. They spent three months getting ready. We didn't really let them take down the power system in the country, but we made them prove that they knew how to do it.

Now, why are we so vulnerable as a country? We're vulnerable because of the enormous productivity improvements that we've sought through information technology in the last 20 years. You're familiar with the SCADA system, Supervisory Control And Data Acquisition Systems? These kinds of systems are used to control physical networks, for example remote switches on a power grid that will open additional switches or bring on new transformers or sensors, valves and pumping stations that are used to regulate the flow of oil through a pipeline.

These systems are used for water irrigation systems in the west. America's infrastructure is being run now through these Supervisory Control And Data Acquisition Systems, SCADA systems. They're commercial systems.

Increasingly, American business, in order to save money and to shed itself of the cost of proprietary networks, is moving these systems onto an Internet-based communications network. So we're finding increasingly, America's business and utilities are controlling the infrastructure through a technology that was never designed with security in mind.

The Defense Department, is surprisingly vulnerable, too. The reason is that over the last 10 years, we have been dramatically shifting our infrastructure over to commercial structure rather than government-owned communications systems. I remember the first time I ever went out to Strategic Air Command 15 years ago.

You'd go out there and there'd be five phones sitting on the desk—a gold phone and a red phone and blue phone and all this kind of stuff. They were all government-owned phones and government switches and government-unique lines. We had our own system. Well, we don't do that anymore. Ninety-five percent of all of our communications now is over commercial systems and networks. And one of the things that surprised us during

Eligible Receiver was the degree to which we had become vulnerable to penetration because we were riding on these networks.

Now, this brings me to my bottom line here. I understand this is a bit controversial, but ultimately you are no different from us. You are going to increasingly do your business over a media that was never designed with security in mind. It was designed as a research tool. We in DoD invented it. It was designed as a research tool. And the protocols are wide open. Everybody knows how to plug in. That's why it's so powerful now in business applications.

So how do you provide security in an environment and a media that inherently is insecure? Many things are required, one of the things that is essential is encryption. Now, I know this is a hot debate and part of the discussion I had with Peter while we were waiting was the issue of encryption. Peter's first question was, "Are you with law enforcement or are you with commerce?" This is the debate that's occurring in Washington. It isn't exactly analogous to Justice versus Commerce. There are law enforcement concerns and Justice and the FBI are responsible for those. We want them responsible for those. Then there are economic response concerns and frankly, civil liberty concerns. Those are contending values of equal importance in our democracy, equal weight, in my mind. I do not believe that it's more important to protect ourselves against terrorists if it means it comes at the expense of civil liberties in the United States.

But I also don't believe that civil libertarians or cyber libertarians have a right to say we as a government have no responsibility to protect American society against criminals or terrorists.

We're going to have to strike a balance here. I personally believe that the debate of whether America's government is threatening our civil liberties is a fraudulent debate. We've never proposed anything that was any different than the mechanism we use every day to balance privacy versus law enforcement and security. Our police don't break into people's houses without a search warrant.

We know how to protect America's privacy, and we know how to safeguard civil liberties. There's a very real reason we fought wars—for these values, these civil liberty values.

We know how to balance them in this country, and we know how we'd balance them as well in this area. And I think that, frankly, the debate that's emerged has been a fraudulent debate, I hope I don't offend people when I say this. We know how to protect the individual and society in America.

Now, you may say that that means I'm siding with law enforcement. I'm not. I think that it's impossible to find a technical solution to this problem.

But I do think it's essential we find technical solution for protection if you're going to operate through the Internet.

Our position in the Department of Defense, and I frankly think it should be your position as well, is that if you're going to operate through these public, insecure modalities, you have to secure yourself And you have to do that through encryption. But I've also got to say the most dangerous thing in the world for us as a war fighter is to get an encrypted message that's a spoofed message. There's an impression of authenticity that comes with an encrypted message. You have to be able to determine the validity of the individual who is sending it to you.

Now, from a business standpoint, I cant' image any of you business people who would turn over to your employees the right to spend your dollars or cut checks or ship technical information and not require those employees to leave an electronic fingerprint on it when they do it. It's a basic premise of internal control.

So your interests and our interests are no different. What it leads me to say is that I'm not picking sides between the law enforcement community and the commerce community, as it were, in this debate. I'm saying we have to go right down through the middle. We have to protect ourselves in this environment and it's got to be with encryption and some form of security management, key recovery in our case. But we're going to make it voluntary. It's our choice and we're going to buy it. We're not going to ask that it be mandated through law on anybody. We're going to pay for it. And we've entered into contracts with a number of companies to help us develop a security infrastructure. We'll get the first one running this fall with Netscape, and hopefully, it'll be operational in October.

But I'm telling you, this is something that you've got to do for own companies and it's something we all have to do, frankly, for the country. It's in your narrow interests as company executives and it's in our broader national interest to do this. And I would ask you to step past this debate that we're having on cyber liberties versus law enforcement. We're going to have to get to a more sophisticated understanding of this problem, and we don't have a lot of time.

I'm going to stop there and I hope that I've stimulated enough interest that there might be some questions.

CHAPTER 19

ENCRYPTION: PROTECTING THE WEB

American Forces Press Service

WASHINGTON—Picture the Internet with millions of bits of your medical and financial information, credit card numbers, legal documents and personal e-mail speeding through cyberspace.

How do we ensure the privacy of our electronic data? How can we prevent hackers from accessing our homes and offices? How will America prevent terrorists from acquiring defense secrets? Federal officials are working to answer these questions.

Electronic communications have been a national security concern ever since private citizens, businesses and the government began relying heavily on computers.

"The Information Age has brought us the Internet, an interconnected global economy and the promise of connecting us all to the same vast world of knowledge, but with that exciting promise comes new challenges." Vice President Al Gore said at the White House Sept. 16 while announcing a new federal encryption policy. He said the policy dramatically increases privacy and security for families and businesses while protecting national security.

Encryption involves scrambling access codes to computer programs and systems to prevent intruders from entering and taking control. Federal officials and private industry are developing strong encryption products that enable law-abiding citizens to protect the privacy of their communications and electronically stored data. At the same time, the programs don't affect law enforcement officials' ability to ensure public safety when criminals use these products.

"We must ensure that the sensitive financial and business transactions that now cruise along the information superhighway are 100 percent safe in cyberspace," Gore said. "We must ensure that new technology does not mean new and sophisticated criminal and terrorist activity which leaves law enforcement outmatched.

We can't allow that to happen.

"We will give law enforcement the ability to fight 21st century crimes with 21st century technology, so our families and businesses are safe, but on-line outlaws are not," Gore said.

Deputy Defense Secretary John Hamre hailed the new policy as an "enormous step forward." He said encryption and computer security programs are vital to the Defense Department. Hackers during the last year successfully accessed several defense systems, he said, and such intrusions will undoubtedly increase in the future.

The new policy covers four main areas:

- American companies will be able to use encryption programs of unlimited strength when communicating between most countries.
- Health, medical and insurance companies will be able to use far stronger electronic protection for personal records and information.
- Law enforcement will still have access to crime-related information under strict and appropriate legal procedures. A federal technical support center will provide federal, state and local law enforcement officials with the resources and technical capabilities needed to fulfill investigative responsibilities.
- The federal government will maintain full ability to fight terrorism and monitor terrorist activity that poses a grave danger to Americans.

Hamre said the military needs encryption because it sends 95 percent of its communications over public networks—public telephone lines and switches, computers and so forth.

Under a proposed new DoD web site policy, DoD plans to review its 1,000 or so Web sites to remove such sensitive information as building and installation plans, operations programs, research and development objectives and personnel data. A Pentagon spokesman said the goal is to provide site visitors useful information without going so far as to aid potential malefactors.

Web security is also a top Defense Department priority because of DoD's electronic commerce activities, Hamre said. "We're the largest company in the world," he noted. "Every month we write about 10 million paychecks. We write about 800,000 travel vouchers. One of our finance centers disburses $45 million an hour. We are a major, major force in business. Electronic commerce is essential for us."

Another department objective is to protect the national infrastructure—power, water and transportation systems, and others. Homeland defense is also an emerging national security challenge, Hamre said. "We must have strong encryption in order to do that, because most of this infrastructure now is being managed through distributed, computer-based management systems.

The new federal policy will also help the Defense Department and law enforcement officials battle terrorism, Hamre said.

"Terrorists and rogue nations are increasingly using [encryption] tools to communicate with each other and to lay their plans. We must have the ability to deal with that."

CHAPTER 20

THE NATIONAL INFORMATION INFRASTRUCTURE PROTECTION ACT OF 1996 LEGISLATIVE ANALYSIS

By
The Computer Crime and Intellectual Property Section
United States Department of Justice

I. INTRODUCTION: The Need for Legislative Reform

Although there has never been accurate nationwide reporting of computer crime, it is clear from the reports which do exist and from anecdotal information that computer crime is on the rise. For example, the Computer Emergency and Response Team at Carnegie-Mellon University reports that from 1991 through 1994, there was a 498% increase in the number of computer intrusions, and a 702% rise in the number of sites affected. *See* CERT Annual Report to ARPA. During 1994, for example, approximately 40,000 Internet computers were attacked in 2,460 incidents. *Id.* Similarly, the FBI's National Computer Crime Squad has opened over 200 hacker cases since the Squad was created in 1991.

That computer crime is on the rise is perhaps a natural result of introducing computers into American society. In an earlier era, the advent of the automobile opened the way for criminals to target the automobile itself (e.g., auto theft) or use it to facilitate traditional crimes (e.g., the bank robbery getaway vehicle). In addition, law enforcement had to learn to seize vehicles to search them for evidence of some offense unrelated to the vehicle itself (e.g., the box of documents in the trunk). In many of the same ways, computers, too, have proven important to criminal investigations. First, a computer may be the target of the offense. In these cases, the criminal's goal is to steal information from, or cause damage to, a computer, computer system, or computer network. Second, the computer may be a tool of the offense. This occurs when an individual uses a computer to facilitate some traditional offense such as fraud (e.g., a bank teller who once stole money from a cash drawer may now use a computer program to skim money directly from depositors' accounts). Last, computers are sometimes incidental to the offense, but significant to law enforcement because they contain evidence of a crime. Narcotics dealers, for example, may use a personal computer to store records pertaining to drug trafficking instead of relying on old-fashioned ledgers.

The different ways in which criminals can use computers have created a philosophical debate among law enforcement experts. Some argue that computer crime is nothing more than traditional crime committed with new, high-tech devices. Others contend that computer crime cannot be analogized to traditional crime and that combatting it requires both innovative law enforcement techniques and new laws designed to address abuses of emerging technologies. In 1984, Congress adopted the latter view and enacted discrete legislation to address crime in electronic environments. Although certain computer crimes appear simply to be old crimes committed in new ways (e.g., the bank teller who uses a computer program to steal money is still committing bank fraud), some computer offenses find their genesis in our new technologies and must be specifically addressed by statute. For example, the widespread damage caused by inserting a virus into a global computer network cannot be prosecuted adequately by relying upon common law criminal mischief statutes. Indeed, it is questionable whether Robert Morris, the individual responsible for launching the Morris worm and crippling 6,000 computers around the world, could have been prosecuted had Congress not had the foresight to enact the Computer Fraud and Abuse Act.

Whether classified as "old" or "new," computer crime creates unique problems for law enforcement and a concomitant threat to the public welfare. The most significant legislative problems stem from technology's shift from a corporeal to an intangible environment. This departure from a physical world (where items are stored in a tangible form that can be carried, such as information written on paper) to an intangible, electronic environment means that computer crimes (and the methods used to investigate them) are no longer subject to traditional rules and constraints. Consider, for example, the way the crimes of theft and criminal mischief have evolved. Before the advent of computer networks, the ability to steal information or damage property was to some extent determined by physical limitations. A burglar could break only so many windows and burglarize only so many homes in a week. During each intrusion, the burglar could carry away only so many items. This does not, of course, make this conduct trivial, but it points out that the amount of property a burglar could steal, or the amount of damage he could cause, had physical limits.

In the information age, of course, these limitations no longer apply. A criminal seeking information stored in a networked computer with dial-in access can acquire that information from virtually anywhere in the world. The quantity of information stolen or the amount of damage caused by malicious programming code may be limited only by the speed of the network and the criminal's computer equipment. Moreover, such conduct can easily occur across state and national borders.

This clear shift to a borderless, incorporeal environment and the increased risk that information will be stolen and transported in electronic form is difficult to address by relying upon older laws written to protect physical property. For example, the statute pertaining to interstate transportation of stolen property, 18 U.S.C. § 2314, speaks of "goods, wares and merchandise," and consequently has been held by at least one court not to apply to intangible property. *See United States v. Brown*, 925 F.2d 1301, 1308 (10th Cir. 1991). Similarly, the long-familiar extortion statute makes it illegal, in some cases, to threaten physical violence to property. 18 U.S.C. § 1951(a). Although a threat to fire bomb a building would clearly satisfy this test, a threat to delete files may not.

II. THE STRUCTURE OF TITLE 18 REFORM

There are two ways, conceptually, to address the growing computer crime problem. The first would be to comb through the entire United States Code, identifying and amending every statute potentially affected by the implementation of new computer and telecommunications technologies. The second would be to focus substantive amendments on the Computer Fraud and Abuse Act to specifically address new abuses that spring from the misuse of new technologies.

The new legislation adopts the latter approach for a host of reasons:

(1) The United States, in a single statute, continues to address the core issues driving computer and information security at both domestic and international levels; that is, protecting the confidentiality, integrity, and availability of data and systems. Indeed, these three themes provide the foundation for the Organization for Economic Cooperation and Development's (OECD) *Guidelines for the Security of Information Systems*. They also serve as the linchpin for emerging domestic works on information privacy. *See, e.g.*, Draft Principles for Providing and Using Personal Information, 60 Fed. Reg. 4362 (January 20, 1995)[hereinafter "Draft Principles"]. By patterning the amended Computer Fraud and Abuse Act on the OECD guidelines, the U.S. is at the forefront of rethinking how information technology crimes must be addressed—simultaneously protecting the confidentiality, integrity, and availability of data and systems. And by choosing this path, we may encourage other countries to adopt a similar framework, thus creating a more uniform approach to addressing computer crime in the existing global information infrastructure.

(2) In most cases, a single point of reference—The Computer Fraud and Abuse Act, 18 U.S.C. § 1030—is provided for investigators, prosecutors, and legislators as they attempt to determine whether a particular abuse of new technology is covered under federal criminal law.

(3) As new technologies are introduced and the criminal law requires reconsideration, fine-tuning § 1030 may well be adequate, and it will not be necessary to continually parse through the entire United States Code.

(4) This statutory scheme will give us a better understanding of the scope of the computer crime problem by enabling more reliable statistics to be generated regarding computer abuse. Under current law, computer crimes can be charged under a host of criminal statutes, and this situation will continue if the U.S. chooses a patchwork approach and amends the various provisions of Title 18 to address new computer crimes. The existence of various computer crime provisions in different parts of Title 18 exacerbates an already obvious problem; i.e., computer crime experts have long admitted that there are no centralized computer crime statistics, not even within the law enforcement community. Indeed, a June 1996 study by the United States Sentencing Commission concluded that there were only 174 cases in which the statute of conviction included 18 U.S.C. § 1030, but conceded that

> . . . pertinent questions remain unanswered. For example, how much criminal behavior that could have been successfully prosecuted under 18 U.S.C. § 1030 was prosecuted under other fraud statutes . . . ?

United States Sentencing Commission, *Report to Congress*: Adequacy of Federal Sentencing Guideline Penalties for Computer Fraud and Vandalism Offenses, pp. 2, 6. By centralizing computer crimes under one statute, we may better measure existing harms, anticipate trends, and determine the need for further legislative reform. Additionally, amendments to the sentencing scheme of 18 U.S.C. § 1030 (and the Federal Sentencing Guidelines—2F1.1—upon which actual sentences are based), will be more effectively determined.

(5) Last, 18 U.S.C. § 1030(f) specifically provides that certain government officials, if engaging in lawfully authorized investigative, protective, or intelligence activities, are not restricted by § 1030. By amending only 18 U.S.C. § 1030 to address new high-tech offenses, this exception clearly continues to apply to any newly defined criminal conduct.

III. Specific Amendments: Protecting the Confidentiality, Integrity, and Availability of Systems and Information

A. Section 1030(a)(1)

Title 18, Section 1030(a)(1) originally provided that anyone who knowingly accesses a computer without authorization or exceeds authorized access and obtains classified information "with the intent or reason to believe that such information so obtained *is to be used* to the injury of the United States, or to the advantage of any foreign nation" is subject to a fine or imprisonment for not more that ten years (for a first offense). 18 U.S.C. § 1030(a)(1)(emphasis added). This scienter element apparently was included when this subsection was originally drafted because it is contained in 18 U.S.C. § 794(a). Section 794(a), however, provides for life imprisonment, whereas § 1030(a)(1) is only a ten-year felony. Therefore, it is more appropriate that the language of § 1030(a)(1) should track the language of 18 U.S.C. § 793(e), which also provides a maximum penalty of ten years' imprisonment for obtaining from any source certain information connected with the national defense and thereafter communicating or attempting to communicate it in an unauthorized manner.

It should be noted that, although there is considerable overlap between 793(e) and § 1030(a)(1) as amended, the two statutes do not reach exactly the same conduct. Section 1030(a)(1) would require proof that the individual knowingly used a computer without authority, or in excess of authority, for the purpose of obtaining classified information or restricted data, and subsequently performed some unauthorized communication or other improper act. In this sense then, it is the use of the computer which is being proscribed, not the unauthorized possession of, control over, or subsequent transmission of the information itself. Existing espionage laws would provide an adequate basis for the prosecution of individuals who attempt to peddle governmental secrets to foreign governments. However, a person who deliberately breaks in to a computer for the purpose of obtaining properly classified or restricted information, or attempts to do so, should be subject to criminal prosecution for this conduct.

B. Section 1030(a)(2)

Subsection (a)(2) is, in the truest sense, a provision designed to protect the confidentiality of computer data. As was noted in 1986 by the Senate Judiciary Committee,

> [t]he premise of 18 U.S.C. § 1030(a)(2) will remain the protection, for privacy reasons, of computerized credit records and computerized information relating to customers' relationships with financial institutions. . . . Because the premise of this subsection is privacy protection, the Committee wishes to make clear that 'obtaining information' in this context includes mere observation of the data.

S. Rep. No. 99-432 at 6.

With the continued evolution of the National Information Infrastructure (NII), however, Congress has come to recognize that not only financial records and credit information warrant federal protection. As noted in the commentary to the Draft Principles, "with the NII, the assumption is that large amounts of sensitive information will be on line, and can be accessed, perhaps without authority by a large number of network users." 59 Fed. Reg. at 27207. Moreover, "the NII will only achieve its full potential if individual privacy is properly protected." *Id.* Therefore, the new subsection 1030(a)(2) is designed to insure that it is punishable to misuse computers to obtain government information and, where appropriate, information held by the private sector. Moreover, the provision has been restructured so that different paragraphs protect different types of information, thus allowing easy additions or modifications to offenses if events require.

Certainly not all computer misuse warrants federal criminal sanctions. The problem is that no litmus test can accurately segregate important from unimportant information, and any legislation may therefore be under- or over-inclusive. For example, a frequent test for determining the appropriateness of federal jurisdiction—a monetary amount—does not work well when protecting information. The theft from a computer of a judge's draft opinion in a sensitive case or the copying of medical records might not meet such a monetary threshold, but clearly such information should be protected. Therefore, the act of taking all of this kind of information is now criminalized. Even so, it is important to remember that the elements of the offense include not just taking the information, but abusing one's computer authorization to do so.

The need to protect information is highlighted by recent studies indicating that people are increasingly misusing computers to obtain information. In 1993, the General Accounting Office (GAO) presented testimony before the House Government Operations Committee, Subcommittee on Information, Justice, Agriculture, and Transportation, on the abuse of National Crime Information Center (NCIC) information. The testimony stated that, following an investigation, GAO determined that (1) NCIC information is valuable, (2) such information has been misused by "insiders" (individuals with authorized access), (3) this misuse included selling NCIC information to outsiders and determining whether friends and relatives had criminal records, and (4) incentives for misuse outweighed potential penalties. Statement of Laurie E. Ekstrand, July 28, 1993, p. 6 [hereinafter "Ekstrand Statement"]. The GAO found that some of this misuse jeopardized the safety of citizens and potentially jeopardized law enforcement personnel. *Id.* at 16. Moreover, because there were no federal or state laws specifically directed at NCIC misuse, most abusers of NCIC were not criminally prosecuted. *Id.* at 17. GAO concluded that Congress should enact legislation with strong criminal sanctions specifically directed at the misuse of NCIC. *Id.* at 20.

Of course, protecting only NCIC data (or, more broadly, criminal history information), would be underinclusive, because other types of sensitive data are clearly at risk. For example, during Operation Desert Storm, it was widely reported that hackers accessed sensitive but unclassified data regarding personnel performance reports, weapons development information, and logistics information regarding the movement of equipment and personnel. *Teen tapped computers of U.S. military*, Chicago Tribune, November 21, 1991 at 3. NASA computers have also been penetrated, *Computer Hacker Charged with Entering NASA System*, Washington Post, September 26, 1991 at A20, as have at least two federal courthouse computer systems. *See*, *e.g., U.S. Says Hackers Scanned Data,* The New York Times, November 15, 1992, at A40. Some Internal Revenue Service employees also improperly used IRS computers to examine tax return information. *I.R.S. Staff Is Cited in Snoopings,* The New York Times, July 19, 1994, at D1, D5.

Clearly, the government should be able to prosecute individuals who obtain government information by misusing computers. Importantly, 18 U.S.C. § 1030(a)(2), as amended, does not punish the mere acquisition of information (which might unduly impede the free flow of ideas), but prohibits intentionally accessing a computer without or in excess of authority *and* then obtaining such information. Moreover, to the extent that the information obtained is or should be available, it should be obtained through legal means (e.g., public sources or FOIA) and not through hacking.

Subsection 1030(a)(2)(C) is designed to protect against the interstate or foreign theft of information by computer. Such a provision is necessary in light of the Tenth Circuit's decision in *United States v. Brown,* 925 F.2d 1301, 1308 (10th Cir. 1991), where the court held that purely intangible intellectual property, such as a computer program, cannot constitute goods, wares, merchandise, securities, or moneys which have been stolen, converted, or taken within the meaning of 2314. "Information" as used in this subsection is meant to be broadly construed and includes information stored in intangible form. Moreover, consistent with Congress's prior construction of § 1030(a)(2), "obtaining information" includes merely reading it; i.e., there is no requirement that the information be copied or transported. This is critically important because, in an electronic environment, information can be "stolen" without asportation, and the original usually remains intact.

Some computers may qualify under more than one subsection of § 1030(a)(2); for example, a particular government computer might be covered by both § 1030(a)(2)(B) and (a)(2)(C). This overlap serves to eliminate legal issues that might have arisen had Congress made the provisions mutually exclusive. Conceivably, in a given case, it may not be clear whether information taken from a government contractor's computer constitutes "information from any department or agency of the United States" under § 1030(a)(2)(B), but the offense might still be chargeable under § 1030(a)(2)(C) if the elements of that subsection are satisfied.

The seriousness of a breach in confidentiality depends, in considerable part, on either the value of the information or the defendant's motive in taking it. Thus, the statutory penalties are structured so that merely obtaining information of minimal value is only a misdemeanor, but certain aggravating factors make the crime a felony. More specifically, the crime becomes a felony if the offense was committed for purposes of commercial advantage or private financial gain, for the purpose of committing any criminal or tortious act in violation of the Constitution or laws of the United States or of any State, or if the value of the information obtained exceeds $5,000.

As for enhancements not based on the value of the property obtained, recent documented cases indicate that individuals misuse information for a variety of unacceptable purposes. The terms "for purposes of commercial advantage or private financial gain" and "for the purpose of committing any criminal or tortious act" are taken from the copyright statute (17 U.S.C. § 506(a)) and wiretap statute (18 U.S.C. § 2511(1)(d)) respectively.

As for the monetary threshold, any reasonable method can be used to establish the value of the information obtained. For example, the research, development, and manufacturing costs, or the value of the property "in the thieves' market," can be used to meet the $5,000 valuation. *See, e.g., United States v. Stegora*, 849 F.2d 291, 292 (8th Cir. 1988).

The relationship between the existing § 1030(a)(3) provision and the newly amended § 1030(a)(2) merits some discussion. Section 1030(a)(3) protects the computer from outsiders, even if the hacker obtains no information. Thus, an intruder who violates the integrity of a government machine to gain network access is nonetheless liable for trespass even when he has not jeopardized the confidentiality of data. Section 1030(a)(2), on the other hand, protects the confidentiality of data, even from intentional misuse by insiders. Additionally, although a first violation of § 1030(a)(3) is always a misdemeanor, a § 1030(a)(2) violation may constitute a felony if the information taken is valuable or sufficiently misused. *See* § 1030(c)(2)(B)(raising the offense to felony level based upon the value or intended use of the improperly acquired data). Although a single act may violate both provisions, the provisions protect against different harms and, in any event, the actor's conduct would be aggregated for the purposes of sentencing.

C. Subsection 1030(a)(3)

Three substantive changes were made to § 1030(a)(3). First, the word "adversely" has been deleted because including this term suggests, inappropriately, that trespassing in a government computer may be benign.

Second, for clarity, the term "the use of the Government's operation of such computer" has been replaced with the term "that use by or for the Government of the United States." When a computer is used for the government, the government is not necessarily the operator, and the old term may have led to confusion. Consistent with this change, a similar change was made to the definition of "federal interest computer" (redesignated as "protected computer") in § 1030(e)(2)(A). Third, Congress inserted "nonpublic" to modify "computer of a department or agency of the United States." This change is intended to reflect the growing use of the Internet by government agencies and, in particular, the establishment of World Wide Web home pages and other public services. Arguably, a person charged under the old subsection (a)(3) might have asserted as a defense that he was not "without authorization to access *any* computer of a department or agency of the United States," because he was authorized to access some publicly available computer of that department or agency, such as a Web site. While this defense would almost have negated the law and thus defied a common-sense interpretation of the former law, Congress added the word "non-public" to make it perfectly clear that a person who has no authority to access any non-public computer of a department or agency may be convicted under (a)(3) even though permitted to access publicly available computers.

D. Subsection 1030(a)(4)

Subsection 1030(a)(4) has been amended to insure that felony level sanctions apply when unauthorized use of the computer (or use exceeding authorization) is significant. At the time the "computer use" exception was originally crafted, the Senate Judiciary Committee noted that:

> [T]he mere use of a computer or computer service has a value all its own. Mere trespasses onto someone else's computer system can cost the system provider a "port" or access channel that he might otherwise be making available for a fee to an authorized user. At the same time, the Committee believes it is important to distinguish clearly between acts of fraud under (a)(4), punishable as felonies, and acts of simple trespass, punishable in the first instance as misdemeanors. That distinction would be wiped out were the Committee to treat every trespass as an attempt to defraud a service provider of computer time.

S. Rep. No. 99-432, 99th Cong., 2d Sess. 10 (1986). *See also* H.R. Rep. No. 99-612, 99th Cong., 2d Sess. 12 (1986).

Although Congress retains the concern about converting every trespass into a felony scheme to defraud, this new amendment clearly recognizes that a blanket exception for computer use may be too broad. Hackers, for example, have broken into Cray supercomputers for the purpose of running password cracking programs, sometimes amassing computer time worth far in excess of $5,000. In light of the large expense to the victim caused by some of these trespassing incidents, it is more appropriate to except from the felony provisions of subsection 1030(a)(4) only cases involving no more than $5,000 of computer use during any one-year period.

E. Subsection 1030(a)(5)

Subsection 1030(a)(5) was completely restructured in 1994, but the 1994 law may have had some unintended consequences. Most notably, certain government and financial institution computers may have been denied previously existing federal protection; some hacking activities may have been inappropriately decriminalized; and certain insider conduct may have been inappropriately criminalized.

In the 1994 amendments, the reach of this subsection was broadened by replacing the term "federal interest computer" with the term "computer used in interstate commerce or communications." The latter term is broader because the old definition of "federal interest computer" in 18 U.S.C. § 1030(e)(2)(B) covered a computer "which is one of two or more computers used in committing the offense, not all of which are located in the same State." This meant that a hacker who attacked other computers in the same state was not subject to federal jurisdiction, even when these actions may have severely affected interstate or foreign commerce. For example, individuals who attack telephone switches may disrupt interstate and foreign calls. The 1994 change remedied that defect.

However, the definition of federal interest computer actually covered more than simply interstate activity. More specifically, 18 U.S.C. § 1030(e)(2)(A) covered, generically, computers belonging to the United States Government or financial institutions, or those used by such entities on a non-exclusive basis if the conduct constituting the offense affected the Government's operation or the financial institution's operation of such computer. By changing § 1030(a)(5) from "federal interest computer" to "computer used in interstate commerce or communications," Congress may have inadvertently eliminated federal protection for those government and financial institution computers not used in interstate communications. For example, the integrity and availability of classified information contained in an intrastate local area network may not have been protected under the 1994 version of 18 U.S.C. § 1030(a)(5), although its confidentiality continued to be protected under 18 U.S.C. § 1030(a)(1). To remedy this situation in the 1996 Act, 18 U.S.C. § 1030(a)(5) was redrafted to cover any "protected computer," a new term defined in § 1030(e)(2) and used throughout the new statute—in § 1030(a)(5), as well as in § 1030(a)(2), (a)(4), and the new (a)(7). The definition of "protected computer" includes government computers, financial institution computers, and any computer "which is used in interstate or foreign commerce or communications."

This broad definition addresses the original concerns regarding intrastate "phone phreakers" (i.e., hackers who penetrate telecommunications computers). It also specifically includes those computers used in "foreign" communications. With the continually expanding global information infrastructure, with numerous instances of international hacking, and with the growing possibility of increased global industrial espionage, it is important that the United States have jurisdiction over international computer crime cases. Arguably, the old definition of "federal interest computer" contained in 18 U.S.C. § 1030(e)(2) conferred such jurisdiction because the requirement that the computers used in committing the offense not all be located in the same state might be satisfied if one computer were located overseas. As a general rule, however, Congress's laws have been presumed to be domestic in scope only, absent a specific grant of extraterritorial jurisdiction. *E.E.O.C.*

v. Arabian American Oil Co., 499 U.S. 244 (1991). To ensure clarity, the statute was amended to reference international communications explicitly.

Another concern with the 1994 version of 18 U.S.C. § 1030(a)(5) involved the overall statutory scheme. Under the 1986 version of subsection § 1030(a)(5), the actor causing the harm must have been *without* authority to access the victim computer. As such, the provision never applied to insiders, although insiders are often responsible for intentionally causing computer damage. Indeed, the Justice Department was forced to decline prosecution in some cases where individuals intentionally inserted malicious programming code into computers, because those individuals were authorized to access the attacked system. The 1994 law, in contrast to the 1986 version, appropriately applied to both insiders and those without authorized access who intentionally caused damage.

Unfortunately, however, by eliminating the trespassing requirement, *and at the same time* requiring the government to prove that the actor either intentionally or recklessly caused damage, the 1994 law no longer punished a person who broke into a federal interest computer and "thereby caused loss." *See* 18 U.S.C. § 1030(a)(5)[1986 version]. Thus, the enactment of the 1994 legislation decriminalized some hacking and inadvertently sent the message that breaking into computers was acceptable so long as the actor neither intended nor recklessly caused damage. However, in these 1996 amendments, criminal liability for such behavior has been restored. This was clearly necessary in light of the increased importance of computer networks in today's society and the nation's considerable interest in creating a trusted national information infrastructure that insures the confidentiality, integrity, and availability of information and systems.

This problem, now corrected, arose because the 1986 and 1994 versions of section 1030(a)(5) defined improper conduct in completely different ways—the former by focusing only on the actor's authority to access the computer; the latter by considering solely the actor's intent. Of course, these two separate litmus tests each cover important aspects of criminal computer damage, but neither measure, taken alone, fully succeeds in describing the acts which should be criminal. For example, although those who intentionally damage a system should be punished regardless of whether they are authorized users, it is equally clear that anyone who knowingly invades a system without authority and causes significant loss to the victim should be punished as well, even when the damage caused is not intentional. In such cases, it is the intentional act of trespass that makes the conduct criminal. To provide otherwise is to openly invite hackers to break into computer systems, safe in the knowledge that no matter how much damage they cause, they commit no crime unless that damage was either intentional or reckless. Rather then send such a dangerous message (and deny victims any relief), it is better to insure that § 1030(a)(5) criminalizes all computer damage done by outsiders, as well as intentional damage by insiders, albeit at different levels of severity.

By using a matrix, it is easy to see that neither the 1986 law nor the 1994 law was adequate, although they fail in different categories.

MATRIX 1: § 1030(a)(5)[1986 Version]
[Based on the defendant's authority to access the computer]

	Trespassers	Authorized Users
Intentional Damage	Felony	No crime
Reckless Damage	Felony	No crime
Negligent Damage	Felony	No crime

MATRIX 2: 18 U.S.C. § 1030(a)(5)[1994 Version]
[Based on the defendant's criminal intent to damage]

	Trespassers	Authorized Users
Intentional Damage	Felony	Felony
Reckless Damage	Misdemeanor	Misdemeanor
Negligent Damage	No crime	No crime

Conceptually, a comprehensive statutory scheme does not treat these two tests—mental state and authority to access—as mutually exclusive. Instead, it integrates them to cover all kinds of serious misconduct. Just as important, it recognizes that some behaviors are less serious, or should not be criminal offenses at all. For example, the 1994 law created a misdemeanor for reckless damage without distinguishing between trespassers and authorized users. Whether authorized users should ever be criminally liable for reckless damage is a debatable question. For example, it could be deemed reckless in today's computer environment to intentionally copy a file from a floppy diskette to a hard drive without first running a virus scan—although imposing criminal sanctions for such conduct is clearly inappropriate, absent other evidence of criminal intent. On the other hand, reckless trespassers warrant felony prosecutions, since they are unauthorized users who pose significant risks to computer systems. Thus, Congress has now chosen an approach that integrates access and authority tests in the following way:

MATRIX 3: 18 U.S.C. § 1030(a)(5) [THE NEW LAW]
[Based on the defendant's authority to access the computer *and* criminal intent to damage]

	Trespassers	Authorized Users
Intentional Damage	Felony	Felony
Reckless Damage	Felony	No crime
Negligent Damage	Misdemeanor	No crime

Essentially, this new statute provides that individuals who access protected computers without authority are responsible for the consequences of their actions, but those accessing with authority are criminally liable only if they intend to cause damage to the victim.

Although subsections § 1030(a)(5)(B) and (a)(5)(C) require that the actor cause damage as a result of his or her unauthorized access, damages are not limited to those caused by the process of gaining illegal entry. Rather, all damage, whether caused while gaining access or after entry, is relevant.

Another concern with the 1994 law was that it required both "damage" and "loss," without clearly articulating what constituted "damage." For example, intruders often alter existing log-on programs so that user passwords are copied to a file which the hackers can retrieve later. After retrieving the newly created password file, the intruder restores the altered log-on file to its original condition. Arguably, in such a situation, neither the computer nor its information has been damaged. Nonetheless, the intruder's conduct allowed him to accumulate valid user passwords to the system, required all system users to change their passwords, and required the system administrator to devote resources to re-securing the system. Thus, although there may be no permanent "damage," the victim does suffer "loss." If the loss to the victim meets the required monetary threshold, the conduct should be criminal, and the victim should be entitled to relief.

It would not have been possible to address all of these concerns by making only minor amendments to subsection 1030(a)(5). Thus, the statutory scheme was altered to both simplify its provisions and adopt the sanctions provided in Matrix 3.

As discussed further below, the term "damage" remains, but is now defined in 18 U.S.C. § 1030(e)(8). Consistent with the view that § 1030(a)(5) protects the integrity and availability of data and systems, "damage" means any impairment of those attributes. The statutory language avoids listing specific acts that can cause such impairment to insure that its coverage is suitably broad. For example, in the 1986 version, the terms "alters, damages or destroys information," were included, inadvertently raising new issues (e.g., whether encrypting data satisfies this test since the underlying original information remains unchanged). Rather than providing a list of prohibited actions and risk being underinclusive, the statute focuses instead on the harms it seeks to prevent.

This harm-based definition of "damage" can now be found in subsections 1030(e)(8)(A) through (D). As in the past, the term "damage" will require meeting one of several significant thresholds. Two of these measures survive from earlier versions of § 1030: the first is significant financial losses—although raised in these amendments from $1000 to $5000— [§ 1030(e)(8)(A)]; the second is potential impact on medical treatment [§ 1030(e)(8)(B)]. In addition, Congress has listed two new threshold harms in its definition of "damage": causing physical injury to any person [18 U.S.C. § 1030(e)(8)(c)] and threatening the public health or safety [18 U.S.C. § 1030(e)(8)(c)]. As the NII and other network infrastructures continue to grow, computers will increasingly be used for access to critical services such as emergency response systems and air traffic control, and will be critical to other systems that we cannot yet anticipate. Thus, any definition of "damage" must broadly encompass the types of harms against which people should be protected.

Having amended the structure of § 1030(a)(5), Congress needed to amend the civil penalty provision under § 1030(g). The subsection as amended provides that victims of computer abuse can maintain a civil action against the violator to obtain compensatory damages, injunctive relief, or other equitable relief, but damages are limited to economic damages for cases where the only damage suffered by the plaintiff is monetary loss as defined by § 1030(e)(8)(A).

F. Subsection 1030(a)(7)

New subsection (a)(7) is designed to respond to a growing problem: the interstate transmission of threats directed against computers and computer networks. Such threats, if accompanied by an intent to extort, may already be covered in some instances by the Hobbs Act, 18 U.S.C. § 1951, which applies to interference with commerce by extortion. They also may be covered in some instances by 18 U.S.C. § 875(d), which applies to interstate communication of a threat to injure the property of another. However, under both of these statutes, it is not absolutely clear that "property" includes the unimpaired operation of a computer or the unrestricted access to the data or programs stored in a computer and its peripheral equipment. Moreover, it is not clear that certain actions (such as encrypting someone's data and then demanding money for the key) constitute a threat to "injure the property of. . .another." *See* 18 U.S.C. § 875(d).

These concerns are not theoretical. In one recent case, for example, an individual threatened to crash a computer system unless he was granted access to the system and given an account. Another case involved an individual who penetrated a city government's computer system and encrypted the data on a hard drive, thus leading the victim to suspect an extortion demand was imminent. (This demand never came, however, and fortunately the victim was able to recover from the incident.) Although the number of such incidents is currently small, the explosion in network access has substantially increased the risk that such conduct will occur, and our nation's increased reliance on computers clearly suggests that such activities, if not deterred, will severely impair our ability to use the NII effectively. Moreover, since such extortion and threats will normally involve interstate and foreign communications, federal law enforcement needed a clear basis to address this new problem quickly.

It is worth noting that subsection (a)(7) covers any interstate or international transmission of threats against computers, computer networks, and their data and programs, whether the threat is received by mail, a telephone call, electronic mail, or through a computerized message service. The provision is worded broadly to cover threats to interfere in any way with the normal operation of the computer or system in question, such as denying access to authorized users, erasing or corrupting data or programs, or slowing down the operation of the computer or system. The extortion element is modeled after that in 18 U.S.C. § 875(b) and (d).

G. Sentencing Provisions: Subsection 1030(c)

The sentencing provisions of § 1030 have been altered to reflect the new statutory scheme and to address an old, technical error. As previously enacted, recidivists were only subject to enhanced penalties if they violated the same *subsection* twice. For example, if an individual violated the Act by committing fraud by computer [subsection (a)(4)] and later committed another computer crime offense by intentionally destroying medical records [subsection (a)(5)], he was not a recidivist because his conduct violated two separate subsections of § 1030. Congress has changed the statutory language to provide that anyone who is convicted twice of committing a computer offense will be subjected to enhanced penalties.

H. Jurisdiction: Subsection 1030(d)

Having created several new crimes in 18 U.S.C. § 1030, Congress needed to consider the jurisdictional grant in 18 U.S.C. § 1030(d). For some time, the Federal Bureau of Investigation and the United States Secret Service have shared concurrent jurisdiction over § 1030 based upon a Memorandum of Understanding. This new Act, by creating certain new crimes, does not alter any existing agreements, nor limit or alter an agency's "traditional" jurisdiction. Thus, there is new language in 18 U.S.C. § 1030(d) to insure that the *status quo* is maintained. For example, the new 18 U.S.C. § 1030(a)(2)(C) addressed gaps in 18 U.S.C. § 2314 (interstate transportation of stolen property), and the new 18 U.S.C. § 1030(a)(7) addressed gaps in 18 U.S.C. § 1951 (the Hobbs Act) and 18 U.S.C. § 875 (interstate communications). All of these statutes are within the traditional jurisdiction of the FBI, therefore 18 U.S.C. § 1030(d) did not extend to the United States Secret Service concurrent jurisdiction over these types of offenses, even when committed by computer. Subsections over which the Secret Service maintains concurrent jurisdiction are § 1030(a)(2)(A) and (B), (a)(3), (a)(4), (a)(5), and (a)(6).

CHAPTER 21

STATEMENT OF SENATOR CARL LEVIN (D-MICH) BEFORE SENATE PERMANENT SUBCOMMITTEE ON INVESTIGATIONS ON DOD'S VULNERABILITY TO INFORMATION WARFARE

May 22, 1996

"Information warfare" is a phrase that sounds like science fiction or a threat from some distant future. But the persons testifying here today know information warfare is not fiction, and it is not a future threat. Information warfare is a reality. It goes on today all across this country, and it poses a current threat to our national security. And we're not paying enough attention to it.

The problem is simple but profound. Today, our national security depends upon computers. Today, we can't move a battleship, communicate battlefield information, develop weapons, deliver supplies, assign personnel, aim missiles, or perform a thousand other military missions without computer systems.

These computer systems use communication pathways and software tools that frequently are not under the control of the Department of Defense. We're told that maybe ten percent of DOD's computer traffic is classified and moves on very secure systems. The other 90 percent of DOD data is unclassified and moves along poorly secured pathways, the majority of which are not government-owned or operated.

These pathways include telephone lines, cable, satellite feeds, even microwaves. Each is susceptible to monitoring, infiltration and manipulation.

Information warfare is based on that fact. Its battlefields are the pathways over which computer data is transmitted. The weapons are software programs that can read, intercept and even alter the data moving from one military computer to another.

The names of these new weapons are colorful. They include sniffers, phone phreaking, worms, Trojan horses, logic bombs and more. The stereotype of someone who breaks into computer systems is the teenage hacker playing games. But this stereotype and the colorful terminology are distractions from the real national security threat.

We will hear today about instances in which computer hackers have sold military information to foreign agents. How hackers attacked military computers to get information during Desert Storm. How hackers have used U.S. computers to lift nuclear information from another country's database, risking international crisis. What happens when hackers learn how to alter battlefield instructions or, during a military confrontation, simply paralyze the computers that move our military supplies and personnel? What if hackers impair our military systems with such subtle software that we don't even know the systems have been hit?

That's not all. An information warfare exercise conducted by the Rand Corporation for DOD looked at attacks on computer systems within the United States and its allies to sabotage domestic infrastructure such as transportation, utilities and finance. These attacks could result in train wrecks, city-wide power outages, banking disruptions, and worse—together generating a domestic chaos that could undermine our national security from within.

DOD is only now establishing the infrastructure needed to detect, assess and counter the threats posed by information warfare. Established three years ago, the Defense Information Systems Agency conducts simulated attacks on individual military computer systems to identify vulnerabilities—succeeding, by the way, in 88% of those simulated attacks. Over the last three years, the military services have each established an information warfare center to detect and counter attacks on their respective computer systems. It is only now that the first DOD wide vulnerability assessments are being made, with results that show we have a long way to go. For example, we will hear today that there are an estimated 250,000 attacks on military computers each year, of which only 4% are detected by the systems under attack and of which only 1 in 500 is reported.

The Defense Department has over 2 million computers, 100 long distance networks and 10,000 local networks. It has 550 installations that operate thousands of active computer systems. Few of these installations have a computer security expert charged with defending the integrity of the installation's computers and data. Filling that gap may be one important step to greater computer security. These hearings will hopefully identify other steps as well, and advance us from describing the problem to designing the solution.

We have years of experience defending our borders and our global interests. Now we have to learn to defend against attacks through cyberspace. I commend Senator Nunn and my colleagues for holding this hearing and look forward to the testimony.

CHAPTER 22

INTELLIGENCE COMMITTEE OFFERS ALTERNATIVE ENCRYPTION LEGISLATION TO ADDRESS SECURITY CONCERNS

September 11, 1997
Contact (202) 225-4121

The House Permanent Select Committee on Intelligence (HPSCI) today approved, by voice vote, legislation proposing an alternative to pending encryption legislation, known as the "Security and Freedom Through Encryption Act" (H.R. 695).

The HPSCI amendment in the nature of a substitute to HR 695 proposes safeguards in the law to meet national security and law enforcement concerns in the debate over the future of United States Encryption policy.

"All members of Congress, and particularly those of us on the Intelligence Committee, have a responsibility to find the proper balance between forward thinking commercial policies and the unquestioned need to protect the security of the American people and America's national interests. We are offering proposals to ensure that we do not plow full steam ahead into the 21st century's information age having seriously weakened our ability to protect the national security, " said HPSCI Chairman Porter J. Goss (Fl-14).

"American citizens have a right to their privacy and their access to the freest possible markets. But they also have a right to their safety and security. Terrorist groups that plot to blow up buildings; drug cartels that seek to poison our children, and those who proliferate in deadly chemical and biological weapons are all formidable opponents of peace and security in the global society. These bad actors must know that the United States' law enforcement and national security agencies, working under the proper oversight, will have the tools to frustrate illegal and deadly activity and bring international criminals to justice," Goss said.

"The bill referred to the Intelligence Committee attempts to deal with complex issues. The substitute adopted by the Committee addresses the legitimate national security and law enforcement concerns that are simply not addressed in H.R. 695. In that respect, the Committee substitute, in my judgment, furthers the debate on these important matters," noted Ranking Democrat Norm Dicks (WA-6).

The main elements of the Intelligence Committee's proposal are:

- Requires exports of encryption products to submit to a one-time review and to include features or functions (that need not be enabled by the manufacturer) allowing for immediate access to plaintext or to decryption information;

- Requires that encryption products manufactured and distributed for sale or use, or import for sale or use, in the United States after January 31, 2000 include features or functions that provide, upon presentment of a court order, immediate access to plaintext data or decryption information from the encryption provider;
- Does not change law enforcement's statutory requirements prior to intercepting oral, wire, or electronic (wireless) communications, or law enforcement's requirements prior to obtaining stored data. Law enforcement will specifically be required to obtain a separate court order to have data, including communications, decrypted;
- Allows for law enforcement access with delayed notification requirements, similar to those allowed in current wiretap statutory provisions:
  - Provide civil remedies and criminal penalties for unlawful access to or disclosure of plaintext or decryption information;
  - Require US government procurement of encryption technology that includes functions or features allowing for immediate access to plaintext or decryption information.

"Our committee has weighed in on these issues in the interest of furthering the important debate now underway about how best to accomplish the multiple goals of a sound encryption policy. Any encryption legislation we consider must take a balanced approach to the national security, law enforcement, public safety and privacy issues at stake. Our action today marks another step in this process, which no doubt will continue to unfold in the days and weeks ahead. I look forward to working with all sides on this debate as we tackle this complex but important issue," Goss said.

The HPSCI, which sought and received sequential referral of H.R. 695, is one of five House committees with jurisdiction on this issue. The Committees on Judiciary, International Relations, National Security and Commerce have also considered this legislation. The HPSCI expects to file its committee report with the House tomorrow, meeting its deadline for action set by the Speaker.

CHAPTER 23

THE UNITED STATES NEEDS A CLEAR AND REALISTIC ENCRYPTION POLICY

STATEMENT OF EDWARD GILLESPIE EXECUTIVE DIRECTOR, AMERICANS FOR COMPUTER PRIVACY

March 4, 1999

Mr. Chairman and members of the Subcommittee,

Thank you for the opportunity to testify before you on H.R. 850, the SAFE Act, sponsored by Representatives Goodlatte and Lofgren and over 200 members of the House. I serve as Executive Director of Americans for Computer Privacy (ACP), a broad coalition of 40 trade associations, over 100 companies and over 3,000 individuals. The coalition includes the financial services, manufacturing, high-tech, and transportation industries as well as law enforcement, civil-liberty, taxpayer and privacy groups. ACP supports policies that allow American citizens continued use of strong encryption without government intrusion, and advocates lifting export restrictions on U.S. made encryption products.

ACP strongly endorses enactment of the SAFE Act, and we appreciate the leadership provided by Representatives Goodlatte and Lofgren very much. We urge your subcommittee to report it promptly for full committee consideration.

As Vice President Gore said in September 1998 when he announced the current administration policy, developing a national encryption policy is one of the most difficult issues facing the country. It requires balancing many competing objectives—all of which are of great importance to the nation.

As ACP noted in our policy paper of May 8, 1998, strong encryption is essential to:

- Protecting the nation's infrastructure and assuring the integrity of information;
- Ensuring the privacy of electronic communications of American citizens and organizations;
- Protecting our national security interests;
- Safeguarding the public; and

- Maintaining U.S. leadership in the development of information technology industry.

As we move into the new millenium, information technology will play an increasingly important role in the way we govern, communicate, conduct commerce, and operate and protect our national infrastructure. Strong encryption is key to the continued vitality and growth of all of these activities. Accordingly, the United States needs a *clear and realistic* national policy to assure that industry is able to develop the products that will help us to meet our national objectives.

Significant progress was made last year with the Administration's policy announced by the Vice President in September and contained in the interim final regulations of December 31, 1998. ACP commends the government for the hard work and thoughtful consideration that went into the development of that policy and those regulations. Last year, ACP had several productive meetings with the Administration's inter-agency task force, including representatives from law enforcement and the Justice Department. Those meetings were conducted in good-faith on both sides and led to a greater understanding on both sides of the needs and concerns of the other. The Clinton Administration incorporated many of our interim recommendations into its updated export policy, including: export relief for encryption products that use symmetric algorithms up to and including 56-bits; products that use asymmetric algorithms up to and including 1024-bits; and relief for various sectors of the business community.

The Clinton Administration, however, has yet to allow U.S. encryption manufacturers to compete on a level playing field in the global marketplace. The Administration policy remains highly problematic and does not represent the clear and realistic national policy that this issue requires.

First, the Administration has entered into an agreement with 32 other countries—the Wassenaar Arrangement—containing certain export controls on encryption. Unfortunately, the Administration's encryption export regulations impose greater restrictions on American companies than those called for under the arrangement. As a first step, we believe the Administration should at least eliminate all controls on encryption software and hardware for products up to 64-bits, and should eliminate all reporting requirements on higher-level encryption exports. Such actions would make U.S. controls consistent with the revised Wassenaar Arrangement.

We also believe that the Administration's efforts to develop a global approach to this issue through the Wassenaar Arrangement are doomed to failure. We recognize that this is a global problem and if it were truly possible to achieve universal agreement that was fairly enforced, industry would no doubt be supportive. But Wassenaar only has 33 members and does not include encryption-producing countries such as China, India,

South Africa, or Israel. Further, the Administration should recognize that the Wassenaar Arrangement is only as effective as the implementing regulations adopted by the member countries. Some of the member nations will promulgate regulations that are less restrictive than those of the United States, thereby providing those nations with a competitive advantage over domestic encryption manufacturers. In short, the Wassenaar Arrangement is a toothless tiger.

Second, the Interim Rule falls short on a number of short-term points. For example, the Interim Rule does not fulfill the mandate promised by Vice President Gore on September 16 to allow all 56-bit encryption products to be eligible for export to all end-users (except terrorist states). In reality, the Interim Rule does *not* allow the export of 56-bit encryption chips, integrated circuits, toolkits, and executable or linkable modules for export under license exception except to U.S. subsidiaries.

Further, the Interim Rule is so complex that a number of the benefits in the new policy are undermined by provisions of the Interim Rule. For example, the reporting requirements are so onerous to companies that reporting costs may exceed the *price* of some products, much less the profit. In the same vein, the Government has shown little understanding of mass-market distribution techniques. It makes no sense that the Government does not expect manufacturers to be able to control mass-market encryption products using 56-bit encryption, but does expect manufacturers to be able to control mass-market encryption products using algorithms higher than 56-bits. Furthermore, it is simply impractical to expect manufacturers to collect reporting data on mass-market encryption products. My personal experience is that I never return registration cards on coffee makers, answering machines, or software products—I expect most people in this room have similar experiences.

And so the Administration's new policy, as grateful as we are for this limited progress, remains flawed even on its own terms.

Beyond this, in the encryption debate in the larger sense, we continue to have good-faith disagreements with the Administration about its current policy, which only Congress and this legislation can address.

Primarily, ACP believes that our current export policy short-changes our long-term national interest in that it puts at jeopardy our current global leadership in this vital technology. Strong, high-quality encryption products are now widely available from foreign makers. That renders our export policy an exercise in futility. We worry that America will lose this critical market to foreign makers. When and if it does, it will be too late to change U.S. policy and too late to preserve U.S. leadership in this vital arena.

If we *do* lose that U.S. leadership position, what will that mean? It will mean that the national security agencies will be confronting ubiquitous encryption made not by U.S. companies, but by foreign companies. Where then will the national security agencies go for technical help on encryption, if the most sophisticated encryption experts and product-makers reside abroad? It could put us in the untenable position of protecting our critical national infrastructure with foreign-made encryption.

We must retain leadership in this vital technology if we are to meet our long-term national security objectives. That is why we must assess our encryption export policies from a long-term, not a short-term, perspective.

In the long run, there can be no doubt that U.S. national security objectives are best served by an IT world in which U.S. companies are market leaders in all aspects, especially encryption. ACP's industrial members have ample evidence of the rapidly growing market share of foreign encryption and examples of U.S. businesses losing out to foreign manufacturers because of the U.S. export regulations. For example, a December 1997 study conducted by Trusted Information System found that 656 non-American encryption products are available from 29 foreign countries. These encryption manufacturers are located as far from the U.S. as China and as close as Mexico. The products in the study were purchased via routine channels, either directly from the foreign manufacturer or from a distributor.

RSA Data Security has lost business opportunities with major foreign conglomerates such as Lloyds TSB PLC, SAP AG, and Siemens Ag because of U.S. export control regulations. U.S. software companies estimate they have lost millions of potential users of their software due to the encryption regulations. It is naïve to believe these foreign customers and entities are forgoing strong encryption to protect their proprietary information because it is not available from U.S. manufacturers, rather than purchasing strong, non-American encryption.

Further, foreign encryption manufacturers are marketing their products by using U.S. encryption regulations against American companies. For example, Baltimore Technologies, an Irish encryption manufacturer that President Clinton visited during his trip to Europe last year, specifically points out the shortcomings of U.S. encryption products in their marketing of their product, WebSecure. Their opening paragraph of their website states that the export versions of U.S. browsers "are limited to 40 bits of encryption, which is not secure enough for most applications." In contrast, WebSecure provides 128-bit encryption for "real security."

Strong encryption is also available for sale and for free on the Internet to anybody in the world with a computer. Here is just one example of the ease with which a person outside the United States can obtain strong encryption with a few clicks on their computer. One, they can visit the international

Pretty Good Privacy site: www.pgpi.com. From that URL, anybody in the world can download strong, 128-bit encryption within 47 seconds. And because any citizen in the U.S. can download encryption legally from the Internet, and anyone in the world has access to those same web sites, the Internet makes controlling encryption exports a very difficult proposition.

ACP also believes it is vital to our national interests that our critical infrastructure is secure and we praise President Clinton for recognizing this vulnerability in his speech earlier this year. We wish, however, that the President recognized the importance of the role of strong encryption produced by U.S. high technology companies.

We do not believe we have all the answers to questions about national security, but ACP strongly believes based on our knowledge of the technology and global markets that our long term national security objectives can only be achieved if the United States realistically acknowledges the inevitability of a world of ubiquitous, strong encryption. Trying to control the proliferation of encryption is like trying to control the proliferation of mathematics. For that is what we are talking about here. Encryption algorithms are nothing but sophisticated mathematics. And while the United States may realistically hope to remain the leader in such a field, it cannot realistically expect to monopolize it.

We are joined in this view by the Center for Strategic and International Studies ("CSIS"). CSIS recently conducted a study of our nation's technical vulnerabilities; the study was chaired by William Webster, the former director of the FBI and Central Intelligence and former U.S. Circuit Judge. The subsequent report, entitled *Cybercrime Cyberterrorism Cyberwarfare Averting an Electronic Waterloo*, calls for the "intelligence gathering communities—law enforcement and foreign intelligence—to examine the implications of the emerging environment and alter their traditional sources and means to address the SIW needs of the twenty-first century. *Continued reliance on limited availability of strong encryption without the development of alternative sources and means will seriously harm law enforcement and national security.*"

For instance, ACP proposed last year the creation of a "NET Center" (and, since then, "Tech Center" has been created) to help law enforcement officials understand how to deal with encryption and other technological advances when encountered in a criminal setting. We have been cooperating on these projects, and we are pleased with the development of this forward-thinking strategy.

On the national security side, Senator Bob Kerrey recently suggested that (1) the President should convene a public-private panel to examine the implications of this new technological age for our national security, and (2) the creation of a new national laboratory for information technology to perform research and to act as a forum for further discussions on technological breakthroughs. These views may deserve further exploration, and ACP wants to play a leading role in crafting industry cooperation.

ACP wishes to emphasize that it recognizes a legitimate governmental need to obtain access to the plain text of communications when authorized by proper legal authority. ACP and its members are responsible citizens of the nation and the globe and have no wish to facilitate the commission of crime, the spread of terrorism or the acquisition and delivery of weapons of mass destruction. Similarly, we are committed to strengthening the nation's infrastructure, enhancing the privacy of American citizens and ensuring the security of electronic commerce. We believe that these sometimes competing objectives can be met, but only if government does not seek to force solutions on the industry that are not compatible with the development of technology and market demands.

ACP has advocated that the U.S. Government should work cooperatively with our nation's hardware and software manufacturers to develop the technical tools and know-how to achieve a policy that effectively responds to society's needs for law enforcement, national security, critical infrastructure protection, privacy preservation, and economic well-being.

In closing, Secretary of Defense William Cohen gave a speech at Microsoft two weeks ago in which he stated: "To maintain peace and stability in this uncertain world, we have mapped out a strategy defined by three words: Shape, Respond, Prepare." ACP and its member companies are willing to do our part in helping the Government prepare for an uncertain 21st century, and we look forward to working with the Government on these projects. But Congress needs to pass the SAFE Act and establish a clear and realistic national policy on encryption. That is the best way to preserve U.S. leadership in encryption technology, upon which the successful protection of our critical infrastructure and achievement of our national security objectives certainly and inevitably depend.

CHAPTER 24

TESTIMONY OF THE CENTER FOR DEMOCRACY AND TECHNOLOGY BEFORE THE HOUSE COMMITTEE ON THE JUDICIARY SUBCOMMITTEE ON COURTS AND INTELLECTUAL PROPERTY

March 4, 1999

Summary

The Center for Democracy and Technology (CDT) is pleased to have this opportunity once again to testify about encryption policy before the House Judiciary Committee. CDT is a non-profit public interest group dedicated to promoting civil liberties and democratic values on the Internet. CDT testified two years ago before this subcommittee in support of the Security and Freedom through Encryption (SAFE) Act, and we are happy to be here supporting the bill once again.

The last two years have made it more clear than ever that Congress should enact SAFE:

- Developments of the last two years have confirmed the need for fundamental revision of U.S. encryption policy. Since this Subcommittee's last encryption hearing in March 1997, 56-bit products have been cracked, key recovery has failed in the marketplace, and overseas the trend continues toward liberalization and away from further controls.
- Over the last two years the Administration has made only incremental changes to a U.S. encryption policy that continues to jeopardize privacy online. The Commerce Department regulations released in December do little to change the fundamental approach of export controls and incentives for key recovery.
- Two years have shown that "key recovery" and "plaintext access" systems are not the solution. Government-driven recovery systems require backdoor access to encrypted data, would impose significant new costs and risks on computer users, and would dramatically increase the surveillance capabilities of law enforcement at the expense of Constitutional liberties.
- Today it is clear that national security and law enforcement are best served by policies supporting the widespread use of strong, unescrowed encryption. Current U.S. policy dangerously impedes the deployment of accessible, easy-to-use, global security systems for

the Internet that are needed to protect our privacy and our critical infrastructures.

Two years ago, there were about 50 million people on the Internet. Today that number has nearly tripled to 140 million people worldwide. Surveys indicate that the number one issue for people as they move online and begin to participate in electronic commerce is privacy and security. The Internet has vast potential to reinvigorate democracy, provide access to information, create new forms of community, and promote economic growth. But the promise of the Internet will not be met unless people can trust it. Widespread availability of strong, encryption without backdoors built in is needed to provide that trust.

It is for all of these reasons that Congress should adopt the SAFE Act of 1999. The Administration has proven unable to change its basic approach to encryption. Congressional action is needed. The SAFE Act of 1999 improves on previous versions of the bill and would help provide Americans with the strong security and privacy products they so badly need. CDT commends Representatives Goodlatte and Lofgren, Chairman Coble, and the other cosponsors of the SAFE Act for their continued commitment to this essential debate about the electronic privacy of Americans.

Developments of the Last Two Years Have Confirmed the Need for Fundamental Revision of U.S. Encryption Policy

Two years ago, this committee held a hearing on encryption strikingly similar to the one being held today. Privacy advocates and industry representatives testified about the need for new encryption policies, and Administration officials argued that new regulations would allow U.S. policy to satisfy the competing interests at hand. In retrospect, the rapid pace of technical and marketplace developments over the last two years have made it clearer than ever before that the U.S. approach to encryption policy remains fundamentally flawed.

A. Exportable encryption has proven increasingly vulnerable

Two years ago privacy advocates argued that 56-bit encryption, the maximum strength exportable for consumers without key recovery, was not secure enough for many applications. The Justice Department disputed this, claiming that "According to the National Security Agency s estimates, the average time needed to decrypt a single message by means of a brute force cryptoanalytic attack on 56-bit DES a strength whose export we are now allowing—would be approximately one year and eighty-seven days using a thirty-million-dollar supercomputer."[1]

---

1 Security and Freedom Through Encryption (SAFE) Act: Hearing on H.R. 695 Before

Technical developments have proven these comments wrong. In the Fall of 1998, a group of researchers sponsored by the Electronic Frontier Foundation built a "DES Cracker" system for less than $250,000 that broke a 56-bit key within 56 hours.[2] Less than six months later, in January 1999, encryption enthusiasts broke a 56-bit code in 22 hours using the DES Cracker and a network of distributed computers. If a non-profit and a group of part-time enthusiasts could develop such a system on a shoestring budget, we are only left to imagine what a foreign government, large corporation, or sophisticated criminal enterprise could do.

The U.S. Government has itself recognized the weakness in 56-bit encryption systems. In a January 1999 draft the National Institute of Standards and Technology (NIST) revised the encryption standard for government use from 56-bit DES to much stronger "Triple DES," citing the vulnerability of DES.[3] Meanwhile, NIST has been leading efforts to create an Advanced Encryption Standard based on the 128-bit (and higher) algorithms that are becoming the world standard for online security. If the government does not trust 56-bit security, why should everyday computer users and companies be expected to rely on this weaker level of security?

B. Key recovery has not been widely accepted

Two years ago before this Subcommittee, Administration witnesses touted key recovery as the compromise that met law enforcement desires and was "going to become the worldwide standard."[4] In fact, since then government-driven key recovery has been greeted with great skepticism and widely discredited.

Research has revealed the vulnerabilities of key recovery systems, which create backdoors to plaintext without the notice or consent of an encryption user. A 1997 report by a group of encryption experts found that "[t]he

---

the Subcomm. on Courts and Intellectual Property of the House Comm. on the Judiciary, 105th Cong., 2nd Sess, No. 9 (1997) (Statement of Robert S. Litt, Deputy Assistant Attorney General, Department of Justice).

2 See ELECTRONIC FRONTIER FOUNDATION, CRACKING DES (1998).

3 "With regard to use of single DES, exhaustion of the DES (i.e. breaking a DES encryption ciphertext by trying all possible keys) has become increasingly more feasible with technology advances. Following a recent hardware based DES key exhaustion attack, NIST can no longer support the use of single DES for many applications." 64 FED. REG. 10, 2625-2628 (1999) (proposed January 15, 1999).

4 "[W]e believe that key recovery encryption is going to become the worldwide standard." Security and Freedom Through Encryption (SAFE) Act: Hearing on H.R. 695 Before the Subcomm. on Courts and Intellectual Property of the House Comm. on the Judiciary, 105th Cong., 2nd Sess, No. 9 (1997) (Statement of Robert S. Litt, Deputy Assistant Attorney General, Department of Justice).

deployment of key-recovery-based encryption infrastructures to meet law enforcements stated specifications will result in substantial sacrifices in security and greatly increased costs to the end-user." A year later, with no substantive response from within the Administration or the technical community, the same group of experts confirmed its findings still held true in June 1998.[5] A copy of their report is being submitted to the Subcommittee along with this testimony.

Despite Administration predictions, the marketplace has shown little interest in even stored data recovery, and there is virtually no demand for key recovery for communications. To CDT's knowledge not one major key recovery encryption product is being widely used by consumers today.[6]

C. The world is not adopting U.S. encryption control policies

Encryption controls are ultimately only effective if other countries control encryption products as well. In 1997, the Administration testified, "We have engaged in extensive international discussions on this topic over the last year, and a consensus is now emerging throughout much of the world that the way to achieve this balance is through the use of a key recovery or trusted third party system. . . . We believe that key recovery will become the worldwide standard for users of the GII."[7] To date, the opposite has been true. The OECD Cryptography Policy Guidelines and the Ministerial Declaration of the European Union, both released in 1997, failed to embrace key recovery despite lobbying by the U.S. government. In the past year, Canada, Ireland and Finland have announced encryption policies allowing free use and export of strong encryption products without key recovery.Even France, a country with sweeping controls on encryption use in the past, recently liberalized its policies.

D. Many in the national security community are now arguing for a change in U.S. policy

Two years ago the national security community seemed to speak with one voice about the danger of strong encryption. Today there has been an

---

5 AN AD-HOC GROUP CRYPTOGRAPHERS AND COMPUTER SCIENTISTS, THE RISKS OF KEY RECOVERY, KEY ESCROW, & TRUSTED THIRD PARTY ENCRYPTION (1997). (Updated 1998 report available at http://www.cdt.org/crypto/risks98/.)

6 Cost may play a role. A recent study by the Business Software Alliance estimated the cost of key escrow systems at $7.7 billion per year and $38.5 billion over a five year period. BUSINESS SOFTWARE ALLLIANCE, THE COST OF GOVERNMENT-DRIVEN KEY ESCROW ENCRYPTION (1998).

7 Security and Freedom Through Encryption (SAFE) Act: Hearing on H.R. 695 Before the Subcomm. on Courts and Intellectual Property of the House Comm. on the Judiciary, 105th Cong., 2nd Sess, No. 9 (1997) (Statement of Robert S. Litt, Deputy Assistant Attorney General, Department of Justice).

increasing recognition of the cost of U.S. encryption policy. The last two years have seen Americans moving their lives online in unprecedented numbers. A Presidential Commission has highlighted the vulnerability of our nation's critical information infrastructure. Together these developments have underscored the importance today of securing the Internet, and deploying strong encryption to do so.

Today many in the national security and law enforcement community have acknowledged the limitations of current U.S. policy. As Sam Nunn testified before the Senate last year, "[I]f the deadlock continues as it is today, building the trust required between the public and private sectors in the broad area of infrastructure protection will be even more difficult."[8] Nunn went on to note that "limiting the power of encryption over the long-haul is simply not going to be feasible. Senator Bob Kerrey, an early proponent of encryption controls, argued in an October 1998 speech that "the encryption debate has hobbled our efforts to write laws that enable our law enforcement and national security agencies to carry out their mission" and argued that it was time to "remove export restrictions on encryption products of any strength."[9]

E. The Administration has proven unable to engage in comprehensive reform

The Department of Commerce has taken a step forward in its recently released encryption regulations, easing exports of 56-bit products and allowing export of strong encryption products to online merchants. However, U.S. policy remains focused on export controls and incentives to use key recovery. The mass market products needed by individual users remain controlled. The special relief for certain industry sectors, while surely welcome by those businesses, does little to change the encryption available to individual computer users or small organizations.

Taken together, these developments argue for a more comprehensive change to U.S. encryption policy, away from export controls and key recovery and towards a view where public safety is best protected by giving people the encryption tools they need to protect themselves on line. The past two years have also shown that such comprehensive reform will most likely only come with the involvement of Congress, as the interests in favor of current policy continue to dominate the Administration's approach to encryption.

---

8 Hearing before the Subcommittee on Technology, Terrorism, and Government Information of the Senate Committee on the Judiciary, 105th Cong., 2nd Sess, (March 17. 1998) (Statement of Sam Nunn, Co-Chair, Advisory Committee to the President's Commission on Critical Infrastructure Protection).

9 144 CONG.REC. S12359 (1998).

U.S. Encryption Policy Continues to Deny Computer Users Essential Tools That Protect Their Privacy

Encryption protects privacy and prevents crime online. In early 1999, it is more clear than ever that the widespread use of encryption is of critical importance for public safety, national security, and law enforcement in the Information Age. The flow of sensitive information over the Internet leaves people increasingly vulnerable to the prying eyes of potential criminals, terrorists, or even foreign governments. Encryption gives people an easy and inexpensive way to protect that information. The need for encryption is becoming ever more acute as sensitive data is finding its way into electronic form:

- Individuals need encryption in order to trust the Internet with private data such as online banking, stock trades, medical records, electronic purchases, or personal communications.
- Businesses need encryption to protect their own proprietary information as it flows across vulnerable global networks.
- The country needs encryption to secure the critical information infrastructure governing such sensitive applications as our utilities, financial markets, or air traffic control networks.

If broad participation in electronic commerce and the information society is to become a reality, the adoption of encryption in most phases of electronic existence will be required.

Encryption is particularly important because of the inherent difficulties of securing the new digital media. The open, decentralized architecture that is the Internets greatest strength also makes it hard to secure. Internet communications often travel "in the clear" over many different computers in an unpredictable path, leaving them open for interception. An email message from Washington to Geneva might pass through New York one day or Nairobi the next—leaving it susceptible to interception in any country where lax privacy standards leave it unprotected. Encryption provides one of the only ways for computer users to guarantee that their sensitive data remains secure regardless of what network—or what country—it might pass through.

Current U.S. policy prevents users from getting the encryption tools they need to protect security online. Today's export controls continue to limit the availability of strong encryption products both domestically and abroad. Such controls directly limit the availability of strong encryption products outside of the U.S., of particular concern to human rights groups and other organizations abroad. Export controls affect people in the U.S. when they communicate abroad, since they may be forced to use the lower levels of encryption available to parties worldwide. Most importantly, export controls have slowed the deployment of strong encryption standards. While some strong encryption products are available to consumers, export controls have largely slowed the seamless integration of good security systems into operating systems, network protocols, and many applications. Encryption should be easy for consumers; because of federal regulations, it is not.

The most recent December 1998 encryption regulations, while a welcome step forward by the Administration, do not change the fundamental premise of U.S. policy: export controls on all but the weakest encryption for mass market consumers, and strong incentives for the use of key recovery and plaintext access systems. The sector relief provided for foreign subsidiaries of U.S. companies, certain industries, and online merchants does little to provide regular consumers with strong encryption. Export controls remain a powerful incentive to adopt key recovery and plaintext access systems. The piecemeal relief offered by the regulations raises the question: When do regular people get to protect their privacy online?

Computer users remain at risk, awaiting the widespread deployment of encryption and facing increasing threats to their unprotected information.

Government-Driven "Key Recovery" and "Plaintext Access" is Not a Solution

The law enforcement community in general has variously endorsed "key escrow," "key recovery," and other forms of "plaintext access" as its favored approach to encryption policy. These variations on the failed "Clipper Chip" policy seek to guarantee third-party access to the keys for all encrypted communications and stored data without the notice or consent of the key owners. Such proposals have been greeted with much skepticism and concern from the global Internet community.

The attempt to institutionalize key recovery worldwide is a fundamental threat to privacy and security both domestically and abroad:

- Global key access systems are vulnerable and unproven—Centralized "back-door" access to the billions of keys used by millions of computer users will introduce new vulnerabilities into a medium that is already difficult to secure. In 1997, eleven renowned computer security experts issued a report on key recovery concluding that, "Building the secure infrastructure of the breathtaking scale and complexity demanded . . . is far beyond the experience and current competency of the field. Even if such an infrastructure could be built, the risks and costs of such a system may ultimately prove unacceptable." In 1998 these experts revisited the question and confirmed that their conclusions remained essentially unchallenged.[10]

---

10 AN AD-HOC GROUP CRYPTOGRAPHERS AND COMPUTER SCIENTISTS, THE

- The Fourth Amendment does not adequately protect key recovery systems both outside and inside of the U.S.—The Administration has been unable to explain what legal standards will internationally protect the communications and data of U.S. individuals and businesses. Moreover, the Administration has indicated that the full Fourth Amendment standards of probable cause and notice would not apply to encryption keys held by third parties, even within the U.S. In a world where personal data is increasingly legally unprotected in the hands of third parties, key recovery systems further erode a person's ability to protect their privacy.
- Recovery will never be appropriate for some applications—For example, the American Association for the Advancement of Science has commented on the sensitive and increasingly important use of encryption by human rights advocates worldwide. "If keys can be recovered by the U.S. government, why should human rights organizations whose entire function is defined by abusive governments trust that their information will remain secure?"[11]

Despite these concerns, current encryption regulations continue to give many encryption producers a Hobbesian choice: accept key recovery or be forced to export lower strength encryption. Moreover, proposals backed by the FBI in the past have sought to further force U.S. encryption users to adopt key recovery through a number of coercive regulations, including outright domestic mandates. While we are encouraged that the Administration appears to have backed away from mandatory domestic controls, we are wary that it has not denounced this approach. And even the current U.S. encryption policy based on key recovery and export controls threatens to leave global Internet users without the technical means to secure their communications or the international legal standards needed to protect their privacy.

National security and law enforcement are best served by policies supporting the widespread use of strong, unescrowed encryption

The state of the emerging information society is making it increasingly clear that the law enforcement benefits of widespread encryption far outweigh the costs. The national security and law enforcement community has begun to recognize the limits of current U.S. policy. As Sam Nunn, Co-Chair of the Advisory Committee to the Presidents Commission on Critical Infrastructure Protection, noted in 1998 Senate testimony, "I do think we

---

RISKS OF KEY RECOVERY, KEY ESCROW, & TRUSTED THIRD PARTY ENCRYPTION (1997). (Updated 1998 report available at http://www.cdt.org/crypto/risks98/.)

11 American Association for the Advancement of Science, Comments on Bureau of Export Administration Interim Rule on Encryption Controls (Feb. 7, 1997).

are in a different era of technology now and I do not think the nostalgia for the old-fashioned wiretap by law enforcement is going to be realistic in this age we are in now. "[12]

The benefits of current U.S. policy to law enforcement are uncertain. U.S. policy will not stop sophisticated criminals from using encryption to evade law enforcement. Strong, non-escrowed encryption is already available both inside and outside of the United States today. Foreign governments and criminals have access to these powerful tools and will be able to encrypt data despite continued export controls or key recovery. Furthermore, nothing in the Administration policies prevents users from "super-encrypting" communications even within a key recovery framework.

The law enforcement problems with encryption are important but more limited than claimed. Law enforcement faces a real, but narrowly focused, problem with encryption. Most encrypted information will still be accessible to law enforcement by legal process even in an encrypted world. For example, businesses will be still be required to produce the plaintext of encrypted business records under proper legal process. Stored information, corporate and business information, and even a great deal of electronic communication will most likely be largely available to law enforcement through legal process similar to that available today.

Important challenges remain for law enforcement interceptions of communications or seizures of data without notice to the party under surveillance. This narrower problem must be put into the context of the benefits provided by encryption and the costs associated with key recovery systems. Moreover, the information economy presents new and powerful tools and opportunities for law enforcement. Online interaction leaves a detailed trail of electronic transactions, credit card purchases, online communications, and Web-based clickstream data presenting new traffic analysis opportunities. In fact, law enforcement is operating today in a Golden Age of surveillance, with online collections of personal data offering unprecedented new tools to obtain evidence of criminal activity and raising important privacy concerns that must be dealt with as well.

U.S. policy is creating a deficit of trust around important issues we could all be working on together. U.S. policy stands in the way of a growing urgent need for strong encryption products that people trust. CDT believes that current U.S. policy dangerously impedes the deployment of accessible, easy-to-use, global security systems for the Internet that are needed to protect our privacy and our critical infrastructure.

---

12 Hearing before the Subcommittee on Technology, Terrorism, and Government Information of the Senate Committee on the Judiciary, 105th Cong., 2nd Sess, (March 17. 1998) (Statement of Sam Nunn, Co-Chair, Advisory Committee to the President's Commission on Critical Infrastructure Protection).

On balance, national security demands strong encryption. CDT agrees with the conclusion of the National Research Council's major study of encryption, which argued in its 1996 encryption study, "On balance, the advantages of more widespread use of cryptography outweigh the disadvantages."[13]

Conclusion

U.S. policy stands in the way of a growing urgent need for strong encryption products that people trust. The past two years have shown that people and businesses are moving more and more of their lives, economic activities, and sensitive data online. The federal government has identified the vulnerability of our nation's critical information infrastructure. Strong encryption, without built-in backdoors, is an essential part of protecting that sensitive data and critical infrastructure.

That is why the SAFE Act is so important. In the current policy standoff between eroding law enforcement arguments and the emerging and acute privacy and security needs of the Information Age, it is Congressional action that is needed. Only Congress is in the position today to change U.S. encryption policy and get Americans the privacy and security tools they need. The private sector cannot do it. The Administration will not do it. The courts may do it, but not without a protracted struggle. Congress must act. CDT believes that immediate liberalization of export controls in the SAFE Act will help provide Americans on the Internet with the strong security and privacy they so badly need.

About the Center for Democracy and Technology

CDT is an independent, non-profit public interest policy organization in Washington, D.C. The Center's mission is to develop and implement public policies to protect and advance individual liberty and democratic values in new digital communications media. The Center achieves its goals through policy development, public education, and coalition building. CDT also coordinates the Digital Privacy and Security Working Group (DPSWG), an ad hoc coalition of more than 50 computer, communications, and public interest organizations and associations working on communications privacy issues. Members of DPSWG assisted in the drafting of the Electronic Communications Privacy Act in 1986 and since have been involved in ongoing policy work regarding privacy and security online.

House Rule XI, clause 2(g)(4) disclosures: Neither Alan Davidson nor the Center for Democracy and Technology have received any federal grant, contract or subcontract in the current or preceding two fiscal years.

For more information contact:

Alan B. Davidson, Staff Counsel<abd@cdt.org>
James X. Dempsey, Senior Staff Counsel<jdempsey@cdt.org>
Center for Democracy and Technology http://www.cdt.org/
202.637.9800 (v)
202.637.0968 (f)

---

13 NATIONAL RESEARCH COUNCIL, CRYPTOGRAPHY'S ROLE IN SECURING THE INFORMATION SOCIETY (1996).

CHAPTER 25

STATEMENT BEFORE THE SUBCOMMITTEE ON COURTS AND INTELLECTUAL PROPERTY COMMITTEE ON THE JUDICIARY, U.S. HOUSE OF REPRESENTATIVES

Dorothy E. Denning
Georgetown University, Computer Science Department
Reiss 225, Washington DC 20057
202-687-5703, denning@cs.georgetown.edu
http://www.cs.georgetown.edu/~denning

March 4, 1999

Thank you for the opportunity to testify on H.R. 850, the "Security and Freedom Through Encryption (SAFE) Act." There are three points that I would like to make.

First, the sad state of security of our country's information infrastructures will not be solved by this bill. This is because the security problems are not the result of using exportable encryption, but rather of not using any encryption at all and of not employing other essential safeguards. Sensitive data, including passwords, is routinely transmitted and stored in the clear. Of the thousands of incidents reported to the Computer Emergency Response Center, I am not aware of any that can be attributed to faulty encryption caused by export controls.

Security also requires much more than encryption. Encryption will not stop insiders from compromising proprietary information, siphoning money from bank accounts, and planting destructive time bombs. It will not stop hackers from exploiting security holes in order to penetrate systems, deface Web pages, and disrupt service. It will not prevent Trojan horses, disguised as appealing software programs, from entering users' computers and stealing passwords and other secrets while they are being typed.

In short, encryption is not a silver bullet. It must be augmented with other security measures, both technical and procedural. These include access controls, authentication, auditing, configuration management, vulnerability testing and repair, intrusion and misuse detection, malicious code detection, and security training and awareness. Cryptographic technologies for authentication, including digital signatures, are not restricted for export and are at least as important as technologies for confidentiality protection.

My second point, which is related to the first, is that high levels of security can be achieved within the context of current export control policy. I'm not just talking about domestic users and U.S. owned companies. An international enterprise can protect its assets by employing fully exportable encryption products that use 128-bit keys or longer and say "Made in USA."

Let me outline one way that can be done. I make two assumptions. First, encryption must be considered within the context of a comprehensive enterprise-wide information security program that encompasses an organization's customers, suppliers, partners, shareholders, consultants, and others who do business with the organization. Second, an organization must be able to protect and retain control over its sensitive information, whether in storage or in transit. These two assumptions lead to an encryption approach that is integrated with an enterprise access control policy. The approach ensures that authorized persons can get the keys needed to decrypt data but that unauthorized persons cannot. It allows for immediate revocation of a user's decryption capabilities. And it provides an audit of every decryption so that policy violations can be detected. It does not require an organization to use third-party key management services, though this would be an option. I recently reviewed a product that offers these protections and is approved for export.

My third and final point is that the current approach of gradually easing export controls may be optimal. If cryptography is over-regulated, our economic competitiveness, technology leadership, and civil liberties are at risk. There can be little doubt that export controls drive some business overseas. Yet if these controls are lifted entirely, law enforcement and national defense are at greater risk. Even though export controls do not prevent domestic or foreign adversaries from getting access to strong encryption, they have influenced major product lines. Many criminals and terrorists use these products rather than going to the trouble of installing add-ons.

Today, Americans enjoy a strong and growing economy and a declining rate in crime. The Administration's encryption policy has imperiled neither our economic competitiveness nor our ability to fight crime and provide for national defense. I am concerned that H.R. 850, either in its current form or with amendments such as those introduced in the last Congress to impose domestic regulations, could upset the delicate balance among our national interests.

In summary, H.R. 850 is not the key to safe electronic commerce or to protecting our critical infrastructures. This is because export controls are not the problem. The bill would help American companies compete in the global marketplace, but it would also remove industry incentives to accommodate law enforcement and national defense interests.

A few years ago, the National Research Council conducted an extensive study of encryption policy at the request of Congress. They made several excellent recommendations, including the progressive relaxation, but not elimination, of export controls. Their proposed course of action is generally consistent with the steps taken by the Administration. This cautious approach to export liberalization may be the best one.

CHAPTER 26

TESTIMONY OF CRAIG MCLAUGHLIN CHIEF TECHNOLOGY OFFICER PRIVADA, INC. BEFORE THE HOUSE JUDICIARY SUBCOMMITTEE ON COURTS AND INTELLECTUAL PROPERTY

March 4, 1999

Good morning, Mr. Chairman and Members of the Subcommittee. Thank you for the opportunity to speak with you this morning about this important topic. I appreciate the efforts of Mr. Goodlatte, Ms. Lofgren and the cosponsors of the SAFE Act for their willingness to address this complex but important issue.

My name is Craig McLaughlin and I am the Chief Technology Officer at Privada, Inc., based in San Jose, CA. I am pleased to be testifying this morning on behalf of the Software & Information Industry Association (SIIA), the result of a merger between the Software Publishers Association and the Information Industry Association. SIIA represents 1400 member companies engaged in every aspect of electronic commerce and strongly supports H.R. 850, the Security and Freedom through Encryption (SAFE) Act.

The Role of Cryptography

Encryption is tremendously important for securing electronic communications and transactions. As the Internet continues to increase, more individuals and businesses "go online," and companies shift their mission-critical operations to the Internet, the need for and importance of cryptography only grows.

As a result, the market for encryption is growing. Users routinely demand robust encryption products, and global sales of encryption products are expected to reach $20 billion by 2002.[1] Companies in every sector are seeking to utilize security products to facilitate online sales, improve their own products, protect their intellectual property and secure their private data and communications.

1 Economic Strategy Institute, 1998.

There can be no question that the demand for encryption products is strong. A 1997 study identified over 1600 encryption products available from more than 900 companies in thirty countries. That same year, Trusted Information Systems found more than 650 encryption products produced abroad; almost 300 of these incorporated DES-level encryption. Encryption is routinely used in software, databases, networking products, telecommunications equipment, computer peripherals, electronic commerce and financial services. Without a doubt, encryption is one of the most important tools companies and individuals have in the digital environment.

While we are beginning to realize the benefits of the global electronic marketplace, we are also realizing some of the challenges that users face. Computer crime, intellectual property theft and privacy fears are some of the issues that the Internet community is being forced to address. By using security products that incorporate robust encryption, companies and individuals can minimize these concerns.

One of the biggest challenges facing us today is the question of online privacy. While users want to take advantage of the Internet's vast resources, many are concerned about the collection and use of their personal information. According to some studies, privacy is the primary concern for online consumers. A Lou Harris poll recently found that 81 percent of Internet users and 79 percent of those who have purchased goods online are concerned about privacy.

My company, Privada, was founded in 1997 on the premise that individuals and organizations should have the ability to control access to and use of their personal information online—that every person should have the freedom to use the Internet responsibly, without sacrificing their privacy. We have developed a suite of products that allow individuals to protect their privacy while using the Internet for browsing, communications or purchases. Our products disassociate one's real-world identity from the online identity, ensuring that individuals can take advantage of the Internet while protecting themselves and their families online.

Companies like mine have worked hard to develop technological solutions that address these concerns, providing both individuals and businesses with the tools needed to assure their privacy. Such efforts help promote a secure online environment and improve user confidence, helping the vibrant electronic commerce market continue to grow.

Current Encryption Policy

The current policy of restricting encryption exports is, I respectfully submit, outdated and counterproductive. The Administration's approach to encryption exports, like others before it, has sought to balance the needs of law enforcement and national security with the needs of Internet users, but instead has only created a situation in which U.S. industry is at a competitive disadvantage to its foreign counterparts, where online communications and transactions may remain vulnerable, where users do not have robust tools to protect their privacy and that ultimately threatens to undermine our technological leadership in this critical area.

Let me address each of these points in some more detail.

Current policy is outdated

The current administration policy has evolved from an era in which encryption was regulated as a munition. Encryption products were largely used to provide a level of secrecy for electronic data and communications. Not widely available, export restrictions on encryption and related products could be relatively effective in limiting the spread of these products around the world.

With the growth of electronic networks, though, the effectiveness of restrictions is seriously compromised. Digital networks cross national borders and reach around the globe. Data flows across the country and around the world in an instant, often without the user knowing where the data is originating or terminating. International networks have made it possible for individual users to take advantage of resources previously unavailable to them and for companies to develop new markets around the world.

More importantly, perhaps, is the fact that encryption is no longer used to simply scramble the text of secret messages. The use of encryption has evolved to include authentication and certification, data integrity and network security applications. These applications are widely used in virtually every industry today and are critical to the further development and use of networks in everyday life.

One example is the protection of sensitive information from misappropriation by unauthorized parties, or misuse by otherwise authorized, but negligent or malicious parties, to a transaction. Encryption is the only practical means by which parties to an online communication can trust that each is who he claims to be. It is the only practical way to guarantee that the communication between those parties remains private.

A further example may be helpful. Many Members of Congress—including yourselves, I'm sure—receive e-mail from their constituents. Some of you choose to reply to your constituents via regular postal mail, but I am sure that many of you choose to use e-mail as a means to communicate with the citizens in your districts. It's effective and inexpensive.

Without technologies like digital signatures, though, your constituents can never really be confident that the message actually came from your office or that the message wasn't modified during the transmission process. Digital signatures, which rely on the enabling technology of encryption, provide users the ability to certify and authenticate the message and therefore trust that the message is authentic. Just as a letter on your stationery with your signature provides a level of confidence, digital signature provide similar assurances for recipients of electronic communications.

Such capabilities are critical for both business and individuals seeking to take advantage and use the Internet. Without robust tools, no one can be assured that their online activities remain private and that their online transactions are trustworthy. Companies are rapidly developing innovative technologies and applications for use on public networks and users are just as rapidly integrating these capabilities into their everyday lives. To ensure that this market continues to grow, consumer concerns like privacy, authentication and security must be addressed. Without encryption, we simply can't do it. We must be able to use and widely deploy encryption if we are to help users protect against the inherent vulnerabilities of public networks.

Current policy puts U.S. companies at a competitive disadvantage

Second, U.S. policy puts U.S. companies like mine at a competitive disadvantage compared to our foreign counterparts. This is an issue that affects us directly at Privada. Because of the current export controls, U.S. companies face restrictions which prevent them from offering competitive products in the global marketplace—restrictions which foreign competitors do not face. Internet users, whether corporate or individual, are sufficiently sophisticated to seek and demand robust encryption tools in the products they use to facilitate their own online activities. Companies that cannot offer these features face an uphill battle in an extremely competitive marketplace.

As a result, companies who choose to incorporate encryption into their products are faced with a Faustian choice. They can either use strong encryption and forgo the lucrative export market, or they can use weaker encryption for their export products, thereby rendering them unattractive to potential customers.

Companies who choose to forgo exports face a significant challenge. In the era of the global electronic marketplace, to have products that cannot be sold on the foreign market is a tremendous disadvantage. For many software and information companies, foreign sales account for a large percentage of their total annual sales; to simply be forced to abandon this market is obstacle that our foreign competitors simply do not face.

Some companies choose an alternative route. They choose to export products that incorporate weaker encryption, placing them at a significant disadvantage to their competitors abroad. Users understand the value of encryption, and simply do not want products that are weak or easily broken. Further, because multiple product lines must be developed, production costs—and thus the cost of products and services—rise. Companies who choose this route often find that their potential international customers go elsewhere to find products that meet their need for robust privacy and security products.

This dichotomy between their foreign counterparts and US companies is so pronounced a foreign competitor of Privada has used it to market its services.

Current policy limits the ability of companies and individuals to protect data and communications

Perhaps the most problematic aspect of current policy is that without strong encryption products, data and communications remain unprotected. With robust encryption, companies and individuals have the tools they need to ensure that their online activities and data are secure, protected and authentic.

Without strong encryption, our products and others cannot provide the level of security that customers are demanding. When forced to use weaker encryption, products and services are vulnerable, undermining the very sense of security and confidence that we seek to instill and foster. In fact, they actually weaken protections by generating a false sense of security—as has been said many times, weak cryptography is worse than none at all.

For my company and others working to develop technological tools to help users protect their privacy, the ability to use and incorporate strong encryption into our offerings is critical. Encryption is the core component of the technologies we develop to help users control how their personal information is collected and used. Without it, our mission is unachievable, and the privacy of millions of individuals is at risk.

Current policy ultimately undermines our technological leadership

Finally, I think that it is critical that we consider what the impact of the current policy will be on our technological leadership in the future. The United States has benefited tremendously from the vibrant technology industries that have seen such rapid innovation and growth in recent years.

As has been widely reported in the press and as implied by our competitor's actions, the methods, algorithms and technologies being discussed today are globally known, understood and published. At its very core, encryption relies on mathematics. And while U.S. manufacturers have developed a wide array of products that incorporate these technologies, there is no reason to believe that competing products developed abroad would not meet users' functionality and performance standards.

As I mentioned earlier, foreign products are widely available. Many have downplayed the quality of these products and services, instead believing that foreign customers automatically assume that U.S.-developed software, information and electronic commerce products are inherently better than their international counterparts. This is simply not true. As I mentioned above, encryption technologies are well understood and available. There is no reason to believe that US products are simply better because they originated here, and it is important not to discount the viability of these foreign products.

The ultimate result, of course, is that companies face restricted markets, unfair competition and reduced sales, resulting in less revenues for research and development of new products. For high-tech industries, especially in software, electronic commerce and information, R&D costs are often quite significant. Without robust sales to fuel additional development, companies cannot afford to innovate or create new products that meet the rapidly changing needs of the electronic marketplace. While it unrealistic to predict that those of us who produce products and services that incorporate encryption will inevitably go out of business or move abroad, it not unreasonable to be very worried about the long-term impact that market restrictions will have on our ability to innovate and lead. Without further research and development, we risk losing the leadership that we have developed in this critical market segment.

The Need for Policy Reform

Clearly, a new approach is needed. It is important that Congress address this issue in a timely manner. We often speak of "Internet time" to refer to the quickly changing electronic environment, and it is critical that our policies remain appropriate to facilitate continued growth.

At the same time, we recognize that there are lingering concerns about the misuse of encryption—the very concerns that have driven the current restrictions. I suggest, though, that a more proactive, forward-looking approach may actually enhance the objectives of the current policy while providing U.S. industry with continued access to robust encryption tools.

How could such a balance be possible? First, let me suggest that maintaining U.S. technological leadership is critical. We must be able to attract and keep those talented individuals and companies that have driven the growth in the industry. If these capabilities move elsewhere or our leadership is compromised, our ability to work with law enforcement and provide assistance will be greatly reduced. As outlined above, we are not going to be able to do so if our companies cannot compete abroad or face unnecessary restrictions on their ability to do so.

In addition, we must provide the tools so that all of our industries can take advantage of new technologies. Economic espionage and computer crime are tremendous threats, and any company that uses computers in any fashion is evaluating ways to make their systems more secure and to protect their data more effectively. To ensure that these organizations, whether they be grocery stores, pharmaceutical research firms or educational facilities, have access to robust tools, we must ensure that our companies are able to develop these products.

The Administration has long recognized the value that encryption has for securing electronic systems. Its recent proposed revisions to the export restrictions, which allow for the export of 56-bit encryption and stronger products for certain sectors, underscore the importance of encryption. I think that it is unrealistic and perhaps a bit short-sighted to assume that the best approach is to regulate which sectors should be able to deploy advanced security products, rather than letting the market and individual users decide what their security requirements are.

Second, companies throughout the industry are developing products that strike the delicate balance between the need for privacy and security with the need to access information. While it may not be feasible for individual users to purchase or deploy many of these products, companies, including those who provide online access to individuals, are beginning to demand these products.

An example may be helpful. Companies, for example, may wish to encrypt their corporate communications to protect their trade secrets or proprietary information. But they also recognize that there may be situations where they need to reconstruct an event or access protected information. The activities of an employee suspected of divulging corporate secrets may need to be investigated, for example. Several products on the market today allow for such access without compromising the security of the original data or communications.

Please do not misconstrue my comments—this is not an endorsement for key recovery. Our products, for example, do not incorporate key recovery but can be used to provide access if needed. That companies are developing such alternatives is simply a recognition that some customers demand such functionality and the market is responding appropriately. Companies must be given the opportunity to respond to market preferences without the intervention of the government because only individual consumers can make decisions regarding what products and protections are appropriate to their unique situation.

Finally, it is important to realize that encryption is widely available from any number of sources, and that maintaining outdated policies will not meet the Administration's objectives. We all know that this genie is out of the bottle, to repeat an oft-used phrase. We cannot simply accept that our export restrictions are effective just because we hope that they are. We must recognize the realities of the market today and adapt our policies before we lose the advantages that we enjoy.

Conclusion

In conclusion, I submit that it is critical that Congress act quickly to remove export provisions on encryption products to ensure that our companies can compete fairly and effectively in the international marketplace and continue to provide users with the tools that they need to protect their privacy and security online. By freeing the market and allowing U.S. companies to take full advantage of the global market for these products, we can ensure that every company and individual has access to technologies that enhance this growing market.

I urge the Members of the Committee to support liberalizing encryption export provisions and to support H.R. 850.

Thank you.

CHAPTER 27

TESTIMONY OF
THOMAS PARENTY
DIRECTOR, DATA AND COMMUNICATIONS SECURITY
SYBASE, INC.
ON BEHALF OF THE
BUSINESS SOFTWARE ALLIANCE

IMMEDIATE NEED FOR RELAXATION OF EXPORT CONTROLS FOR SOFTWARE AND HARDWARE WITH ENCRYPTION CAPABILITIES

BEFORE THE
COURTS AND INTELLECTUAL PROPERTY SUBCOMMITTEE
OF THE
COMMITTEE ON THE JUDICIARY
U.S. HOUSE OF REPRESENTATIVES
WASHINGTON, D.C.

March 4, 1999

SUMMARY OF
STATEMENT OF THOMAS PARENTY
DIRECTOR, DATA AND COMMUNICATIONS SECURITY, SYBASE, INC.

BSA strongly supports the Security and Freedom through Encryption ("SAFE") Act (H.R. 850) because it ensures that all Americans may use and sell any encryption domestically and provides badly needed export control relief.

Congress must immediately relax export controls on software and hardware with encryption capabilities. Widespread deployment of American products with encryption capabilities will help to accelerate dramatically the growth of electronic commerce by protecting consumers' privacy and preventing electronic crime. Export control relief also is vital for protecting America's critical infrastructures and ensuring that American software and hardware companies remain not only internationally competitive but also the market leaders in both software and hardware products.

The Administration took the first step towards developing a sensible long-term encryption policy by permitting exports of select products to

select users, but they still have not gone far enough. A successful encryption policy must be based on technological and market realities. It must recognize that:

- The worldwide standard is 128-bit encryption;
- Mass market software and hardware is uncontrollable; and
- It is in America's national and economic security interests to have American designed and manufactured encryption products deployed worldwide.

Without relaxation of export controls, U.S. manufacturers remain at a competitive disadvantage, and foreign consumers will purchase encryption products from foreign suppliers.

Foreign products are comparable in capabilities and quality. When a foreign purchaser cannot obtain an American product they simply purchase it from a foreign supplier. Unfortunately, not only are American companies losing a sale of an encryption item, but they are also losing the sale of the program or hardware such as an Internet server or an application browser that uses the encryption capability. In fact, companies risk losing sales of entire systems because of their inability to provide necessary security features. The only impact of the Administration's export policy is widespread deployment of foreign designed and manufactured software and hardware.

The SAFE Act recognizes that the United States should not try to control uncontrollable exports of mass market and public domain software and hardware. It also permits exports of 128-bit level custom software and hardware. At the same time, the SAFE Act prohibits the government from mandating the use of key escrow, key recovery or recoverable encryption or requiring Americans to use key escrow, key recovery or recoverable encryption if they want to use an electronic signature. Ultimately, the SAFE Act will help Americans to use encryption to protect privacy, prevent crime and protect our national security.

Introduction

Good Morning. My name is Thomas Parenty, and I am the Director of Data and Communications Security for Sybase, Inc. In this capacity, I am responsible for all security-related product development for one of the ten largest software companies in the world. I have been active in the cryptography and computer security field for over 15 years, including my tenure at the National Security Agency (NSA) in the early to mid-eighties.

While at the NSA, I advised the Director of the NSA on internal NSA computer security issues and worked on the security of global nuclear command and control networks, focusing on the formal verification of cryptographic protocols and internal computer security controls. Because of my specialized security knowledge, I worked on national security-related, compartmentalized programs at other government agencies during my service at the NSA. In addition, I have worked on the security design of operating systems, networks and database management systems for government agencies, including the Central Intelligence Agency, the Defense Intelligence Agency, the Air Force, as well as many U.S. computer vendors. Most recently, I served as an advisor to the President's Commission on Critical Infrastructure Protection, specifically addressing the needs of the telecommunications and banking infrastructures. I am also a member of the National Research Council's panel on information technology.

Headquartered in Emeryville, California, Sybase, Incorporated, is a worldwide leader in distributed, open computing solutions with revenues in 1998 of over $ 800 million. We provide customers and partners with the software and services to create business solutions for strategic, competitive advantage. These high-performance, end-to-end solutions encompass client/server, Internet and intranet transaction processing and data mart and data warehousing applications. Sybase's Adaptive Component Architecture™ enables rapid design, development and deployment of distributed multi-tier business applications. Our product lines include Sybase high-performance database servers, EnterpriseConnect™ distributed data access and connectivity products, and Powersoft open business development tools.

I greatly appreciate the opportunity to appear today before this Committee on behalf of Sybase and the Business Software Alliance (BSA). Since 1988, BSA has been the voice of the world's leading software developers before governments and with consumers in the international marketplace. BSA promotes the continued growth of the software industry through its international public policy, education and enforcement program in 65 countries throughout North America, Europe, Asia and Latin America. Its members represent the fastest growing industry in the world. BSA worldwide members include Adobe, Attachmate, Autodesk, Bentley Systems, Corel Corporation, Lotus Development, Microsoft, Network Associates, Novell, Symantec and Visio. Additional members of BSA's Policy Council include Apple Computer, Compaq, Intel, Intuit and my company, Sybase. BSA websites: *www.bsa.org; www.nopiracy.com*.

But we really are here today to speak on behalf of the tens of millions of *users* of American software and hardware products. The American software and hardware industries have succeeded because we have listened and responded to the needs of computer users worldwide. We develop and sell products that users want and for which they are willing to pay.

One of the most important features computer users *are* demanding is the ability to protect their electronic information and to interact securely worldwide. American companies have innovative products which can meet this demand and compete internationally. But there is one thing in our way—the continued application of overbroad, unilateral, export controls by the U.S. Government.

The Security and Freedom through Encryption (SAFE) Act, H.R. 850, modernizes U.S. export laws regarding software and hardware with encryption capabilities to permit American companies to compete on a level international playing field and to provide computer users with their choice of adequate protection for their confidential information and critical infrastructures.

For these reasons, BSA strongly supports the SAFE Act. We urge the Committee to report the SAFE Act unamended and look forward to its passage by the House early this year.

We want to pay tribute to the tremendous efforts of Representatives Goodlatte and Lofgren in championing this legislation, as well as thank both you, Mr. Chairman, and Mr. Frank and the other Subcommittee members who were among the 205 original cosponsors of the SAFE Act.

This morning I want to make three points:

- Widespread deployment of encryption is not only desirable, it is critical;
- America's export policy should promote widespread deployment of products with encryption capabilities in the worldwide market; and
- BSA strongly supports the SAFE Act because it provides freedom for Americans to use and sell any encryption domestically and provides greatly needed export control relief.

Widespread Deployment Of Encryption Is Not Only Desirable, It Is Critical

Confidential Information And Secure Networks In The Internet Age Are The Key To Privacy And Commerce

American individuals and companies are rapidly becoming networked together through private local area networks (LANs), wide area networks (WANs) and public networks such as the Internet. Combined, these private and public networks are the economic engine driving electronic commerce, transactions and communications. This engine is being choked by the lack of availability of strong encryption products.

Traffic on the Internet doubles every 100 days. Predictions of business-to-business Internet commerce for the year 2000 range from $66 billion to $171 billion, and by 2002, electronic commerce between businesses is expected to reach $300 billion. During 1997, one leading manufacturer of computer software and hardware sold $3 million per day online for a total of $1.1 billion for the year.

More and more individual consumers also are going on line and spending. More than 10 million people in North America alone have purchased something over the Internet and at least 40 million have obtained product and price information on the Internet only to make the final purchase off-line. Imagine the boost in volume of e-commerce if all of these consumers had enough confidence in the security of the Internet to purchase on-line.

Yet in 1996 the Computer Security Institute/FBI Computer Crime Survey indicated that our worldwide corporations will be increasingly under siege: over half from within the corporation, and nearly half from outside of their internal networks.

Network users *must* have confidence that their communications and data—whether personal letters, financial transactions or sensitive business information—are secure and private. Electronic commerce is transforming the marketplace—eliminating geographic boundaries and opening the world to buyers and sellers. Companies, governments and individuals now realize that they can no longer protect data and communications from others by relying on limiting physical access to computers and maintaining stand-alone centralized mainframes. Instead, users expect to be able to pick up their e-mail or modify a document from any computer anywhere in the world simply by using their Internet browsers. Thus, consumers worldwide are demanding to be able to protect their electronic information and interact securely worldwide, and access to products with strong encryption capabilities has become critical to providing them with confidence that they will have this ability.

Full Deployment Of Strong Encryption Is Vital For Protecting America's Critical Infrastructures

Governments also are recognizing that without encryption, the electronic networks that control such critical functions as airline flights, health care functions, electrical power and financial markets remain highly vulnerable. The U.S. General Accounting Office in its report issued in May of 1996 entitled "Information Security: Computer Attacks at Department of Defense Pose Increasing Risks" found that computer attacks are an increasing threat, particularly through connections on the Internet, such attacks are costly and damaging, and such attacks on Defense and other U.S. computer systems pose a serious threat to national security.

As the President said on January 22, 1999, before the National Academy of Sciences, "[w]e must be ready—ready if our adversaries try to use computers to disable power grids, banking, communications and transportation networks, police, fire and health services—or military assets. More and more, these critical systems are driven by, and linked together with, computers, making them more vulnerable to disruption."

The President has been so concerned that he established a Commission on Critical Infrastructure Protection to provide him with guidance and issued two Presidential Directives based on the Commission's recommendations.

In the Report of the President's Commission on Critical Infrastructure Protection entitled Critical Foundations: Protecting America's Infrastructures (October 1997), the Commission emphasized that "Strong encryption is an essential element for the security of the information on which critical infrastructures depend." In fact "[p]rotection of the information our critical infrastructures are increasingly dependent upon is in the national interest and essential to their evolution and full use. A secure infrastructure requires the following:

- Secure and reliable telecommunications networks.
- Effective means for protecting the information systems attached to those networks. . . .
- Effective means of protecting data against unauthorized use or disclosure.
- Well-trained users who understand how to protect their systems and data."

An earlier blue ribbon National Research Council (NRC) Committee similarly concluded in its (May 1996) CRISIS Report ("Cryptography's Role in Securing the Information Society") that encryption *promotes* the national security of the United States by protecting "nationally critical information systems and networks against unauthorized penetration."

Thus, the NRC Committee found that on balance the advantages of widespread encryption use outweighed the disadvantages and that the U.S. Government has "an important stake in assuring that its important and sensitive . . . information . . . is protected from foreign government or other parties whose interests are hostile to those of the United States."

Information security is critical to the integrity, stability and health of individuals, corporations and governments. While cryptography is but one element of security, it is the keystone of secure, distributed systems. Frankly, there is no substitute for good, widespread, strong cryptography when attempting to prevent crime and sabotage through these networks. The security of any network, however, is only as good as its weakest link. America's infrastructures cannot be protected if they are networked with foreign infrastructures using weak encryption.

U.S. unilateral export controls have also had a significant impact on the availability of strong American encryption domestically, which is ultimately harming the American consumer. The U.S. software and hardware industry makes at least one-half of its revenues through exports. For this reason and due to the significant difficulties companies encounter selling foreign purchasers a weaker version of an encryption product, some software companies have offered products with the same encryption capabilities both domestically and abroad. Therefore, the American consumer has fewer strong American encryption products to choose from than they would without U.S. export controls. The American consumer is ultimately left with an unfortunate choice: they may either buy strong encryption which they cannot use internationally, or they may simply purchase strong foreign encryption products that are not subject to U.S. export controls. Neither choice is the best for protecting America's critical infrastructures.

In the long-term, we believe it is in America's best interest to have America's critical infrastructures and national security be protected by widespread reliance on strong American encryption products both here and abroad.

Relaxed Export Controls On Encryption Products Is Vital For Ensuring America's Global Competitiveness

American companies *do* have exciting and innovative products that can meet the demand for 128-bit encryption and compete internationally. But unless the current unilateral U.S. export restrictions are changed to allow the use of strong encryption, American individuals and businesses will not be active participants in this new networked world of commerce—let alone continue to be the leaders in its development. Furthermore, American companies will no longer be providing the world, and its critical infrastructures, with the answers to their security problems. Instead foreign companies will. It is unclear how U.S. national security or law enforcement will be aided or how our critical infrastructures will be secure when foreign encryption products dominate the world market.

The computer software and hardware industries are American success stories, but they are being threatened. America's software and hardware industries are important contributors to U.S. economic security—now and in the future. Information technology industries are now directly responsible for over one-third of real growth of the U.S. economy. Between 1980 and 1992, the computing and software industry grew at an annual rate of over 28%, while overall domestic growth was less than 3%. From 1990 through 1996, the software industry grew at a rate of 12.5%, nearly 2.5 times faster than the overall U.S. economy.

More than 7 million people work in IT industries. In 1996, the software industry provided a total of over 619,000 direct jobs and $7.2 billion in tax revenues for the U.S. economy. The software industry is expected to create an average of 45,700 new jobs each year through 2005. If piracy were to be eliminated in the United States, the number of new software jobs created would double to an average of 93,000 a year.

Moreover, the computer software industry has achieved tremendous success in the international marketplace with global sales of packaged (*i.e.*, non-custom) software reaching over $118.4 billion in 1996, and rising to $135.4 billion in 1997. American produced software accounts for 70% of the world market, with exports of U.S. programs constituting half of the industry's output.

The incredible growth of the industry and its exporting success benefits America through the creation of jobs here in the United States. Many of these jobs are in highly skilled and highly paid areas such as research and development, manufacturing and production, sales, marketing, professional services, custom programming, technical support and administrative functions. In the U.S. software industry, workers enjoy more than twice the average level of wages across the entire economy—$57,319 versus $27,845 per person.

All of these revenues and jobs are dependent upon American software and hardware producers remaining the market leaders around the world, especially as the major growth markets continue to be outside the United States. Strong export controls on products with encryption capabilities are crippling the ability of these companies to compete with foreign providers.

America's Export Policy Should Promote Widespread Deployment Of American Products with Encryption Capabilities In The Worldwide Market

As embodied in the SAFE Act, the most successful encryption policy will ensure that Americans can use and sell any encryption that they want domestically, prohibit both Federal and State governments from imposing encryption standards or techniques, and relax export controls on products with encryption capabilities in a manner that is based on technological and market realities. Just because law enforcement and national security interests wish that they could turn back the clock and limit consumers access to strong encryption approved by the government, it will not happen, especially on a worldwide basis. This is especially true for mass market software and hardware, which by its inherent nature is uncontrollable.

The Administration Took The First Step Towards Developing A Sensible Long-Term Encryption Policy, But They Still Have Not Gone Far Enough

The BSA members welcomed the Administration's efforts to relax export controls on select products used by select users. However, the Administration's actions are merely a first step. A truly successful, sensible encryption policy would be based on technological and market realities, and would not create winners and losers in the encryption marketplace on a sector-by-sector basis. It would recognize that:

- The worldwide encryption standard is 128-bit encryption;

- Mass market software and hardware is inherently uncontrollable; and
- It is in America's national and economic security interests to have American designed and manufactured encryption products deployed worldwide.

Moreover, we believe it is preferable for Congress to put encryption policy on a statutory basis—sending a strong message around the world that encryption is important for a strong defense, for protecting the privacy of citizens and for preventing crime.

Unilateral U.S. Export Controls Harm American Interests

Currently, there are no restrictions on the use of cryptography within the United States. However, the U.S. Government maintains strict *unilateral* export controls on computer products that offer strong encryption capabilities.

American companies are forced to limit the strength of their encryption to the 56-bit key length level set late in 1998. The recently announced regulations will also permit companies to export stronger encryption on a sector-by-sector, user-by-user basis. However, this policy ignores the fact that:

- The minimum strength now required by new Internet applications is 128-bit encryption;
- The most widely used encryption program, PGP, with over two million users worldwide, uses the Swiss developed IDEA encryption algorithm, with a 128-bit key;
- American companies cannot export encryption products to a vast majority of non-U.S. commercial entities. Foreign manufacturers provide 128-bit encryption alternatives and add-ons—filling the market void created by U.S. export controls;
- Providing sector-by-sector relief is unworkable for mass market products and does not reflect commercial realities for sales of custom products; and
- 56 bit encryption has been demonstrated to be vulnerable to commercial let alone governmental attack. (In the beginning of this year at the RSA Encryption Conference, a 56-bit DES encoded message was broken by private companies and individuals working together in 22 hours and 15 minutes—imagine what a hostile government with serious resources could do.)

Export controls also have made American companies less competitive and opened the door for foreign software and hardware developers to gain significant market share—decreasing our national and economic security.

I want to take one minute to discuss the Wassenaar Arrangement at this point. Please do not be fooled by any claims from the Administration that the Wassenaar Arrangement is the multilateral agreement on encryption that they have been touting was just around the corner for the past several years.

The Wassenaar Arrangement was an agreement among only 30 countries, and it actually decontrolled encryption products. Many countries, such as Israel and South Africa, who export strong encryption are not signatories to the Arrangement. The Wassenaar Arrangement eliminated controls of any sort on 56-bit encryption and permits exports of up to 64-bit encryption in mass-market software and hardware. It also removed any reporting requirements—the sole official means for actually monitoring what countries are doing. Although the Arrangement left open the possibility that countries might individually control 128-bit encryption, we are skeptical that they will do so. There is no penalty for failing to control 128-bit encryption, and most countries are actually moving towards *encouraging* the use of stronger encryption. Finally, a country could technically comply with the Arrangement, while still permitting easy exports of strong encryption.

Even France, traditionally the country which placed the greatest restrictions on its own citizens by limiting them to the easily broken 40-bit level of encryption, has recognized that technology has progressed. Near the end of 1998, France relaxed controls on the domestic use of encryption and is now permitting, and in fact encouraging, the use of 128-bit encryption by its citizens.

Without Export Relief, Foreign Consumers Will Purchase Their Products From Foreign Suppliers, Keeping U.S. Manufacturers At A Competitive Disadvantage

As a result of U.S. unilateral export controls, encryption expertise is being developed off-shore by foreign manufacturers who now provide hundreds of encryption alternatives and add-ons. The Administration's export controls are in no way preventing foreigners, let alone those with criminal intent, from obtaining access to encryption products. In fact, foreign software and hardware manufacturers have seized the opportunity to create sophisticated encryption products and to capture sales.

As long ago as 1995, the General Accounting Office confirmed that sophisticated encryption software is widely available to foreign users on foreign Internet sites. In 1996, a Department of Commerce study again confirmed the widespread availability of foreign manufactured encryption programs and products. An on-going industry study by Trusted Information Systems (TIS Study) highlights the ever-increasing availability of foreign developed and manufactured products as it discovered there were 656 foreign programs and products available from 29 countries as of December 1997.

Further demonstrating the worldwide availability, use and sophistication of encryption abroad is the Department of Commerce's National Institute of Standards and Technology (NIST) efforts to work with the private sector to develop an Advanced Encryption Standard (AES). Individuals and companies from eleven different countries proposed 10 out of the 15 candidate algorithms submitted to NIST: Australia's LOKI97; Belgium's RIJNDAEL; Canada's CAST-256 and DEAL; Costa Rica's FROG; France's DFC; Germany's MAGENTA; Japan's E2; Korea's CRYPTON; and the United Kingdom, Israel and Norway's SERPENT algorithms. Only 5 out of the 15 candidate algorithms were submitted by U.S.-based individuals or companies.

The impact of lost sales is enormous. If an encryption product is combined with other applications such as Internet browsers and application servers, U.S. companies will generally lose both sales. In fact, companies risk losing sales of entire systems because of inability to provide necessary security features. This permits foreign manufacturers to gain entry into companies as well as gain credibility—providing the foreign manufacturers with further opportunity to take away future sales in the same and other product lines.

I would like to mention a few specific examples with respect to foreign availability of encryption products. The Apache Group, based in the U.K., announced in April 1997 that its Apache Unix Internet Server software with very strong encryption had a 29% market share of Web server software. Today the Apache web server serves over half—50%—of the domains on the Internet.

Companies such as Brokat Informationssysteme, a German company, are developing products that are more than simply add-ons to American products. Brokat's modular e-services platform, Twister, which companies use to offer their customers secure and simple electronic services via various electronic channels, such as the Internet or mobile communications networks, is already being used by more than 1,500 companies worldwide. Brokat's sales outside of Germany, including to the United States, have now increased to be 56 percent of the company's total sales. The American market research institute Meridien Research described BROKAT as the leading company worldwide for Internet banking solutions.

The merger of two foreign companies, Zergo Holdings (U.K.) and Baltimore Technologies (Ireland), into a new company called Baltimore only further illustrates that foreign companies are flourishing solely because there is no U.S. competition. According to the Gartner Group in a Research Note dated January 28, 1999, the new company is "a competitive participant in providing e-commerce and enterprise security, with 11 international offices and a global partner network . . . with customers in 40 countries."

U.S. Encryption Export Controls Hurt American Companies Without Helping Law Enforcement Or National Security

U.S. export controls have had the effect of creating an encryption expertise outside the United States that is gathering momentum. Unfortunately, every time research and development of an encryption technique or product moves off-shore, U.S. law enforcement and national security agencies lose. We believe that continuing down this path will be ultimately more harmful to our national security and law enforcement efforts as American companies will no longer be the world leaders in creating and developing encryption products.

In fact, as long ago as 1996, the NRC Committee concluded that as demand for products with encryption capabilities grows worldwide, foreign competition could emerge at levels significant enough to damage the present U.S. world leadership in information technology products. The Committee felt it was important to ensure the continued economic growth and leadership of key U.S. industries and businesses in an increasingly global economy, including American computer, software and communications companies. Correspondingly, the Committee called for an immediate and easy exportability of products meeting general commercial requirements—which is currently 128-bit level encryption!

To summarize:

Foreign competitors not subject to outdated U.S. export controls are ready to take sales and customers from U.S. companies today.

Complex and cumbersome U.S. export controls make American companies less competitive. They significantly increase the costs of developing, marketing and selling products with encryption capabilities, delay the introduction of new products or features, and encourage foreign customers to purchase from foreign suppliers due to the uncertainty and delay in obtaining a comparable American product.

Current export controls do not keep strong encryption out of the hands of foreign customers; they just keep U.S. products out of their hands.

BSA Strongly Supports The SAFE Act Because It Provides Freedom For Americans To Use And Sell Any Encryption Domestically And Provides Greatly Needed Export Control Relief

The SAFE Act Preserves Americans' Domestic Encryption Freedom

The SAFE Act ensures that Americans may use and sell whatever kind of encryption they want domestically. It ensures that the U.S. government may not require or provide other incentives for Americans to use encryption products "approved" by the government or meeting certain standards.

Also, the Act does not permit the government to link electronic signatures to the use of certain types of encryption products.

The SAFE Act Provides Law Enforcement With Important Safeguards

Importantly, the SAFE Act does permit the Secretary of Commerce to continue preventing exports to countries of terrorist concern or other embargoed countries pursuant to the Trading With The Enemy Act or the International Emergency Economic Powers Act.

The bills also contain safeguards when relaxing export controls for strong encryption products—the Secretary of Commerce is not required to permit such exports if there is substantial evidence that the software or hardware will be diverted or modified for military or terrorist use or re-exported without requisite U.S. authorization.

The SAFE Act Recognizes That Mass Market Products Are Uncontrollable And Should Be Exportable

U.S. export controls still ignore the realities of mass-market software and hardware distribution. Mass-market hardware manufacturers and software publishers sell products through multiple distribution channels such as OEMs (*i.e.*, hardware manufacturers that pre-load software onto computers), value-added resellers, retail stores and the emerging channel of on-line distribution. Thus, mass market products are available to the general public from a variety of sources.

The mass-market distribution model presupposes that hardware manufacturers and software publishers will take full advantage of these multiple channels to ship identical or substantially similar products worldwide (allowing only for differences resulting from localization) irrespective of specific customer location or characteristics.

Uncontrollable products at 56-bits cannot suddenly become controllable products at 128-bits. The SAFE Act recognizes as a fundamental proposition that the United States should *not* try to control the export of something that is, by its very nature, uncontrollable. Trying to control the uncontrollable squanders the limited resources of companies trying to comply with unrealistic export controls as well as the resources of the government as it tries to enforce unenforceable export controls, undermining the credibility of the entire system of export controls.

The SAFE Act Permits Exports Of Custom Hardware And Software

The SAFE Act ensures that if strong encryption products have been permitted to be exported to foreign banks, then custom software and hardware with comparable encryption capabilities should be exportable to other foreign commercial purchasers in that country. The U.S. should not control exports of competitive custom products embodying world encryption standards. Note that the type of software and hardware we are talking about here is a "custom" product (if it were generally available it would not need an individual license under the bill's other provisions).

The Time For Action Is Now

To keep American vendors on a level international playing field and American computer users adequately protected, U.S. export controls must be immediately updated to reflect technological and international market realities.

Thank you.

CHAPTER 28

TESTIMONY OF WILLIAM A. REINSCH UNDER SECRETARY FOR EXPORT ADMINISTRATION DEPARTMENT OF COMMERCE BEFORE THE HOUSE COMMITTEE ON THE JUDICIARY SUBCOMMITTEE ON COURTS AND INTELLECTUAL PROPERTY ON THE SECURITY AND FREEDOM THROUGH ENCRYPTION ACT

March 4, 1999

Thank you, Mr Chairman, for the opportunity to testify on the direction of the Administration's encryption policy. We have made a great deal of progress since my last testimony on this subject in September 1997.

Even so, encryption remains a hotly debated issue. The Administration continues to support a balanced approach which considers privacy and commerce as well as protecting important law enforcement and national security equities. We have been consulting closely with industry and its customers to develop a policy that provides that balance in a way that also reflects the evolving realities of the market place.

The Internet and other electronic media are becoming increasingly important to the conduct of international business. One of the many uses of the Internet which will have a significant affect on our everyday lives is electronic commerce. According to a recent study, the value of e-commerce transactions in 1996 was $12 million. The projected value of e-commerce in 2000 is $2.16 billion. Many service industries which traditionally required face to face interaction such as banks, financial institutions, and retail merchants are now providing cyber service. Customers can now sit at their home computers and access their banking and investment accounts or buy a winter jacket with a few strokes of their keyboard.

Furthermore, most businesses maintain their records and other proprietary information, such as health records or sales strategies, electronically. They now conduct many of their day-to-day communications and business transactions via the Internet and E-mail. An inevitable byproduct of this growth of electronic commerce is the need for strong encryption to provide the necessary secure infrastructure for electronic communications, transactions and networks. The disturbing increase in computer crime and electronic espionage has made people and businesses wary of posting their private and company proprietary information on electronic networks if they believe the infrastructure may not be secure. A robust secure infrastructure can help allay these fears, and allow electronic commerce to continue its explosive growth.

Developing a new encryption policy has been complicated because we do not want to hinder its legitimate use—particularly for electronic commerce; yet at the same time we want to protect our vital national security, foreign policy and law enforcement interests. We have concluded that the best way to accomplish this was to continue a balanced approach: to promote the development of strong encryption products that would allow lawful government access to plaintext under carefully defined circumstances; to promote the legitimate uses of strong encryption to protect confidentiality; and continue looking for additional ways to protect important law enforcement and national security interests.

During the past three years, we have learned that there are many ways to assist in lawful access. There is no one-size-fits-all solution. The recovery encryption plans we received showed that different technical approaches to recovery of plaintext exist. In licensing exports of encryption products under individual licenses, we also learned that, while some products may not meet the strict technical criteria of our regulations, they are nevertheless consistent with our policy goals.

Additionally, we learned that the use of strong non-recovery encryption within certain trusted industry sectors is an important component of our policy in order to protect private consumer information and allow our US high tech industry to maintain its lead in the information security market while minimizing risk to national security and law enforcement equities.

Taking into account all that we have learned and reviewing international market trends and realities, in 1998 we made several changes to our encryption policy that I will summarize for you.

On September 22, 1998, we published a regulation implementing our decision to allow the export, under a license exception, of unlimited strength encryption to banks and financial institutions located in countries that are members of the Financial Action Task Force or have effective anti-money laundering laws. The regulation also allows exports, under a license exception, of encryption products that are specially designed for financial transactions. This new policy recognizes the fact that we need to secure and safeguard our financial networks, and the banking and financial communities cooperate with government authorities when information is required to combat financial and other crimes. The direct result of this policy change is that over 100 of the world's largest banks and almost 70% of the international financial institution market is now eligible for strong American-made encryption.

As I mentioned earlier, we have been looking for ways to make our policy consistent with both market realities and national security and law enforcement concerns. Since last March, the Administration has been engaged in a dialogue with U.S. industry, law enforcement, and privacy groups on how our policy might be improved to find technical solutions, in addition to key recovery, that can assist law enforcement in its efforts to combat crime. At the same time, we wanted to find ways to assure U.S. technology leadership, promote secure electronic commerce, and protect important privacy concerns. The purpose of this dialogue was to find cooperative solutions that could assist law enforcement, while protecting national security, plus assuring continued U.S. technology leadership and promoting the privacy and security of U.S. firms and citizens in electronic commerce. We believed then and now that the best way to make progress on this issue is through a constructive cooperative dialogue, rather than seeking legislative solutions. Through our dialogue, there has been increased understanding among the parties. And we have made progress.

The result of this dialogue was an update to our encryption policy which Vice President Gore unveiled last September 16. The regulations implementing the update were published on December 31. This will not end the debate over encryption controls, but we believe the regulation addresses some private sector concerns by opening large markets and further streamlining exports.

The policy update liberalizes controls on 56-bit products and on products of unlimited bit length, whether or not they contain recovery features, to certain industry sectors. Many of the new reforms permit the export of encryption to certain end-users under a license exception. That is, after the product receives a one-time review, it can be exported by the manufacturer, resellers and distributors without the need for a license or other additional review. In developing our policy we identified the key sectors that will form the basis of creating a reliable secure infrastructure for communicating and storing critical personal information: banks, financial institutions, insurance companies, on-line merchants, and health facilities.

Specifically, the new policy allows for:

- the export of 56-bit hardware and software worldwide to any end user under a license exception;
- exports of strong encryption, including technology, to U.S. companies and their subsidiaries under a license exception to protect important business proprietary information;
- exports of strong encryption to the insurance and medical/health sectors in 46 countries under a license exception for use in securing proprietary medical and health information;

- exports of strong encryption to secure on-line transactions between on-line merchants and their customers in 46 countries under a license exception.
- "recovery capable" or "recoverable encryption products of any key length, such as the so-called "doorbell" products, can now be approved under a kind of bulk license called an "encryption licensing arrangement" to recipients in 46 countries. Examples of such products are systems that are managed by a network or corporate security administrator.

I would note that these provisions apply to products with or without key recovery features. One of the aspects of our policy update is to permit exports of strong encryption with or without key recovery to protect electronic commerce while also minimizing the risk to national security and law enforcement. For example, in some cases we have limited our approval policy to a list of countries or a set of end users, rather than permit exports on a global basis, to help protect national security interests.

We have also expanded our policy to encourage the marketing of a wider variety of "recoverable" products that may not be key recovery in a narrow sense but which may be helpful to law enforcement pursuant to strict authorities. Again, these are typically systems managed by a network or corporate administrator. We also further streamlined exports of key recovery products by no longer requiring a review of foreign key recovery agents and no longer requiring companies to submit business plans.

This past year, we also made progress on developing a common international approach to encryption controls through the Wassenaar Arrangement, which was established in 1996 as the successor to COCOM. It is an international export control arrangement among 33 countries whose purpose is to prevent destabilizing accumulations of arms and civilian items with military uses in countries or regions of concern. It is the multilateral basis for many of our export controls.

In December, through the hard work of Ambassador Aaron, the President's special envoy on encryption, the Wassenaar Arrangement members agreed on several changes relating to encryption controls. These changes go a long way toward increasing international security and public safety by providing countries with a stronger regulatory framework for managing the spread of robust encryption.

Specific changes to multilateral encryption controls include removing controls on all encryption products at or below 56 bit and certain consumer items regardless of key length, such as entertainment TV systems, DVD products, and on cordless telephone systems designed for home or office use.

Most importantly, the Wassenaar members agreed to remove encryption software from the General Software Note and replace it with a new cryptography note. First drafted in 1991, the General Software Note allowed countries to export mass market encryption software without restriction. It was essential to modernize the GSN and close the loophole that permitted the uncontrolled export of encryption with unlimited key length. Under the new cryptography note, mass market hardware has been added and a 64-bit key length or below has been set as an appropriate threshold. This will enable governments to review the dissemination of 64-bit and above encryption.

I want to be clear that this does not mean encryption products of more than 64 bits cannot be exported. Our own policy permits that. It does mean, however, that such exports must be reviewed by governments consistent with their national export control procedures.

Export control policies without a multilateral approach have little chance of success. Agreement, by the Wassenaar members, to close the loophole for mass market encryption products is a strong indication that other countries are beginning to share our public safety and national security concerns. Contrary to what many people thought two years ago, we have found that most major encryption producing countries are interested in developing a harmonized international approach to encryption controls.

At the same time, we recognize that this is an evolutionary process, and we intend to continue our dialogue with industry. Our policy should continue to adapt to technology and market changes. We will review our policy again this year with a view toward making further changes. An important component of our review is input from industry, which we are receiving through our continuing dialogue.

With respect to H.R.850, the Administration opposes this legislation as we did its predecessor in the last Congress. The bill proposes export liberalization far beyond what the Administration can entertain and which would be contrary to our international export control obligations. Despite some cosmetic changes the authors have made, the bill in letter and spirit would destroy the balance we have worked so hard to achieve and would jeopardize our law enforcement and national security interests. I defer to other witnesses to describe the impact of the bill on their equities, but let me describe two of its other problems

First, I want to reiterate that this Administration is not seeking controls or restraints on domestic manufacture or use of encryption. We continue to believe the best way to make progress on ways to assist law enforcement is through a constructive dialogue. As a result, we see no need for the statutory prohibitions contained in the bill.

Second, once again we must take exception to the bill's export control provisions. In particular, the references to IEEPA as I understand them would preclude controls under current circumstances and in any future situation where the EAA had expired, and the definition of general availability, as in the past, would preclude export controls over most software.

In addition, whether intended or not, we believe the bill as drafted could inhibit the development of key recovery even as a viable commercial option for those corporations and end users that want it in order to guarantee access to their data. The Administration has repeatedly stated that it does not support mandatory key recovery, but we endorse and encourage development of voluntary key recovery systems, and, based on industry input, we see growing demand for them, especially corporate key recovery, that we do not want to cut off.

As this Committee knows better than most, public debate over encryption policy has been spirited. Many on both sides of the debate have had difficulty grasping their counterparts' views or realizing that there is a middle ground. Our dialogue with industry has gone a long way toward bridging that gap and finding common ground. We will continue this policy of cooperative exchange as it is clearly the best way to pursue our policy objectives of balancing public safety, national security, and the competitive interests of US companies.

CHAPTER 29

STATEMENT OF THE HONORABLE HOWARD COBLE CHAIRMAN SUBCOMMITTEE ON COURTS & INTELLECTUAL PROPERTY ON H.R. 850, THE "SECURITY AND FREEDOM THROUGH ENCRYPTION (SAFE) ACT"

March 4, 1999

The Subcommittee will come to order.

Today the Subcommittee will hear testimony on H.R. 850, the "Security and Freedom Through Encryption (SAFE) Act." As all of you well know, encryption is the process of encoding data or communications in a form that only the intended recipient can understand. Once the exclusive domain of the national security agencies, encryption has become increasingly important to persons and companies in the private sector concerned with the security of the information they transmit. H.R. 850 seeks to provide a means of ensuring protection for confidential communications transmitted in this information age. It also seeks to lift restrictions on the exportation of advanced encryption so U.S. information companies will remain the world leader.

This matter has consumed abundant time and energy of this Subcommittee and the Full Committee over the last few years, and it is time make sure it becomes law. While I am ever mindful of the concerns of law enforcement and national security agencies, I believe the reforms contained in H.R. 850 are critical to ensure that the United States will continue as the leader in information technologies, as well as guaranteeing protection from intrusions to free speech as new information technology develops.

I once again commend the gentleman from Virginia, Mr. Goodlatte, for his work in this area and for introducing this very important piece of legislation which I support. I might mention that this bipartisan bill currently has over 200 cosponsors including both Republicans and Democratic Leadership. I would like to ask the Ranking Member, Mr. Berman, to make an opening statement and will then ask Mr. Goodlatte to further explain his bill.

CHAPTER 30

TESTIMONY OF DAVE MCCURDY PRESIDENT ELECTRONIC INDUSTRIES ALLIANCE BEFORE THE HOUSE JUDICIARY COMMITTEE SUBCOMMITTEE ON COURTS AND INTELLECTUAL PROPERTY U.S. HOUSE OF REPRESENTATIVES REGARDING ENCRYPTION POLICY FOR THE INFORMATION AGE

March 4, 1999

Thank you, Mr. Chairman for the opportunity to testify today on U.S. encryption policy for the Information Age. With over 2000 member companies, EIA is the premier alliance of trade associations for the high-technology industry. Our mission is to promote an economic and political environment, in this country and around the world, in which our industry can thrive.

I am also a former member of Congress from Oklahoma. During my 14 year tenure in this body, I served as Chairman of the House Intelligence Committee, as well as subcommittee chairman on the Armed Services Committee and the Science Committee. I continue to serve as a member of the Weapons of Mass Destruction Commission, a group of experts investigating how this country can combat proliferation. So I am well aware of how dual-use civilian technologies like encryption can be used for illicit purposes, and the important role of export controls to our national security. But I also recognize the severely limited effectiveness of export controls, as well as the vital importance of a strong and innovative high-technology sector to keep our armed forces a step ahead of any adversary.

Mr. Chairman, it is now clear that 1998 was the year the Internet became mainstream, with more than 100 million people worldwide using the network, up from only three million just four years earlier. Furthermore, shoppers are likely to spend upwards of $10 billion online this year, and the figure could quadruple by 2002. But as the Net proves itself to be a powerful means for commerce, with potentially hundreds of billions of dollars moving across the network, it also becomes a tempting target for criminals seeking to steal consumers' credit card data or eavesdrop on their online communications. Similarly, as large and small companies turn to the Net to exchange information with their various divisions, or communicate with their suppliers and customers around the world, those electronic transmissions become targets for hackers or business rivals. Other examples of electronic information needing protection are the intellectual property on DVD's or posted on websites; voice conversations moving across wireless networks; or product designs, employee records, and other sensitive data stored on companies' internal computers.

In each of these cases, individuals and companies need encryption to protect themselves. Additionally, with the exponential advances in computing power available to ordinary people and criminals alike, ever stronger encryption is needed to defend against illegal attempts to unscramble sensitive electronic information. Yet, the Administration continues to impose regulations which threaten the security and privacy of millions of Internet users.

The net effect of this policy is to damage the global competitiveness of the U.S. high-tech industry, as well as to jeopardize the security of individuals and companies which operate internationally. Meanwhile, if the so-called "bad actors" want to use encryption, they have the choice of buying it off the shelf of any software store in America, or downloading it from the Internet, or buying it from a producer overseas no questions asked. Because of these overly strict and burdensome, yet futile export controls, one of this country's most dynamic and competitive industries is ceding marketshare to foreign competitors. Even the Commerce Department, which administers the encryption rules, has acknowledged that there are over 600 encryption products being produced in 29 countries outside the United States, and they are competitive with anything the U.S. produces. It would be tragic if these export rules forced U.S. high-tech companies, and the high-paying jobs associated with them, to move offshore. Such a development would also have the ironic effect of compelling our law enforcement and national security officers to confront encryption made by more non-U.S. companies, with which they have no cooperative relationships.

Mr. Chairman, I have the privilege of representing the most dynamic and competitive industry in the U.S. economy today—actually, I should say, in the world economy today. The companies we represent operate globally, they think and plan in global terms, and they face intense international competition. The fact is, the days when U.S. companies dominated the high-technology industry are over. Similarly, the days when the domestic U.S. market could sustain the industry are also over. It has become almost cliche, but the global economy is a fact of doing business for us, and is a critically important concept to keep in mind as we formulate public policy in this area.

As any successful CEO will tell you, competing—indeed, surviving—in the global economy means exporting. The phenomenal success of the U.S. technology industry comes from its entreprenurialism, its aggressiveness, its willingness to compete—all those free market forces that drive innovation. In this kind of business environment, tapping new markets before the competition does is the key to success. In 1997, more than one-third of what the U.S. electronics industry produced was exported overseas, over $150 billion in goods. That means more than a third of the 1.8 million employees who work for U.S. electronics companies depend on exports for their jobs, and the percentage goes up every year.

We must also recognize that our high-tech companies are the engine for technological innovation and economic growth in the world today. The U.S. economy is the most competitive in the world due in no small part to the amazing advancements our companies have achieved. Technologies which, not long ago, had only military or limited civilian applications are now pervasive in our society, and the greater economic efficiency stemming from this diffusion of technology has been the driving force for the remarkable prosperity so many Americans are experiencing.

The impact of export controls on how this industry competes in the global economy can hardly be overstated. They hold us back from competing. Unilateral export controls essentially force us to cede the playing field to our overseas competitors, or at least burden us to the point that we cannot compete effectively. The case of export controls on encryption is perhaps the best example. No amount of government subsidies could do more to develop the European encryption industry than U.S. export controls have.

We assert that a more balanced policy is needed—one which recognizes the interests of government, the high-tech industry, and corporate and individual users. While we in the business community recognize the importance of keeping potentially dangerous technologies out of the wrong hands, the government must similarly recognize the importance of a dynamic and innovative high-tech industry to our economy, and not incidentally, to our national security. We believe that the national security vs. economic arguments present a false choice, and that a well-balanced and realistic compromise is within reach.

The Security and Freedom Through Encryption Act (SAFE) is a vital aspect of the strategy to develop a meaningful Information Age encryption policy, and we sincerely appreciate the tremendous efforts of Congressmen Goodlatte and Lofgren as leaders of effort. Among other things, this bill would reaffirm the right of all Americans to use whatever encryption they choose to protect themselves, their digital property, and their electronic communications. Furthermore, it would prevent the government from requiring businesses to use only certain types of encryption in their global operations. EIA is excited to be part of the coalition working to enact this important legislation, and we look forward to working with this committee towards that end.

EIA has also put forward a proposal which we urge the Administration to consider, and which is attached to my written testimony. In brief, our compromise proposal has four basic elements. First, the government needs to significantly ease the restrictions on low-level and mass market encryption software. It was not very long ago that encryption was a solely military application, and therefore easily controlled, but to continue imposing onerous controls on software which anyone can purchase at the local shopping mall just does not make sense.

Second, the law enforcement and national security agencies could better define their access requirements, thereby allowing industry to develop a variety of marketable solutions, as well as enabling the Clinton Administration to finally abandon its key recovery policy.

Third, our new policy needs to differentiate between the increasingly numerous uses for the technology, such as for voice communications, data transmission, and in consumer electronics, with appropriate controls on those applications that clearly present problems for government, and decontrolling the rest.

Finally, U.S. policymakers need to recognize the futility of unilateral export restrictions, which serve only to damage our domestic industries while doing little to protect our national security. Only when we encourage our allies to develop meaningful multilateral controls can we hope to prevent the bad actors from acquiring these technologies.

The bottom line is, we in the high-tech industry believe this is a controversy which has dragged on for too long, and that a reasonable solution is possible if all sides are willing to compromise. Our industry is willing to accept restrictions on exports if the controls are reasonable, if they are effective at addressing the problem they are meant to solve, and if they do not impose unnecessary, overly burdensome requirements. We believe that by implementing these basic proposals, the Administration's legitimate concerns can be addressed, the U.S. high-tech industry will be allowed to compete globally, and users will have the security they need.

I would be happy to take your questions.

CHAPTER 31

STATEMENT OF KEITH A. RHODES
TECHNICAL DIRECTOR FOR COMPUTERS AND TELECOMMUNICATIONS
ACCOUNTING AND INFORMATION MANAGEMENT DIVISION

BEFORE THE
SUBCOMMITTEE ON TECHNOLOGY,
COMMITTEE ON SCIENCE,
HOUSE OF REPRESENTATIVES

INFORMATION SECURITY—
THE MELISSA COMPUTER VIRUS DEMONSTRATES URGENT NEED FOR STRONGER PROTECTION OVER SYSTEMS AND SENSITIVE DATA

April 15, 1999

Madam Chairwoman and Members of the Subcommittee:

Thank you for inviting me to participate in today's hearing on the "Melissa" computer virus. Although it did disrupt the operations of thousands of companies and some government agencies, this virus did not reportedly permanently damage systems and did not compromise sensitive government data. Nevertheless, it has shown us just how quickly computer viruses can spread and just how vulnerable federal information systems are to computer attacks. Moreover, Melissa has clearly highlighted the urgent and serious need for stronger agency and governmentwide protection over sensitive data. Today, I will discuss the immediate effects of the Melissa virus and variations of it as well as its broader implications. I will also discuss some critical measures that should be taken to help ensure that federal departments and agencies are better prepared for future viruses and other forms of attack.

The Melissa Virus and Its Immediate Impact

Melissa is a "macro virus" that can affect users of Microsoft's Word[1] 97 or Word 2000. Macro viruses are computer viruses that use an application's

---

1 Word processing software. The virus can also infect Word 98 for Macintosh and documents created by this application. However, in the Macintosh environment, the virus will not automatically send the infected document to others.

own macro programming language[2] to reproduce themselves. Macro viruses can inflict damage to the document or to other computer software.

Melissa itself is delivered in a Word document. Once the Word document is opened, and the virus is allowed to run, Melissa:

- Checks to see if Word 97 or Word 2000 is installed.
- Disables certain features of the software, which makes it difficult to detect the virus in action.
- Generally sends copies of the infected document to up to 50 other addresses using compatible versions of Microsoft's Outlook electronic mail program.[3]
- Modifies the Word software so that the virus infects any document that the user may open and close. If these documents are shared, the virus is spread.

Under some circumstances, Melissa could cause confidential documents to be disclosed without the user knowing it.

If addresses in an electronic mail address book are within the same organization, Melissa can quickly overload electronic mail servers and result in a denial of service. According to Carnegie Mellon University's CERT Coordination Center,[4] for example, one site reported receiving 32,000 copies of mail messages containing Melissa on its systems within 45 minutes.

In fact, what made Melissa different from other macro viruses was its ability to take advantage of the Microsoft e-mail application and the speed at which it spread. According to the CERT Coordination Center, the first confirmed reports of the virus were received on Friday, March 26, 1999. By Monday, March 29, it had reached more than 100,000 computers at more than 300 organizations.

In the course of spreading, variations of the Melissa virus also surfaced, including the "Papa" virus—a Microsoft Excel 97[5] or Excel 2000 macro

---

2 Macros are tools for customizing computer applications so that often-used commands can be automatically executed.

3 Outlook is a desktop information manager that also provides e-mail support. If any of the first 50 addresses in Outlook's address book represents a mailing list, then everyone on that list also receives a copy of the virus. In addition, if the user has more than one address book, the first 50 addresses in each book are used.

4 Originally called the Computer Emergency Response Team, the center was established in 1988 by the Defense Advanced Research Projects Agency. It is charged with establishing a capability to quickly and effectively coordinate communication among experts in order to limit the damage associated with and respond to incidents and building awareness of security issues across the Internet community.

5 Spreadsheet software.

virus that can also be delivered by e-mail. According to the Microsoft Corporation, this virus could generate commands that result in significant network traffic congestion without the user's knowledge.

Fortunately, aside from shutting down e-mail systems, Melissa did not reportedly permanently damage government and private sector information systems and did not compromise sensitive government data. However, because the federal government does not have a process for reporting and analyzing the effects of such attacks, quantitative analysis is difficult.

In its information bulletin on the virus, the Department of Energy (DOE) reported that Melissa had been detected at multiple DOE sites and had spread widely within the department. According to DOE, the risk of damage was low because most users did not have macros in files and would be alerted by Word's macro detector.[6] However, at the time it issued its advisory, DOE believed the risk of lost productivity and lost mail messages was high as mail servers might need to be shut down and purged of infected messages.

Broader Implications of the Melissa Virus

Although the Melissa virus reportedly did not compromise sensitive government data or damage systems, it demonstrated the formidable challenge the federal government faces in protecting its information technology assets and sensitive data.

First, Melissa showed just how quickly viruses can proliferate due to the intricate and extensive connectivity of today's networks–in just days after the virus was unleashed, there were widespread reports of infections across the country. Worse yet, as the virus made its way through the Internet, variations appeared that were able to bypass security software designed to detect Melissa. These two factors alone made it extremely difficult to launch countermeasures for the infection.

Second, Melissa showed how hard it is to trace any virus back to its source. At first, it was widely assumed that Melissa was created by a writer, known by the computer handle "VicodinES," who was distributing the virus from an America Online account known as "Sky Roket." But later, after receiving a tip from America Online, investigators discovered

---

6 As noted elsewhere, Melissa is a macro virus and requires the host program, such as Word 97, to allow it to execute. By taking advantage of Word's ability to notify the user whenever a macro is going to be executed, a user can prevent the virus from executing in the first place.

that this account was allegedly stolen by the suspect arrested for creating the virus. Without this level of cooperation, the suspect might not have ever been identified.

Third, Melissa demonstrated that vulnerabilities in widely adopted commercial-off-the-shelf (COTS) products can be easily exploited to attack all their users. This is alarming because agencies are increasingly turning to COTS products to support critical federal operations. Because they are built to appeal to a broad market and not to satisfy a particular organization's unique functional and security requirements, agencies must thoroughly analyze the vulnerabilities and threats associated with COTS products before acquiring them. It is estimated that Microsoft's Office suite, which includes Word and Excel, represented 89 percent of the revenues for this market in 1997.

Fourth, Melissa illustrated that there are no effective agency and governmentwide processes for reporting and analyzing the effects of computer attacks. There is not complete information readily available on what agencies were hit and only partial data on the Department of Defense and DOE. Moreover, there are no data available at this time that quantify the impact of the virus, for example, productivity lost or the value of data lost.

Fifth, Melissa proved that computer users can do a good job of protecting their systems when they know the risks and dangers of computing and when they are alerted to attacks. Reports from the media revealed that organizations that trained their employees and warned them of the attack fared much better than those that did not.

More important, Melissa is a symptom of broader information security concerns across government. Over the past several years, we and inspectors general have identified significant information security weaknesses in each of the largest 24 federal agencies.[7] These include inability to detect, protect against, and recover from viruses such as Melissa; inadequately segregated duties which increase the risk that people can take unauthorized actions without detection; and weak configuration management processes, which cannot prevent unauthorized software

---

7 Information Security: Serious Weaknesses Place Critical Federal Operations and Assets at Risk (GAO/AIMD-98-92, September 23, 1998); Information Security; Strengthened Management Needed to Protect Critical Federal Operations and Assets (GAO/T-AIMD-98-312, September 23, 1998); and Financial Audit: 1998 Consolidated Financial Statements of the United States Government (GAO/AIMD-99-130, March 31, 1999).

from being implemented. Examples of significant security lapses that have been reported follow.

- In November 1997, the Social Security Administration Inspector General reported that security weaknesses subjected sensitive information to potential unauthorized access, modification, or disclosure. The Inspector General reported that 29 convictions involving agency employees were obtained during fiscal year 1997, most of which involved creating fictitious identities, fraudulently selling social security cards, misappropriating funds, or abusing access to confidential information.
- In May 1998, we reported that (1) the Department of State's information systems and the sensitive data they maintain were vulnerable to access, change, disclosure, and disruption by unauthorized individuals[8] and (2) weak computer security practices at the Federal Aviation Administration jeopardized flight safety.[9]
- In October 1998, we reported that weaknesses at Treasury's Financial Management Service placed billions of dollars of payments and collections at risk of fraud.[10]
- Over the past 7 years the U.S. Department of Agriculture's (USDA) Inspector General reported that USDA's National Finance Center, which annually makes over $21 billion in payroll disbursements to about 434,000 employees, had not ensured that (1) systems security adequately prevented misuse or unauthorized modifications, (2) access to data was needed or appropriate, and (3) modifications made to software programs were properly authorized and tested.
- In September 1998, we reported that general computer control weaknesses placed critical Department of Veterans Affairs (VA) operations, such as financial management, health care delivery, benefit payments, and life insurance services, at risk of misuse and disruption. In addition, sensitive information contained in VA systems, including financial transaction data and personal information on veteran medical records and benefit payments were vulnerable to inadvertent or deliberate misuse, fraudulent use improper disclosure, or destruction–possibly occurring without detection.[11]

---

8 Computer Security: Pervasive, Serious Weaknesses Jeopardize State Department Operations (GAO/AIMD-98-145, May 18, 1998).

9 Air Traffic Control: Weak Computer Security Practices Jeopardize Flight Safety (GAO/AIMD-98-155, May 18, 1998).

10 Financial Management Service: Areas for Improvement in Computer Controls (GAO/AIMD-99-10, October 20, 1998).

11 VA Information Systems: Computer Control Weaknesses Increase Risk of Fraud, Mis-

In view of these and other pervasive security weaknesses, we designated information security as a new governmentwide high-risk area in February 1997. In performing audits at selected individual agencies, we and the inspectors general have also developed hundreds of specific recommendations aimed at improving the effectiveness of information security programs.

Since our 1997 High-Risk Report, the recognition of the importance of addressing information security problems has greatly increased and led to significant actions. In late 1997, for example, in response to our recommendations, the Chief Information Officers (CIO) Council designated information security a priority area and established a Security Committee. During 1998, the committee sponsored a security awareness seminar and developed plans for improving incident response services.

Also, in May 1998, Presidential Decision Directive 63 (PDD 63) was issued. This established entities within the National Security Council, the Department of Commerce, and the Federal Bureau of Investigation to address critical infrastructure issues. It required each major department and agency to develop a plan for protecting its own critical infrastructure. Other provisions include (1) enhanced analysis of information on threats, (2) assessments of government systems' susceptibility to exploitation, and (3) incorporation of infrastructure assurance functions in agency strategic planning and performance measurement frameworks.

Melissa and other recent incidents demonstrate, however, that still much more needs to be done to ensure that systems and data supporting critical federal operations are adequately protected.

Measures That Can Help Ensure Agencies Are Better Prepared for Future Viruses and Computer Attacks

To help strengthen computer security practices, we issued an executive guide in May 1998 entitled Information Security Management: Learning From Leading Organizations (GAO/AIMD-98-68). It describes a framework for managing risks through an ongoing cycle of activity coordinated by a central focal point. The guide, which is based on the best practices of organizations noted for superior information security programs, has been endorsed by the CIO Council, and distributed to all major agency heads, CIOs, and inspectors general. By adopting the following 16 practices recommended by the guide, agencies can be better prepared to *protect* their systems, *detect* attacks and *react* to security breaches.

---

use, and Improper Disclosure (GAO/AIMD-98-175, September 23, 1998).

Principles	Practices
Assess risk and determine needs	1. Recognize information resources as essential organizational assets 2. Develop practical risk assessment procedures that link security to business needs 3. Hold program and business managers accountable 4. Manage risk on a continuing basis
Establish a central management focal point	5. Designate a central group to carry out key activities 6. Provide the central group ready and independent access to senior executives 7. Designate dedicated funding and staff 8. Enhance staff professionalism and technical skills
Implement appropriate policies and related controls	9. Link policies to business risks 10. Distinguish between policies and guidelines 11. Support policies through a central security group
Promote awareness	12. Continually educate users and others on risks and related policies 13. Use attention-getting and user-friendly techniques
Monitor and evaluate policy and control effectiveness	14. Monitor factors that affect risk and indicate security effectiveness 15. Use results to direct future efforts and hold managers accountable 16. Be alert to new monitoring tools and techniques

Just as it is important for agencies to implement comprehensive security programs, it is important that a comprehensive governmentwide strategy emerge from current efforts to implement PDD 63 and strengthen the CIO Council's focus on security. As we recently recommended to the Director of the Office of Management and Budget (OMB) and the Assistant to the President for National Security Affairs, such a strategy should:[12]

- Clearly delineate the roles of federal organizations with responsibilities for information security.
- Rank the greatest risks.
- Promote the use of proven security tools and best practices.
- Ensure the adequacy of workforce skills.

12 Information Security: Serious Weaknesses Place Critical Federal Operations and Assets at Risk (GAO/AIMD-98-92, September 23, 1998).

- Provide for evaluating systems on a regular basis.
- Identify long-term goals, as well as time frames, priorities, and annual performance goals.

OMB, the CIO Council, and the National Security Council agree that such a strategy should be implemented and are working collaboratively on a plan to (1) assess agencies' security postures, (2) implement best practices, and (3) establish a process of continued maintenance.

Conclusions

Federal agencies were fortunate that the worst damage done by Melissa was to shut down e-mail systems and temporarily disrupt operations. Because of the increasing reliance on the Internet and standard COTS products as well as the increasing improvements in computer attacker tools and techniques, (as evidenced in the additional capability and techniques employed in the Melissa attack), it is likely that the next virus will propagate faster, do more damage, and be more difficult to detect and to counter. It is imperative, therefore, that federal agencies and the government as whole swiftly implement long-term solutions to protect systems and sensitive data. It is also critical that the federal government establish reporting mechanisms that facilitate analyses of viruses and other forms of computer attacks and their impact. Our Information Security Best Practice guide offers a good framework for agencies to follow, but sustained governmentwide leadership is needed to ensure that executives understand their risks, monitor agency performance, and resolve issues affecting multiple agencies.

Madam Chairwoman, this concludes my testimony. I will be happy to answer any questions you or Members of the Subcommittee may have.

CHAPTER 32

TESTIMONY OF ALAN B. DAVIDSON, STAFF COUNSEL CENTER FOR DEMOCRACY AND TECHNOLOGY BEFORE THE HOUSE PERMANENT SELECT COMMITTEE ON INTELLIGENCE

June 9, 1999

Summary

Encryption is the essential tool for protecting privacy, security, public safety, and free expression online. The Center for Democracy and Technology (CDT) is pleased to have this opportunity to testify about U.S. encryption policy before the House Intelligence Committee. CDT is a public interest group dedicated to promoting civil liberties and democratic values on the Internet. CDT has long supported the Security and Freedom through Encryption (SAFE) Act, which is badly needed to reform a U.S. encryption policy that does not serve the interests of the United States.

1. On balance, the widespread use of strong encryption is in the long term national security interests of the United States.

Encryption is vital to the security of electronic commerce, personal communications, and the increasingly important information infrastructure. While encryption creates new challenges for law enforcement and national security, no U.S. export control or key recovery policy can stop determined criminals, terrorist groups, or rogue governments from getting access to strong encryption. Moreover, the law enforcement problem is narrower than it appears: Even without controls, in many cases the plaintext of encrypted data will be available where it is decrypted or stored, or through lawful access to existing decryption information. Meanwhile, encryption export controls are slowing down the use of strong encryption which would allow us to protect proprietary and personal information and our critical infrastructures.

For all of these reasons, we agree with the conclusion of the National Research Council's 1996 report *Cryptography's Role in Securing the Information Society*. Its distinguished panel of technology, policy, and national security experts, who received classified briefings, concluded that "*On balance, the advantages of more widespread use of cryptography outweigh the disadvantages.*"

2. Government-mandated key recovery or plaintext access threatens security and privacy

Two years ago, in June 1997, a group of eminent computer security experts and cryptographers issued a damning critique of government proposals to require plaintext access features in encryption systems. Their report, *The Risks of Key Recovery, Key Escrow, and Trusted Third Party Encryption* concluded that:

> "The deployment of key-recovery-based encryption infrastructures to meet law enforcement's stated specifications will result in substantial sacrifices in security and greatly increased costs to the end-user. . . . These basic requirements [for plaintext access by government] make the problem of general key recovery difficult and expensive—and potentially too insecure and too costly for many applications and many users."

Two years later, the assertions of the experts remain unrebutted. Mandatory government access has been rejected by the marketplace and by policy makers around the world. Mandatory plaintext access raises serious constitutional concerns about non-particularized searches. And the government remains unable to explain how the security of keys will be protected globally when U.S. legal protections stop at the border.

3. The world is rejecting U.S. policy of export controls and key recovery.

Contrary to Administration predictions, foreign governments are not embracing export controls and key recovery. Since November, three major U.S. allies have moved in the opposite direction. France, long held out as a supporter of domestic controls, has abandoned mandatory key recovery and liberalized controls. The U.K., viewed as the Administration's staunchest ally, in May abandoned a domestic key recovery proposal. And just last week Germany announced that it would affirmatively promote the export of strong encryption.

4. This Committee should demand responses to the unanswered questions about U.S. encryption policy.

The Administration has been unable to provide satisfactory answers to basic questions about the effectiveness and costs of encryption export controls and plaintext access systems. This Committee should seek answers to these questions:

- How can export controls be sustained in the long run, when the basic technology underlying good encryption is well understood outside of the U.S., good foreign products exist, and foreign governments are rejecting the U.S. policy approach?
- Don't plaintext access requirements increase the vulnerability of encryption systems? Law enforcement is demanding nearly real-time ac-

cess to plaintext. Aren't the same systems that provide fast government access susceptible to mistake or misuse by criminals? Don't they create a new hole in the encryption that otherwise would never exist?

- How does plaintext access controlled by third parties meet the needs of the intelligence agencies? Are intelligence agencies prepared to identify themselves to foreign service providers or foreign governments in order to gain access to keys or plaintext?

  Scenario: Suppose U.S. intelligence agencies surreptitiously intercept encrypted communications from a suspected terrorist organization. The group uses key recovery, with keys recoverable by lawful process in the Middle Eastern country the group operates from. Will U.S. agencies go through that lawful process?

- How will the U.S. government respond to foreign demands for reciprocal sharing of keys? What standards will protect the keys of U.S. companies and citizens accessible abroad?

  Scenario: The government of China seeks decryption keys held in the U.S. for the encrypted communications of a pro-democracy group in China. Under what circumstances will the keys be turned over?

  Scenario: Following the U.S. lead, France requires that U.S. companies make keys available to the French government for all communications in France. What legal standards will protect the trade secrets or other sensitive data sent with these systems, in a country without a Fourth Amendment and with a track record for government industrial espionage?

The Administration has been unable to answer these fundamental questions about U.S. policy. It has been unable to do so because export controls and key recovery are not a workable policy. Congressional action is needed. While CDT remains concerned with the criminal provisions in the SAFE Act, overall the SAFE Act would help provide Americans with the strong security and privacy products they so badly need. We urge this committee to support SAFE, and to seek answers to the hard questions about a U.S. policy that, in its current state, does not add up.

1. Developments of the Last Two Years Confirm the Need for a New U.S. Encryption Policy

Two years ago, this committee and four others in the House held hearings and voted on the SAFE Act in the context of a debate on encryption strikingly similar to the one going on today. Privacy advocates and industry representatives testified about the need for new encryption policies, and Administration officials argued that new regulations would allow U.S. policy to satisfy the competing interests at hand. In retrospect, the rapid pace of technical and marketplace developments over the last two years has made it clearer than ever before that the U.S. approach to encryption policy remains fundamentally flawed.

A. Exportable encryption has proven increasingly vulnerable.

Two years ago, the Justice Department disputed arguments by privacy advocates that 56-bit encryption, the maximum strength exportable for consumers without key recovery, was not secure enough for many applications. The Justice Department claimed that "According to the National Security Agency s estimates, the average time needed to decrypt a single message by means of a brute force cryptoanalytic attack on 56-bit DES—a strength whose export we are now allowing—would be approximately one year and eighty-seven days using a thirty-million-dollar supercomputer."[1]

Technical developments have proven these comments wrong. In the Fall of 1998, a group of researchers sponsored by the Electronic Frontier Foundation built a "DES Cracker" system for less than $250,000 that broke a 56-bit key within 56 hours.[2] Less than six months later, in January 1999, encryption enthusiasts broke a 56-bit code in 22 hours using the DES Cracker and a network of distributed computers. If a non-profit and a group of part-time enthusiasts could develop such a system on a shoestring budget, we are only left to imagine what a foreign government, large corporation, or sophisticated criminal enterprise could do.

The U.S. Government has itself recognized the weakness in 56-bit encryption systems. In a January 1999 draft the National Institute of Standards and Technology (NIST) revised the encryption standard for government use from 56-bit DES to much stronger "Triple DES," citing the vulnerability of DES.[3] Meanwhile, NIST has been leading efforts to create an Advanced Encryption Standard based on the 128-bit (and higher) algorithms that are becoming the world standard for online security. If the government does not trust 56-bit security, why should everyday computer users and companies be expected to rely on this weaker level of security?

B. Key recovery has not been widely accepted.

Two years ago, Administration witnesses touted key recovery as the compromise that met law enforcement desires and was going to "become

---

1 Security and Freedom Through Encryption (SAFE) Act: Hearing on H.R. 695 Before the Subcomm. on International Economic Policy and Trade of the House Comm. on International Relations, 105th Cong., 2nd Sess, 57-73 (1997) (Statement of Robert S. Litt, Deputy Assistant Attorney General, Department of Justice) (emphasis added).

2 See ELECTRONIC FRONTIER FOUNDATION, CRACKING DES (1998).

3 "With regard to use of single DES, exhaustion of the DES (i.e. breaking a DES encryption ciphertext by trying all possible keys) has become increasingly more feasible with technology advances. Following a recent hardware based DES key exhaustion attack, NIST can no longer support the use of single DES for many applications." 64 FED. REG. 10, 2625-2628 (1999) (proposed January 15, 1999).

the worldwide standard."[4] In fact, since then government-driven key recovery has been greeted with great skepticism and widely discredited.

Research has revealed the vulnerabilities of key recovery systems, which create backdoors to plaintext without the notice or consent of an encryption user. A 1997 report by a group of encryption experts found that "[t]he deployment of key-recovery-based encryption infrastructures to meet law enforcements stated specifications will result in substantial sacrifices in security and greatly increased costs to the end-user." A year later, with no substantive response from within the Administration or the technical community, the same group of experts confirmed its findings still held true in June 1998.[5] A copy of their report is being submitted to the Committee along with this testimony.

Despite Administration predictions, the marketplace has shown little interest in even stored data recovery, and there is virtually no demand for key recovery for communications. To CDT's knowledge, not one major key recovery encryption product is being widely used by consumers today.[6]

C. The world is not adopting U.S. encryption control policies.

Encryption controls are ultimately only effective if other countries control encryption products as well. In 1997, the Administration testified, "We have engaged in extensive international discussions on this topic over the last year, and a consensus is now emerging throughout much of the world that the way to achieve this balance is through the use of a key recovery or trusted third party system. . . We believe that key recovery will become the worldwide standard for users of the GII."[7]

---

4 "[W]e believe that key recovery encryption is going to become the worldwide standard." Security and Freedom Through Encryption (SAFE) Act: Hearing on H.R. 695 Before the Subcomm. on International Economic Policy and Trade of the House Comm. on International Relations, 105th Cong., 2nd Sess, 57-73 (1997) (Statement of Robert S. Litt, Deputy Assistant Attorney General, Department of Justice).

5 AN AD-HOC GROUP CRYPTOGRAPHERS AND COMPUTER SCIENTISTS, THE RISKS OF KEY RECOVERY, KEY ESCROW, & TRUSTED THIRD PARTY ENCRYPTION (1997). (Updated 1998 report available at http://www.cdt.org/crypto/risks98/.)

6 Cost may play a role. A recent study by the Business Software Alliance estimated the cost of key escrow systems at $7.7 billion per year and $38.5 billion over a five year period. BUSINESS SOFTWARE ALLIANCE, THE COST OF GOVERNMENT-DRIVEN KEY ESCROW ENCRYPTION (1998).

7 Security and Freedom Through Encryption (SAFE) Act: Hearing on H.R. 695 Before the Subcomm. on International Economic Policy and Trade of the House Comm. on International Relations, 105th Cong., 2nd Sess, 57-73 (1997) (Statement of Robert S. Litt, Deputy Assistant Attorney General, Department of Justice).

To date, the opposite has been true. The OECD Cryptography Policy Guidelines and the Ministerial Declaration of the European Union, both released in 1997, failed to embrace key recovery despite lobbying by the U.S. government. In the past year, Canada, Ireland and Finland have announced encryption policies allowing free use and export of strong encryption products without key recovery. Recent Administration claims of a new encryption control regime through the Wassenaar Arrangement have been overstated. In fact, many countries, including Wassenaar signatories, still allow encryption exports consistent with Wassenaar (See below for a greater explanation of the Wassenaar changes.)

Perhaps most importantly, major U.S. trading partners have recently moved away from U.S. policy. The U.K., the staunchest supporter of U.S. key recovery policy, this Spring rejected a major plaintext access proposal. France, long held out as a country with sweeping controls on encryption use in the past, recently liberalized its domestic limits. And this June the German government rejected domestic controls on encryption and committed to affirmatively promote German encryption exports.

D. The Administration has proven unable to engage in comprehensive reform.

The Department of Commerce has taken a step forward in its recently released encryption regulations, easing exports of 56-bit products and allowing export of strong encryption products to online merchants. However, U.S. policy remains focused on export controls and incentives to use key recovery. The mass market products needed by individual users remain controlled. The special relief for certain industry sectors, while surely welcome by those businesses, does little to change the encryption available to individual computer users or small organizations.

Taken together, these developments argue for a more comprehensive change to U.S. encryption policy, away from export controls and key recovery and towards a view where public safety is best protected by giving people the encryption tools they need to protect themselves on line.

2. Encryption Export Controls Jeopardize Privacy and Free Expression While Providing Few Real Benefits

Encryption export controls continue to threaten the privacy and security of Internet users and businesses both abroad and within the United States. They also threaten free expression: a Federal appeals court recently held that controls on encryption source code violate the First Amendment, and controls in general keep people from getting the security tools they need to protect free expression and free association. At the same time, the rationale for export controls is increasingly eroding due to the availability of good foreign encryption products. Finally, contrary to some claims, export controls are not required by the Wassenaar Arrangement.

A. Encryption protects privacy

Export controls limit the availability of strong encryption products to law-abiding consumers both domestically and abroad. Such controls directly limit the availability of strong U.S. products abroad, of particular concern to human rights groups and other international organizations. Export controls affect people in the U.S. when they communicate abroad, since they may be forced to use the lower levels of encryption available to the parties they communicate with. Export controls have also slowed the deployment of strong encryption standards. While some strong encryption products are available to consumers, export controls have largely slowed the seamless integration of good security systems into operating systems, network protocols, and many applications. Encryption should be easy for consumers; because of federal regulations, it is not.

The importance of encryption to computer users is real. Some examples include:

- Human rights workers now routinely use encryption to protect the highly sensitive information and sources they collect in their investigations of human rights abuses by governments around the world. According to Human Rights Watch, encryption is used by its workers in countries including Bosnia, China, Lebanon, Rwanda, or Kashmir. Failure to use encryption can literally have deadly consequences: in 1998 the unencrypted notes of a UN human rights investigator in Kinshasa led to a manhunt of informants.[8]
- Journalists communicating sensitive information abroad increasingly need encryption to protect the stories and notes they communicate internationally. Journalists increasingly use the public Internet to communicate, and according to the Center for Public Integrity,

---

8 Security and Freedom Through Encryption (SAFE) Act: Hearing on H.R. 695 Before the Subcomm. on International Economic Policy and Trade of the House Comm. on International Relations, 106th Cong., 1st Sess. (May 18, 1999) (Statement of Dinah PoKempner, Deputy General Counsel, Human Rights Watch).

many are only just becoming aware of the need for regular use of strong encryption.[9]

- Ordinary computer users in the United States are threatened by export controls when they communicate abroad. In the seamless global network, someone might be forced to reveal personal information insecurely when say, purchasing flowers for a loved one abroad, buying a book from a foreign website, or sending papers to a colleague abroad.

In these cases, encryption export controls have a real impact on security. As Human Rights Watch testified in Congress earlier this Spring, "Strong encryption, such as the PGP program, is available worldwide, but restraints have inhibited both the sharing of compatible programs throughout the human rights community and the development of new encryption products for the mass market. Were mass-market strong encryption readily available worldwide, protection in times of crisis would be a reality for many more users." Weaker encryption may not be sufficient for sensitive applications; for example, human right workers believe the Yugoslavian government has access to Russian surveillance technology allowing them the readily crack the 40-bit encryption most readily available to potential informants there.[10]

The most recent December 1998 encryption regulations, while a welcome step forward by the Administration, do not change the fundamental premise of U.S. policy: export controls on all but the weakest encryption for mass market consumers, and strong incentives for the use of key recovery and plaintext access systems. The sector relief provided for foreign subsidiaries of U.S. companies, certain industries, and online merchants does little to provide regular consumers with strong encryption. Export controls remain a powerful incentive to adopt key recovery and plaintext access systems. The piecemeal relief offered by the regulations raises the question: When do regular people get to protect their privacy online?

Computer users remain at risk, awaiting the widespread deployment of encryption and facing increasing threats to their unprotected information.

---

9 See Regina Joseph, "The Encryption Imperative", Editor & Publisher, January 1999.

10 Security and Freedom Through Encryption (SAFE) Act: Hearing on H.R. 695 Before the Subcomm. on International Economic Policy and Trade of the House Comm. on International Relations, 106th Cong., 1st Sess. (May 18, 1999) (Statement of Dinah PoKempner, Deputy General Counsel, Human Rights Watch).

B. Export controls hinder free expression and have been ruled unconstitutional

Encryption export controls violate the First Amendment prohibitions on prior restraints of speech. Courts have ruled that the ideas expressed in encryption source code make source code a form of protected speech, and the courts continue to find encryption export controls unconstitutional. In May, the Ninth Circuit Court of Appeals ruled that current export controls on encryption source code violated the First Amendment. While the decision is likely to be appealed—leaving many still in need of immediate relief—it is highly significant that yet another court has now found that the Administration's urgent national security claims do not override the constitutional infirmity of such regulations vesting "boundless discretion in government officials."

C. Good Encryption is Increasingly Produced Abroad

Strong encryption products are widely available abroad, undercutting the basic rationale of U.S. export controls. Ideas cannot be stopped at the border, and so it is no surprise that strong encryption products are increasingly being produced abroad. Surveys have documented literally hundreds of encryption products produced outside of the United States, many of them stronger than the limits imposed on U.S. exports.[11] The open, global, decentralized nature of the Internet makes transfer of encryption software around the world a simple matter. For example, a short online search reveals a sample of Web sites from around the world (see Figure 1 below) distributing free versions of the popular 128-bit encryption software "Pretty Good Privacy" (PGP). PGP is distributed on Web sites around the world ranging from those run by well-known organizations, to those provided by Internet Service Providers, to individual home pages.

While some have argued that foreign encryption is easily broken, there is much reason to believe that much of the internationally-produced encryption is just as secure as American encryption. Cryptography has become a global science. Many cryptography researchers live outside of the U.S. Many of the important cryptography conferences are held outside of the U.S. each year. Some of the most important advances in cryptography have been made by researchers outside of the U.S. For

---

11 Encryption software can be purchased from many other countries across the globe, and easily downloaded over the Internet. According to a recent study by the Economic Strategy Institute, 1,601 encryption products were available as of September 1997 from 941 firms in thirty countries. Of this total, 653 are made outside the United States by 472 foreign firms.

example, just this May an important new method for attacking encryption systems was released by Israeli cryptographer Adi Shamir.

Figure 1: Some Foreign Sites Supplying Pretty Good Privacy™ (popular 128-bit encryption product)

Country	Site Name	URL
Norway	The International PGP Page	http://www.pgpi.com/
Austria	Arges Tempo Internet Service Provider	http://www.arges.tempo.at/pgp.download/
Brazil	The Best of Internet Software (Personal Home Page)	http://www.intensa.com.br/pesquisas/down2.htm
Germany	(Personal Homepage)	http://www.westfalen.de/hugo/index.html
Singapore	Centre for Internet Research	http://www.irdu.nus.sg/pgp/download.htm
Sweden	ÇoMa's PGPClick Download Page	http://www.torget.se/users/c/CoMa/pgp/pgdownload.html

Source: Center for Democracy and Technology, May 1999.

Other examples of high quality foreign cryptography can be found in the ongoing efforts of the National Institute for Standards and Technology (NIST) to create a new Advanced Encryption Standard (AES), due to be completed by 2002. NIST solicited submissions for the AES algorithms, expected to become the world-class standard for the strongest (128-bit and higher) encryption products. Of the 15 submissions NIST received, ten were from industry and university researchers outside of the United States—including Australia, Belgium, Canada, Costa Rica, England, France, Germany, Israel, Japan, and Korea.

D. The Export of Strong Encryption Does Not Violate the Wassenaar Arrangement

The Clinton Administration has long argued that the world community would imminently agree to limit the spread of encryption. In December 1998, the Administration claimed victory when 32 other nations agreed under the Wassenaar Arrangement to create a separate category for encryption products, removing encryption of 64 bits or greater from the General Software Note, a license exception for software products that are generally available and in the public domain.[12] Administration officials claimed that the amendments to Wassenaar "closed a loophole" by

12 General Software Note, Wassenaar Arrangement (http://www.wassenaar.org/list/gtngsn.pdf)

creating an international ceiling on bit-length.[13] However, the Wassenaar Arrangement does not impose multilateral controls on encryption products as the Administration claims.

Administration officials have argued that the export liberalization provisions in SAFE would violate the Wassenaar Arrangement.[14] However, Wassenaar does not impose multilateral export controls, but rather provides a set of non-binding guidelines for participating countries to follow in the spirit of international stabilization. Compliance with standards set by Wassenaar is entirely at the discretion of each participating country: "All measures undertaken with respect to the arrangement will be in accordance with national legislation and policies will be implemented on the basis of national discretion." The member countries are not required to adopt Wassenaar standards, and there are no penalties for exercising national discretion. Several Wassenaar signatories such as Canada, Finland, and Ireland readily allow the export of strong encryption products.

Moreover, several important countries that are not members of Wassenaar—Israel, China, and India—allow export of strong encryption.

The SAFE Act is completely consistent with the letter, and the spirit, of Wassenaar. It eases export controls on encryption products that are already generally available on the international market. The bill also prohibits mandatory key escrow, a system that is being rejected by the international community because it imposes serious privacy risks on the encryption user. SAFE also includes provisions that allow the Secretary of Commerce to prohibit export of specific encryption products to specific countries if "such encryption products will be used for military or terrorist end-use."

3. Government-Driven "Key Recovery" or "Plaintext Access" is Not a Solution

The law enforcement community in general has variously endorsed "key escrow," "key recovery," and other forms of "plaintext access" as its favored approach to encryption policy. These variations on the failed "Clipper Chip" policy seek to guarantee third-party access to the keys for

---

13 Security and Freedom Through Encryption (SAFE) Act: Hearing on H.R. 695 Before the Subcomm. on Courts and Intellectual Property of the House Comm. on the Judiciary, 106th Cong., 1st Sess. (March 4, 1999) (Statement of William A. Reinsch, Under Secretary for Export Administration, Department of Commerce).

14 At a March 1999 meeting of the President's Export Subcommittee on Encryption, Under Secretary Reinsch identified provisions of the SAFE bill that he asserted would violate the Wassenaar Arrangement. Summary of Open Session, President's Export Subcommittee on Encryption (March 12, 1999).

all encrypted communications and stored data without the notice or consent of the key owners. Such proposals have been greeted with much skepticism and concern from the global Internet community.

The attempt to institutionalize key recovery worldwide is a fundamental threat to privacy and security both domestically and abroad:

- Global key access systems are vulnerable and unproven—Centralized "back-door" access to the billions of keys used by millions of computer users will introduce new vulnerabilities into a medium that is already difficult to secure. In 1997, eleven renowned computer security experts issued a report on key recovery concluding that, "Building the secure infrastructure of the breathtaking scale and complexity demanded . . . is far beyond the experience and current competency of the field. Even if such an infrastructure could be built, the risks and costs of such a system may ultimately prove unacceptable." In 1998 these experts revisited the question and confirmed that their conclusions remained essentially unchallenged.[15]
- The Fourth Amendment does not adequately protect key recovery systems both outside and inside of the U.S.—The Administration has been unable to explain what legal standards will internationally protect the communications and data of U.S. individuals and businesses. Moreover, the Administration has indicated that the full Fourth Amendment standards of probable cause and notice would not apply to encryption keys held by third parties, even within the U.S. In a world where personal data is increasingly legally unprotected in the hands of third parties, key recovery systems further erode a person's ability to protect their privacy.
- Recovery does not address national security needs—The recovery of a key, under lawful authority and with the assistance of a third party, can be of value to domestic law enforcement. It is of little use for national security signals intelligence purposes, where eavesdropping often occurs abroad without seeking local legal authority and where the assistance of third parties may not be possible.
- Recovery will never be appropriate for some applications—For example, the American Association for the Advancement of Science has commented on the sensitive and increasingly important use of encryption by human rights advocates worldwide. "If keys can be recovered by the U.S. government, why should human rights organizations whose entire

---

15 AN AD-HOC GROUP CRYPTOGRAPHERS AND COMPUTER SCIENTISTS, THE RISKS OF KEY RECOVERY, KEY ESCROW, & TRUSTED THIRD PARTY ENCRYPTION (1997). (Updated 1998 report available at http://www.cdt.org/crypto/risks98/.)

function is defined by abusive governments trust that their information will remain secure?"[16]

Despite these concerns, current encryption regulations continue to give many encryption producers a Hobbesian choice: accept key recovery or be forced to export lower strength encryption. Moreover, proposals backed by the FBI in the past have sought to further force U.S. encryption users to adopt key recovery through a number of coercive regulations, including outright domestic mandates. While we are encouraged that the Administration appears to have backed away from mandatory domestic controls, we are wary that it has not denounced this approach. And even the current U.S. encryption policy based on key recovery and export controls threatens to leave global Internet users without the technical means to secure their communications or the international legal standards needed to protect their privacy.

4. National Security and Law Enforcement are Best Served by the Widespread Use of Strong Encryption

It is increasingly clear that the benefits of widespread encryption far outweigh the costs. The last two years have seen Americans moving their lives online in unprecedented numbers. A Presidential Commission has highlighted the vulnerability of our nation's critical information infrastructure. Together these developments have underscored the importance of securing the Internet, and deploying strong encryption to do so.

Two years ago the national security community seemed to speak with one voice about the danger of strong encryption. Today there has been an increasing recognition of the cost of U.S. encryption policy. As Sam Nunn, Co-Chair of the Advisory Committee to the Presidents Commission on Critical Infrastructure Protection, noted in 1998 Senate testimony, "I do think we are in a different era of technology now and I do not think the nostalgia for the old-fashioned wiretap by law enforcement is going to be realistic in this age we are in now."[17] Senator Bob Kerrey, an early proponent of encryption controls, argued in an October 1998 speech that "the encryption debate has hobbled our efforts to write laws that enable our law enforcement and national security agencies to carry out their mis-

---

16 American Association for the Advancement of Science, Comments on Bureau of Export Administration Interim Rule on Encryption Controls (Feb. 7, 1997).

17 Hearing before the Subcommittee on Technology, Terrorism, and Government Information of the Senate Committee on the Judiciary, 105th Cong., 2nd Sess, (March 17. 1998) (Statement of Sam Nunn, Co-Chair, Advisory Committee to the President's Commission on Critical Infrastructure Protection).

sion" and argued that it was time to "remove export restrictions on encryption products of any strength."[18]

The benefits of current U.S. policy to law enforcement are uncertain. U.S. policy will not stop sophisticated criminals from using encryption to evade law enforcement. Strong, non-escrowed encryption is already available both inside and outside of the United States today. Foreign governments and criminals have access to these powerful tools and will be able to encrypt data despite continued export controls or key recovery. Furthermore, nothing in the Administration policies prevents users from "super-encrypting" communications even within a key recovery framework.

The law enforcement problems with encryption are important but more limited than claimed. Law enforcement faces a real, but narrowly focused, problem with encryption. Most encrypted information will still be accessible to law enforcement by legal process even in an encrypted world. For example, businesses will be still be required to produce the plaintext of encrypted business records under proper legal process. Stored information, corporate and business information, and even a great deal of electronic communication will most likely be largely available to law enforcement through legal process similar to that available today.

Important challenges remain for law enforcement interceptions of communications or seizures of data without notice to the party under surveillance. This narrower problem must be put into the context of the benefits provided by encryption and the costs associated with key recovery systems. The information economy presents new and powerful tools and opportunities for law enforcement surveillance. Online interaction leaves a detailed trail of electronic transactions, credit card purchases, online communications, and Web-based clickstream data presenting new traffic analysis opportunities. In fact, law enforcement is operating today in a Golden Age of surveillance, with online collections of personal data offering unprecedented new tools to obtain evidence of criminal activity (and raising important privacy concerns that must be dealt with.)

U.S. policy is creating a deficit of trust around important issues we could all be working on together. U.S. policy stands in the way of a growing urgent need for strong encryption products and better computer security in general. As Sam Nunn testified before the Senate last year, "[I]f the deadlock continues as it is today, building the trust required between the public and private sectors in the broad area of infrastructure protection will be even more difficult."[19] Nunn went on to note that "limiting the power of encryption

---

18 144 CONG.REC. S12359 (1998).

19 Hearing before the Subcommittee on Technology, Terrorism, and Government Information of the Senate Committee on the Judiciary, 105th Cong., 2nd Sess, (March 17.

over the long-haul is simply not going to be feasible." Current U.S. policy dangerously impedes the deployment of accessible, easy-to-use, global security systems for the Internet that are needed to protect our privacy and our critical infrastructure.

On balance, national security demands strong encryption. CDT agrees with the conclusion of the National Research Council's major study of encryption, which argued in its 1996 encryption study, "On balance, the advantages of more widespread use of cryptography outweigh the disadvantages."[20]

5. Conclusion: Congress Should Support the SAFE Act

U.S. policy stands in the way of a growing urgent need for strong encryption products that people trust. The past two years have shown that people and businesses are moving more and more of their lives, economic activities, and sensitive data online. The federal government has identified the vulnerability of our nation's critical information infrastructure. Strong encryption, without built-in backdoors, is an essential part of protecting that sensitive data and critical infrastructure.

That is why the SAFE Act is so important. SAFE would take the commonsense step of allowing export of the mass-market encryption software most difficult to control, and most important to everyday computer users. It is a measured step; it does not remove all export controls, retaining for example restrictions on exports to countries like Iran or Libya. And CDT remains particularly concerned about provisions criminalizing the use of encryption to obstruct law enforcement in the course of a federal felony. While this provision is narrowly crafted, we remain concerned that it not be interpreted broadly enough to chill the routine use of encryption. On balance though, CDT believes SAFE would be a dramatic step forward for the security of computer users.

In the current policy standoff between an unsustainable control policy and the emerging and acute privacy and security needs of the Information Age, Congressional action is needed. Only Congress is in the position today to change U.S. encryption policy and get Americans the privacy and security tools they need. The private sector cannot do it. The Administration will not do it. The courts may do it, but not without a protracted struggle. Congress must act. CDT believes that immediate liberalization of export controls in the SAFE Act will help provide Americans on the Internet with the strong security and privacy they so badly need.

1998) (Statement of Sam Nunn, Co-Chair, Advisory Committee to the President's Commission on Critical Infrastructure Protection).

20 NATIONAL RESEARCH COUNCIL, CRYPTOGRAPHY'S ROLE IN SECURING THE INFORMATION SOCIETY (1996).

About the Center for Democracy and Technology

CDT is an independent, non-profit public interest policy organization in Washington, D.C. The Center's mission is to develop and implement public policies that protect and advance individual liberty and democratic values in new digital communications media. The Center achieves its goals through policy development, public education, and coalition building. CDT also coordinates the Digital Privacy and Security Working Group (DPSWG), an ad hoc coalition of more than 50 computer, communications, and public interest organizations and associations working on information privacy and security issues.

House Rule XI, clause 2(g)(4) disclosures: Neither Alan Davidson nor the Center for Democracy and Technology have received any federal grant, contract, or subcontract in the current or preceding two fiscal years.

For more information contact:

Alan B. Davidson, Staff Counsel — adb@cdt.org
James X. Dempsey, Senior Staff Counsel — jdempsey@cdt.ort
Center for Democracy and Technology — http://www.cdt.org/
202.637.9800 (v)
202.637.0968 (f)

CHAPTER 33

TESTIMONY OF WILLIAM A. REINSCH UNDER SECRETARY FOR EXPORT ADMINISTRATION DEPARTMENT OF COMMERCE BEFORE THE SENATE COMMERCE COMMITTEE

June 10, 1999

Thank you, Mr Chairman, for the opportunity to testify on the direction of the Administration's encryption policy. We have made a great deal of progress since my last testimony before this Committee on this subject.

Even so, encryption remains a hotly debated issue. The Administration continues to support a balanced approach which considers privacy and commerce as well as protecting important law enforcement and national security equities. We have been consulting closely with industry and its customers to develop a policy that provides that balance in a way that also reflects the evolving realities of the market place.

One of the many uses of the Internet which will have a significant affect on our everyday lives is electronic commerce. The Internet and other digital media are becoming increasingly important to the conduct of international business. There were 43.2 million Internet hosts worldwide last January compared to only 5.8 million in January 1995. According to a recent study, the value of e-commerce transactions in 1996 was $12 million. The projected value of e-commerce in 2000 is $2.16 billion. To cite one example, travel booked on Microsoft's Website has doubled every year since 1997, going from 500,000 to an estimated 2.2 million this year. Many service industries which traditionally required face-to-face interaction such as banks, financial institutions and retail merchants are now providing cyber service. Customers can now sit at their home computers and access their banking and investment accounts or buy a winter jacket with a few strokes of their keyboard.

Furthermore, most businesses maintain their records and other proprietary information digitally. They now conduct many of their day-to-day communications and business transactions via the Internet and E-mail. An inevitable byproduct of this growth of electronic commerce is the need for strong encryption to provide the necessary secure infrastructure for digital communications, transactions and networks. The disturbing increase in computer crime and electronic espionage has made people and businesses wary of posting their private and company proprietary information on electronic networks if they believe the infrastructure may not be secure. A robust secure infrastructure can help allay these fears, and allow electronic commerce to continue its explosive growth.

Developing an encryption policy has been complicated because we do not want to hinder its legitimate use—particularly for electronic commerce; yet at the same time we want to protect our vital national security, foreign policy and law enforcement interests. We have concluded that the best way to accomplish this is to continue a balanced approach: to promote the development of strong encryption products that would allow lawful government access to plaintext under carefully defined circumstances; to promote the legitimate uses of strong encryption to protect confidentiality; and continue looking for additional ways to protect important law enforcement and national security interests.

During the past three years, we have learned that there are many ways to assist lawful access. There is no one-size-fits-all solution. The plans for recovery encryption products we received from more than 60 companies showed that a number of different technical approaches to recovery exist. In licensing exports of encryption products under individual licenses, we also learned that, while some products may not meet the strict technical criteria of our regulations, they are nevertheless consistent with our policy goals.

Additionally, we decided that the use of strong non-recovery encryption within certain trusted industry sectors is an important component of our policy to protect private consumer information and allow our U.S. high-tech industry to maintain its lead in the information security market. Taking into account all that we have learned and reviewing international market trends and realities, we made several changes in 1998 to our encryption policy that I will now summarize.

In September 1998, we published a regulation allowing the export, under a license exception, of unlimited strength encryption to banks and financial institutions located in 46 countries which allows U.S. companies new opportunities to sell encryption products to the world's leading economy. This policy recognizes the need to secure our financial networks, and the history of cooperation which the banking and financial communities have with government authorities when information is required to combat financial and other crimes.

More importantly, on September 16th, Vice President Gore unveiled an update to our encryption policy. This Policy Update was the result of a dialogue with U.S. industry, law enforcement, and privacy groups on how our policy might be improved to find technical solutions, in addition to key recovery, that can assist law enforcement in its efforts to combat crime. At the same time, we wanted to find ways to assure continued U.S. technology leadership, promote secure electronic commerce, and protect privacy concerns. We believed then and now that the best way to make progress on this issue is through a constructive, cooperative dialogue, rather than by legislative solutions. Through dialogue lasting more than a year, there has been increased understanding among the parties and we have made progress.

On December 31, we published regulations implementing the Vice President's policy announcement. These regulations will not end the debate over encryption controls, but we believe the regulation addresses some private sector concerns by opening large markets and further streamlining exports.

The Update permits the export of 128-bit encryption products and higher (with or without key recovery) to several important industry sectors. Now, banks, financial institutions, health facilities, and on-line merchants can secure their sensitive financial, medical, and on-line transactions in digital form. This update also allows U.S. companies to export 128-bit or greater encryption products, including technology to subsidiaries around the world, to protect its proprietary information and to develop new products. Further, this update allows the export of 128-bit or greater "recovery capable" or "recoverable" encryption products under an encryption licensing arrangement. Such products include those that are readily available in the marketplace such as general purpose routers, firewalls, and virtual private networks. These recoverable products are usually managed by a network or corporate security administrator without any involvement by a third party. Since the Update announcement, Industry has been taking advantage of this new liberalization and the streamlined process awarded to such products.

Many of the updates permit the export of encryption to these end-users under a license exception. That is, after the product receives a technical review, it can be exported by manufacturers, resellers and distributors without the need for a license or other additional review. These license exceptions currently apply to a list of countries or a set of end users. We also have a general policy of approval for exports to those sectors through encryption licensing arrangements (ELA), a kind of bulk license, to allow unlimited shipments of strong encryption to the sectors worldwide.

We also further streamlined exports of key recovery products by no longer requiring a review of foreign key recovery agents and no longer requiring companies to submit business plans.

We recognize that the development of our policy is an evolutionary process, and we intend to continue our dialogue with industry. Our policy will continue to adapt to technology and market changes. We will review our policy again this year with a view toward making further changes. An important component of our review is input from industry, which we are receiving through our continuing dialogue.

This past year, we also made progress on developing a common international approach to encryption controls through the Wassenaar Arrangement. Established in 1996 as the successor to COCOM, it is a multilateral export control arrangement among 33 countries whose purpose is to prevent destabilizing accumulations of arms and industrial equipment with military uses in countries or regions of concern. Wassenaar provides the basis for many of our export controls.

In December, through the hard work of Ambassador David Aaron, the President's special envoy on encryption, the Wassenaar Arrangement members agreed on several changes relating to encryption controls. These changes go a long way toward increasing international security and public safety by providing countries with a stronger regulatory framework for managing the spread of robust encryption. Specific changes to multilateral encryption controls include removing multilateral controls on all encryption products at or below 56 bit and certain consumer items regardless of key length, such as entertainment TV systems, DVD products, and on cordless telephone systems designed for home or office use.

Most importantly, the Wassenaar members agreed to remove encryption software from Wassenaar's General Software Note and replace it with a new cryptography note. Drafted in 1991, when banks, government and militaries were the primary users of encryption, the General Software Note allowed countries to export mass market encryption software without restriction. The GSN was created to release general purpose software used on personal computers, but it inadvertently also permitted countries to release encryption. It was essential to modernize the GSN and close the loophole that permitted the uncontrolled export of encryption with unlimited key length. Under the new cryptography note, mass market hardware has been added and a 64-bit key length or below has been set as an appropriate threshold. This will lead governments to review the dissemination of 64-bit and above encryption.

I want to be clear that this does not mean encryption products of more than 64 bits cannot be exported. Our own policy permits that, as does the policy of most other Wassenaar members. It does mean, however, that such exports now can be reviewed by governments consistent with their national export control procedures.

Export control policies without a multilateral approach have little chance of success. Agreement among the Wassenaar members on the treatment of mass market encryption products is a strong indication that other countries share our public safety and national security concerns. Contrary to what many people thought two years ago, we have found that most major encryption producing countries are interested in developing a common approach to encryption controls.

The PROTECT Act

With respect to S.789, the Administration opposes this legislation for a number of reasons. Overall the bill does not promote the balance that this Administration has worked so hard to achieve over the past several years. Let me now discuss some of the more problematic sections.

Under section 505, the removal of export controls on publicly or generally available encryption is in effect left to an advisory board composed of private sector and government representatives, with the concurrence of the Secretary. We believe such a board would be unworkable. Although availability is one of the factors we use to decide whether an encryption product may be exported, it is not the only factor and should not be elevated above the others. We need to be able to take all factors, including national security and public safety, into account when making export control decisions. Disallowing or downgrading important considerations will only serve to weaken our export control system. The broad definitions used in the bill would give the Board wide latitude in making its findings on what is available. This could place the Secretary in the position of having to routinely object to the removal of export controls when important national security and law enforcement interests are at stake. The bill makes this decision subject to judicial review. The Administration does not think it is wise public policy for the courts to adjudicate Executive Branch decisions on these matters.

Section 501 removes the Department of Justice from the encryption export license consultation process. Since law enforcement interests are an important consideration in regard to encryption, we cannot support this provision.

We support the provisions in the bill that require a technical review for eligibility to export encryption under a license exception. In fact, this is consistent with current regulations. What we cannot support, however, is the portion of section 504 that would provide automatic eligibility after 15 days if the exporter has not received a decision from the government. In all cases, a very careful technical review is completed in order to determine that a product is technically eligible for a particular license exception. Although we try to perform these reviews as quickly as possible, a 15-day automatic approval will severely limit our ability to do a careful review.

Section 504 also proposes control parameters and export liberalizations beyond what the Administration can entertain and which would be contrary to our international export control obligations. For example, Wassenaar agreed to decontrol encryption products up to 56-bits whereas this bill would decontrol encryption products using a key length at 64-bits or less. Section 504 also expands the set of products, end users, and countries eligible to receive encryption under a license exception beyond what we believe is prudent.

Another troubling part of this bill is section 102, which would permit a U.S. person located anywhere in the world to develop, manufacture, sell or use any type of encryption. If this provision were construed to permit U.S. citizens to develop, manufacture and sell encryption products overseas, even with the use of non-public controlled technology that they had acquired in the United States, it would, in effect, prevent the government from requiring a license for U.S. persons to develop and manufacture encryption abroad. As a result, U.S. companies would likely move all development and manufacture of encryption out of the United States in order to take advantage of this loophole. This is not in our country's economic or national security interest.

Section 103 contains a provision that would prohibit the U.S. Government from conditioning any approval on the fact that a product is recoverable. A fundamental feature of our encryption policy is that we provide incentives for companies to develop products that provide strong security and also meet the needs of national security and law enforcement. The bill would eliminate this laudable feature of our policy that industry wanted us to include in last year's update. In addition, this provision of the bill is inconsistent with section 504 which allows license exception treatment for recoverable products.

Section 506 would eliminate any export controls on products using the forthcoming Advanced Encryption Standard (AES). We oppose the removal of export controls on encryption products simply because they implement a government standard. Products incorporating the AES should be exportable to the same extent as any other product incorporating encryption of similar strength. Under our current policy, AES-based products could be exported to banks, large corporations, on-line merchants without restriction and to many other safe endusers depending on the nature of the product. We do not think it is wise to link development of the AES to export controls. Such a linkage might bring undue pressure on NIST to complete the AES process faster than planned, and may therefore not allow prudent study of the security features of the candidate algorithms before selection.

With respect to the provisions of the bill that do not relate to export controls, we have a number of questions and concerns.

One such provision in Section 202 requires that encryption products used by the Government must interoperate with other commercial encryption products. The extent to which interoperability is required is unclear in the bill, but we believe the practical result of this requirement is that the Government could not use encryption because no single encryption product interoperates with all other products. It also appears that this provision could prohibit the use of encryption developed by the government for its own internal use in "closed" systems that are purposefully designed not to interoperate with other systems.

Section 202 also appears to prevent mandatory use of recoverable encryption when communicating with U.S. Federal, state and local governments. This would appear to preclude an agency from requiring key recovery or recoverable products for business purposes. We believe the effect of this provision may be much broader than simply preventing government from using recoverable encryption when dealing with the public. The practical effect would be that Government sites would have to be capable of supporting secure communications using all encryption methodologies on the market. This is absurd.

We are concerned that section 302 of the bill may preclude NIST's work with voluntary standards organizations because it prohibits the Secretary of Commerce from carrying out any policy that establishes an encryption standard for use by businesses or other entities other than for computer systems operated by the United States Government. The Secretary of Commerce is prohibited from establishing standards for business; however, when invited by standards organizations to do so, NIST does, as a matter of policy, work together with those organizations. Cooperation between NIST and standards organizations is important for both NIST and industry, and it is consistent with government policy to use voluntary standards and to purchase commercial off-the-shelf products. If the government cannot have input to the standards process, we may end up with less secure products available for government agencies. We want to encourage, to the extent possible, the development of voluntary standards that meet the needs of the government. This reduces costs for both government and industry.

In regard to section 401 dealing with the "Information Technology Laboratory," we have two concerns. First, we do not think it is appropriate for NIST to undertake research and development of new technologies to facilitate lawful access to communications and electronic information. This activity is more appropriately done by the FBI. Second, we are concerned that the bill will provide NIST with new tasks but no new funding to carry out this work. We have similar concerns with section 402. The advisory board, whose correct statutory name is " Computer System Security and Privacy Advisory Board," is made up of 13 volunteers. Again, any additional tasks assigned to this board would require necessary funding.

The Administration does not seek encryption export control legislation, nor do we believe such legislation is needed. The current regulatory structure provides for balanced oversight of export controls and the flexibility needed to adjust to our economic, foreign policy and national security interests to advances in technology. This is the best approach to an encryption policy that promotes secure electronic commerce, maintains U.S. lead in information technology, protects privacy, and protects public safety and national security interests.

As you know, public debate over encryption policy has been lively and often acrimonious. Some of those on both sides of the debate are not interested in searching for a middle ground that can meet all of our needs. Our dialogue with industry has gone a long way toward bridging that gap and finding common ground. We will continue this policy of cooperative exchange, which is clearly the best way to pursue our policy objectives of balancing public safety, national security, and the competitive interests of U.S. companies.

CHAPTER 34

TESTIMONY OF
D. JAMES BIDZOS
VICE CHAIR
SECURITY DYNAMICS TECHNOLOGIES, INC.
PARENT COMPANY OF RSA DATA SECURITY, INC.
ON BEHALF OF
AMERICANS FOR COMPUTER PRIVACY

IMMEDIATE NEED FOR RELAXATION OF EXPORT CONTROLS FOR SOFTWARE AND HARDWARE WITH ENCRYPTION CAPABILITIES

BEFORE THE
COMMERCE, SCIENCE AND TRANSPORTATION COMMITTEE
OF THE
U.S. SENATE

Washington, D.C.
June 10, 1999

SUMMARY OF STATEMENT OF D. JAMES BIDZOS VICE CHAIR OF SECURITY DYNAMICS TECHNOLOGIES, INC.

Congress must immediately relax export controls on software and hardware with encryption capabilities. Widespread deployment of American products with encryption capabilities will help to accelerate dramatically the growth of electronic commerce by protecting consumers' privacy and preventing electronic crime.

Without relaxation of export controls, U.S. manufacturers remain at a competitive disadvantage, and foreign consumers will purchase encryption products from foreign suppliers. Foreign products are comparable in capabilities and quality. When a foreign purchaser cannot obtain an American product they simply purchase it from a foreign supplier. Unfortunately, not only are American companies losing a sale of an encryption item, but they are also losing the sale of the program or hardware such as an Internet server or an application browser that uses the encryption capability. In fact, companies risk losing sales of entire systems because of their inability to provide necessary security features. The only impact of the Administration's export policy is widespread deployment of foreign designed and manufactured software and hardware.

The Administration took the first step towards developing a sensible long-term encryption policy by permitting exports of select products to select users, but they still have not gone far enough.

The PROTECT Act is an improvement over current Administration policy. It affirms that Americans may use and sell any type of encryption domestically, and ensures that the U.S. Government may not use its full powers and capabilities to compel Americans to use or sell a certain type of encryption. The PROTECT Act also provides a broader range of export relief for American encryption products and provides a certain timeframe for the export review process. Also, the Act provides Congressional support for, and sets a 5-year limit on the selection of, the 128-bit Advanced Encryption Standard.

The PROTECT Act should be further improved to reflect market and technological realities. The PROTECT Act does not permit individual foreign consumers to obtain strong, non-recoverable encryption, making it impossible for them to securely purchase products from American companies. Also, the Act does not provide immediate export relief for encryption sales to small businesses—one of the fastest growing worldwide business sectors.

Unfortunately, the PROTECT Act limits easy exportability of mass market products with strong 128-bit encryption until NIST adopts the Advanced Encryption Standard. This means individual consumers and small businesses will have to wait three years to obtain strong American encryption, and foreign companies will have had three more years to market their products. Exportability in the meantime is dependent on an unwieldy complex bureaucracy that will determine whether American products are generally available or compete with comparable foreign products. We believe the evidence already is overwhelming regarding these facts.

Introduction

Good Morning. My name is Jim Bidzos, and I am Vice Chair of Security Dynamics Technologies, Inc., a Massachusetts-based security firm that is also the parent company of RSA Data Security, located in San Mateo, California. For over 13 years, until earlier this year, I was the President and CEO of RSA Data Security, the world's leading encryption company.

RSA's technology is embedded in both Netscape and Microsoft browsers, and in over 500 other products, all used by hundreds of millions of people around the world to secure Internet transactions and digital data of many types. Over many years, I have personally negotiated hundreds of licenses to RSA encryption technology, including licenses with companies such as IBM, Microsoft, ATT, Netscape, Oracle, and Motorola. These negotiations almost always involve discussions about encryption needs, end-user requirements, and export policy. I have thus gained unique insights into the needs and concerns of both industry and users with respect to encryption.

I am also founder and chairman of Verisign, Inc., the leader in Internet authentication. Verisign is the world's largest Internet security products and services company as measured by both customers and market capitalization.

I am a member of the board of directors of several other security companies. One specializes in virtual private networks. Another is a manufacturer of security tokens. Another offers cryptographically secure digital time stamping services. I am also a director of a UK-based encryption hardware company, a Dublin-based secure electronic payments company, and two Japanese security companies.

I have been deeply involved in the debate over encryption, from many aspects, including US policy on the export of this technology. Over the last 13 years, I have testified many times before both the House and Senate on encryption policy, and I have participated in numerous US and international standards activities.

I believe that my long and unique history in the encryption area allows me to offer testimony today that may help the committee better understand industry's concerns over US encryption policy.

On behalf of Americans for Computer Privacy ("ACP"), thank you for the opportunity to testify on S.798, the PROTECT Act, sponsored by Chairman McCain and cosponsored by four other committee members Senators Burns, Wyden, Abraham, and Kerry.

ACP is a coalition of over 3,500 individuals, 40 trade associations and over 100 companies representing financial services, manufacturing, high-tech, and transportation industries as well as law enforcement, civil-liberty, taxpayer and privacy groups. ACP supports policies that allow American citizens to continue using strong encryption without government intrusion, and advocates the lifting of export restrictions of U.S. made encryption products.

But we really are here today to speak on behalf of the tens of millions of users of American software and hardware products. The American software and hardware industries have succeeded because we have listened and responded to the needs of computer users worldwide. We develop and sell products that users want and for which they are willing to pay.

One of the most important features computer users are demanding is the ability to protect their electronic information and to interact securely worldwide. American companies have innovative products which can meet this demand and compete internationally. But there is one thing in our way—the continued application of overbroad, unilateral, export controls by the U.S. Government.

At the outset, I want to say that the PROTECT Act definitely moves us in the right direction and is a significant improvement over the Administration's current policy—but it could be further improved in several important respects (along the lines of the SAFE Act).

ACP recognizes a legitimate governmental need to obtain access to information and communications when authorized by proper legal authority. ACP and its members are responsible citizens. We have no wish to facilitate the commission of crime or the spread of terrorism. Similarly, we are committed to strengthening the nation's infrastructure and promoting national security, enhancing the privacy of American citizens and ensuring the security of electronic commerce.

But we believe that the best way of meeting all these objectives is promote the widespread use of encryption!

Ultimately, any truly successful, sensible encryption policy that has America's best interests at heart must be based on technological and market realities, and should not create winners and losers in the encryption marketplace on a sector-by-sector basis. It would recognize that:

- The worldwide encryption standard is 128-bit encryption;
- Mass market software and hardware is inherently uncontrollable; and
- It is America's national and economic security interests to have American designed and manufactured encryption products deployed worldwide.

We believe it is preferable for Congress to put encryption policy on a statutory basis rather than continuing to leave it up to inconsistent Administration regulations—sending a strong message around the world that encryption is important for protecting the privacy of citizens, for promoting e-commerce, preventing crime and protecting our critical infrastructures and national defense.

The American Computer Software and Hardware Industries—An American Success Story

The computer software and hardware industries are American success stories, but they are being threatened. America's software and hardware industries are important contributors to U.S. economic security. Information technology industries now are directly responsible for over one-third of real growth of the U.S. economy, and both the computer and software industries are continuing to grow. From 1990 through 1996, the software industry grew at a rate of 12.5%, nearly 2.5 times faster than the overall U.S. economy.

More than 7 million people work in IT industries. In 1996, the software industry provided a total of over 619,000 direct jobs and $7.2 billion in tax revenues for the U.S. economy. The software industry is expected to create an average of 45,700 new jobs each year through 2005. If piracy were to be eliminated in the United States, the number of new software jobs created would double to an average of 93,000 a year.

Moreover, the computer software industry has achieved tremendous success in the international marketplace with global sales of packaged (*i.e.*, non-custom) software reaching over $118.4 billion in 1996, and rising to $135.4 billion in 1997. American produced software accounts for 70% of the world market, with exports of U.S. programs constituting half of the industry's output.

The incredible growth of the industry and its exporting success benefits America through the creation of jobs here in the United States. Many of these jobs are in highly skilled and highly paid areas such as research and development, manufacturing and production, sales, marketing, professional services, custom programming, technical support and administrative functions. In the U.S. software industry, workers enjoy more than twice the average level of wages across the entire economy—$57,319 versus $27,845 per person.

All of these revenues and jobs are dependent upon American software and hardware producers remaining the market leaders around the world, especially as the major growth markets continue to be outside the United States. Strong export controls on products with encryption capabilities are crippling the ability of these companies to compete with foreign providers and are only ensuring that foreign products are securing worldwide critical infrastructures, not American products.

Secure Networks And Confidential Information In The Internet Age Are The Key To Privacy And Commerce.

American individuals and companies are rapidly becoming networked together through private local area networks (LANs), wide area networks (WANs) and public networks such as the Internet. Combined, these private and public networks are the economic engine driving electronic commerce, transactions and communications. This engine is sputtering and threatens to stall.

Traffic on the Internet doubles every 100 days. Predictions of business-to-business Internet commerce for the year 2000 range from $66 billion to $171 billion, and by 2002, electronic commerce between businesses is expected to reach $300 billion. During 1997, one leading manufacturer of computer software and hardware sold $3 million per day online for a total of $1.1 billion for the year.

More and more individual consumers also are going on line and spending. Five years from today, we anticipate nearly 60 percent of all Americans to be using the Internet. More than 10 million people in North America alone have already purchased something over the Internet, and at least 40 million have obtained product and price information on the Internet only to make the final purchase off-line. Altogether last year, consumers spent nearly $8 billion online. Nearly 1.5 million Americans join the online population every month, and the number of worldwide online users is expected to reach 248 million by 2002.

The incredible participation by American consumers in the Internet phenomenon clearly demonstrates that the need for strong encryption is no longer merely the purview of our national security agencies concerned about securing data and communications from interception by foreign governments. Today, every American even merely dabbling on the Internet requires access to strong encryption. Imagine the boost in volume of e-commerce if all of these consumers had enough confidence in the security of the Internet to purchase on-line. Yet in 1996 the Computer Security Institute/FBI Computer Crime Survey indicated that our worldwide corporations will be increasingly under siege: over half from within the corporation, and nearly half from outside of their internal networks.

Network users *must* have confidence that their communications and data—whether personal letters, financial transactions or sensitive business information—are secure and private. Electronic commerce is transforming the marketplace—eliminating geographic boundaries and opening the world to buyers and sellers. Companies, governments and individuals now realize that they can no longer protect data and communications from others by relying on limiting physical access to computers and maintaining stand-alone centralized mainframes. Instead, users expect to be able to pick up their e-mail or modify a document from any computer anywhere in the world simply by using their Internet browses. Thus, consumers worldwide are demanding to be able to protect their electronic information and interact securely worldwide, and access to products with strong encryption capabilities has become critical to providing them with confidence that they will have this ability.

Unilateral U.S. Export Controls Harm American Interests

Currently, there are no restrictions on the use of cryptography within the United States. However, the U.S. Government maintains strict *unilateral* export controls on computer products that offer strong encryption capabilities.

American companies are forced to limit the strength of their encryption to the 56-bit key length level set late in 1998. The recently announced regulations will also permit companies to export stronger encryption on a sector-by-sector, user-by-user basis. However, this policy ignores the fact that:

- The minimum strength now required by new Internet applications is 128-bit encryption;
- American companies cannot export encryption products to a vast majority of non-U.S. commercial entities. Foreign manufacturers provide 128-bit encryption alternatives and add-ons-filling the market void created by U.S. export controls;
- Providing sector-by-sector relief is unworkable for mass market products and does not reflect commercial realities for sales of custom products;
- 56-bit encryption has been demonstrated to be vulnerable to commercial let alone governmental attack. (In the beginning of this year at the RSA Encryption Conference, a 56-bit DES encoded message was broken by private companies and individuals working together in 22 hours and 15 minutes—imagine what a hostile government with serious resources could do); and
- New developments in technology are introduced everyday that speed up decryption time. Adi Shamir, the Israeli computer scientist who is the "S" in RSA, recently announced "Twinkle:, which is a proposed method for quickly unscrambling computer-generated codes that have until now been considered secure, at the International Association for Cryptographic Research's latest meeting in Prague.

The Wassenaar Arrangement Is Not A Multilateral Agreement To Control Encryption

I want to take one minute to discuss the Wassenaar Arrangement at this point. Please do not be fooled by any claims from the Administration that the Wassenaar Arrangement is the multilateral agreement on encryption that they have been touting was just around the corner for the past several years.

The Wassenaar Arrangement replaced the old COCOM regime with a non-binding agreement among 30 countries to report on their sensitive exports. The December 1998 Wassenaar Arrangement agreement actually *decontrolled* encryption products. Many countries, such as Israel and South Africa, who export strong encryption are not signatories to the Arrangement. The Wassenaar Arrangement eliminates controls of any sort on 56-bit encryption and permits exports of up to 64-bit encryption in mass-market software and hardware. It also removed any reporting requirements—the sole official means for actually monitoring what countries are doing. Although the Arrangement left open the possibility that countries might individually control 128-bit encryption, we are skeptical that they will do so. There is no penalty for failing to control 128-bit encryption, and most countries are actually moving towards *encouraging* the use of stronger encryption. Finally, a country could technically comply with the Arrangement, while still permitting easy exports of strong encryption.

Ironically, the U.S. government is a good example of the lack of effect of the Wassenaar Arrangement. In its new encryption regulations, the Administration is still controlling encryption products with greater than 56, not 64, bit keys, and they have imposed reporting requirements on mass market products even if they are using 64-bit encryption.

Recently, on June 2, 1999, the German government established a new encryption policy seeking to improve protection of German users of global information networks and clarifying that any encryption product may be developed, produced marketed and used without restrictions in Germany. The German government declared its intention to simplify their export review process and to strengthen the performance and ability of German manufacturers to compete internationally. The German government will monitor abuses of encryption for illegal purposes and attempt to further improve the technical capabilities of German law enforcement and security agencies to handle advances in encryption technology.

Even France, traditionally the country which placed the greatest restrictions on its own citizens by limiting them to the easily broken 40-bit level of encryption, has recognized that technology has progressed. Near the end of 1998, France relaxed controls on the domestic use of encryption and is now permitting, and in fact encouraging, the use of 128-bit encryption by its citizens.

Without Export Relief, Foreign Consumers Will Purchase Their Products From Foreign Suppliers, Keeping U.S. Manufacturers At A Competitive Disadvantage

Export controls also have made American companies less competitive and opened the door for foreign software and hardware developers to gain significant market share—decreasing our national and economic security.

As a result of U.S. unilateral export controls, encryption expertise is being developed off-shore by foreign manufacturers who now provide hundreds of encryption alternatives and add-ons. The Administration's export controls are in no way preventing foreigners, let alone those with criminal intent, from obtaining access to encryption products. In fact, foreign software and hardware manufacturers have seized the opportunity to create sophisticated encryption products and to capture sales.

As long ago as 1995, the General Accounting Office confirmed that sophisticated encryption software is widely available to foreign users on foreign Internet sites. In 1996, a Department of Commerce study again confirmed the widespread availability of foreign manufactured encryption programs and products. Professor Hoffman today releases the results of his latest survey which shows the continuing growth in foreign encryption products in the face of U.S. export controls.

If an encryption product is combined with other applications such as Internet browsers and application servers, U.S. companies generally will lose both sales. In fact, companies risk losing sales of entire systems because of inability to provide necessary security features. This permits foreign manufacturers to gain entry into companies as well as gain credibility—providing the foreign manufacturers with further opportunity to take away future sales in the same and other product lines.

U.S. Encryption Export Controls Hurt American Companies Without Helping Law Enforcement Or National Security

U.S. export controls have had the effect of creating an encryption expertise outside the United States that is gathering momentum. Unfortunately, every time research and development of an encryption technique or product moves off-shore, U.S. law enforcement and national security agencies lose. We believe that continuing down this path will be ultimately more harmful to our national security and law enforcement efforts as American companies will no longer be the world leaders in creating and developing encryption products.

In fact, as long ago as 1996, the NRC Committee concluded that as demand for products with encryption capabilities grows worldwide, foreign competition could emerge at levels significant enough to damage the present U.S. world leadership in information technology products. The Committee felt it was important to ensure the continued economic growth and leadership of key U.S. industries and businesses in an increasingly global economy, including American computer, software and communications companies. Correspondingly, the Committee called for an immediate and easy exportability of products meeting general commercial requirements—which is currently 128-bit level encryption!

To summarize:

- Foreign competitors not subject to outdated U.S. export controls are ready to take sales and customers from U.S. companies today.
- Complex and cumbersome U.S. export controls make American companies less competitive. They significantly increase the costs of developing, marketing and selling products with encryption capabilities, delay the introduction of new products or features, and encourage foreign customers to purchase from foreign suppliers due to the uncertainty and delay in obtaining a comparable American product.

- Current export controls do not keep strong encryption out of the hands of foreign customers; they just keep U.S. products out of their hands.
- In the future, if export controls on encryption are not relaxed, both American and foreign infrastructures will be secured by foreign encryption products, creating a significant problem for American law enforcement and national security agencies.

American companies do have exciting and innovative products that can meet the demand for 128-bit encryption and compete internationally. But unless the current unilateral U.S. export restrictions are changed to allow the use of strong encryption, American individuals and businesses will not be active participants in this new networked world of commerce—let alone continue to be the leaders in its development. Furthermore, American companies will no longer be providing the world, and its critical infrastructures, with the answers to their security problems. Instead foreign companies will. It is unclear how U.S. national security or law enforcement will be aided or how our critical infrastructures will be secure when foreign encryption products dominate the world market.

The Bernstein Case

The absurdity of the existing export control regime is further highlighted by the recent decision of the 9th Circuit Court of Appeals in *Bernstein v. DOJ*. In that case, the court held that the existing restrictions on the export of source code, the language in which programmers communicate their ideas to one another, are an unconstitutional prior restraint on first amendment rights of free speech. So now we have a situation where it is permissible to export jobs (because one can export source code to teach foreign programmers), but not American products (because one cannot embody that source code in a product)!

More generally, Judge Fletcher's opinion raises some very valid, more general questions and points out how important encryption is to the mainstream life of Americans rather than merely to obscure technologists. Judge Fletcher states:

> In this increasingly electronic age, we are all required in our everyday lives to rely on modern technology to communicate with one another. This reliance on electronic communication, however, has brought with it a dramatic diminution in our ability to communicate privately. Cellular phones are subject to monitoring, email is easily intercepted, and transactions over the Internet are often less than secure. Something as commonplace as furnishing our credit card number, social security number, or bank account number puts each of us at risk. Moreover, when we employ electronic methods of communication, we often leave electronic "fingerprints" behind, fingerprints that can be traced back to us. Whether we are surveilled by our government, by criminals, or by our neighbors, it is fair to say that never has our ability to shield our affairs from prying eyes been at such a low ebb. The

> availability and use of secure encryption may offer an opportunity to reclaim some portion of the privacy we have lost. Government efforts to control encryption thus may well implicate not only the First Amendment rights of cryptographers intent on pushing the boundaries of their science, but also the constitutional rights of each of us as potential recipients of encryption's bounty. Viewed from this perspective, the government's efforts to retard progress in cryptography may implicate the Fourth Amendment, as well as the right to speak anonymously, . . ., the right against compelled speech, . . . , and the right to informational privacy. While we leave for another day the resolution of these difficult issues, it is important to point out that Bernstein's is a suit not merely concerning a small group of scientists laboring in an esoteric field, but also touches on the public interest broadly defined.

The Administration Took a Small First Step Towards Developing A Sensible Long-Term Encryption Policy, But They Still Have Not Gone Far Enough

Progress was made last year in the new Administration policy announced by the Vice President in September and contained in the interim final regulations of December 31, 1998.

ACP welcomed the Administration's efforts to relax export controls on select products used by select users. We especially appreciated the Administration's apparent abandonment of its key escrow policy that would have required all encryption exports (except for 40-bit and less encryption) to be capable of providing third parties with immediate access to the plaintext of stored data or communications without the knowledge of the user. Foreign companies and consumers simply would not purchase such products as a multitude of foreign products without key escrow are readily available.

However, the Administration's actions are merely a first step. U.S. export controls still ignore the realities of mass-market software and hardware distribution. Mass-market software publishers and hardware manufacturers sell products through multiple distribution channels such as OEMs, (*i.e.*, hardware manufacturers that pre-load software onto computers), value-added resellers, retail stores and the emerging channel of on-line distribution. Thus, mass market products are available to the general public from a variety of sources. (It also is why continued reporting requirements about end-uses and end-users make no sense.)

The mass-market distribution model presupposes that software publishers and hardware manufacturers will take full advantage of these multiple channels to ship identical or substantially similar products worldwide (allowing only for differences resulting from localization) irrespective of specific customer location or characteristics. As mass market products are uncontrollable, ACP believes U.S. companies should be able to export the current market standard of 128-bit encryption. Unfortunately, the Administration has only proposed permitting easy exports of 56-bit encryption even if foreign products exist in the marketplace.

ACP also believes that encryption hardware and software should be treated identically. However, contrary to the Administration's original announcement regarding export relief which included export relief for hardware, the new regulations still do not permit 56-bit encryption chips, integrated circuits, toolkits and executable or linkable modules to be easily exported except to subsidiaries of U.S. companies or otherwise relax export controls on stronger mass market hardware.

In addition, ACP believes that the new regulations are so complex and contain unrealistic requirements that they undermine many of the benefits of the Administration's export relief for stronger encryption, especially for mass market hardware and software. U.S. companies are now required to meet a number of new, unilateral reporting requirements. For example, exporters now are required to report the name and address of end-users, a virtual impossibility for mass-market exporters because registration of end-users is customarily voluntary. A system to obtain the names and addresses of each of the millions of potential health care end-users, for example, would cost more than the profits yielded from many products.

ACP also is disappointed that the Administration's regulations do not clearly provide online merchants with the level of export control relief originally envisioned as they do not permit ISPs to provide "services" as a permissible end-use. This could chill the use by ISPs located abroad of U.S.-origin encryption products for billing, payment, and delivery purposes, despite the widespread foreign availability of such products.

The PROTECT Act Is An Improvement Over Current Administration Policy

The PROTECT Act Establishes The Correct Domestic Encryption Policy

The PROTECT Act affirms that Americans may use and sell any type of encryption domestically. Even more importantly, the PROTECT Act ensures that the U.S. Government may not use its full powers and capabilities to compel, directly or indirectly, Americans to use or sell a certain type of encryption. This will prevent the U.S. Government from attempting to achieve domestic controls on encryption through regulations or "incentives".

For example, the Act prohibits the U.S. Government from linking the ability to electronically sign a document to a requirement that the consumer use a particular encryption methodology for ensuring confidentiality. Thus, the U.S. Government cannot require Americans to use a certain type of encryption (such as key escrow) to engage in electronic commerce.

Also, the PROTECT Act specifically restricts the government from requiring any American to use a particular encryption product or methodology to communicate with or transact business with the government. The U.S. Government may only specify technologies for its own internal uses.

The PROTECT Act Provides Additional Export Relief For Encryption Products

The PROTECT Act provides a broader range of export relief for American encryption products than the Administration. We are pleased that the PROTECT Act provides immediate export relief after a one-time review by the government for:

- All encryption products using key lengths of 64-bits or less rather than the less secure 56-bit key lengths proposed by the Administration;
- All recoverable encryption products regardless of key length, including telecommunications related products; and
- All encryption products using key lengths greater than 64-bits to certain legitimate and responsible commercial users, including publicly traded firms, firms subject to government regulation, U.S. companies' foreign subsidiaries, affiliates and strategic partners, on-line merchants who use encryption products to support electronic commerce, and foreign governments who are members of NATO, OECD and ASEAN.

We are also pleased that the PROTECT Act recognizes the need for a quicker and more certain timeframe for the export review process. Businesses simply cannot live with the U.S. Government taking between 3 to 6 months to determine whether a product is exportable when many Internet products have 90 day product cycles and most businesses do not want to wait through one or two business quarters to update their computer systems.

The PROTECT Act Begins To Recognize Mass Market Product Realities

We also are encouraged that the PROTECT Act recognizes the difficulties in complying with reporting requirements for mass market encryption products and eliminates such reporting requirements. It is virtually impossible for mass-market exporters to report the name and address of each end-user. Millions of these products are sold through multi-level distribution channels (*e.g.*, VAR's and chain stores). Moreover, as registration of mass market products is customarily voluntary. This is a vast improvement over the Administration's proposed regulations which effectively require companies to develop a system to obtain the names and addresses for each health and medical end-user of stronger encryption products and all foreign online merchants.

The PROTECT Act also provides Congressional support for, and sets a 5-year limit on the selection of, the 128-bit Advanced Encryption Standard which is being developed under the auspices of the National Institute of Standards and Technology. The 2002 deadline will provide impetus for NIST to finish developing the standard in a timely manner while providing NIST with sufficient time to study the final standard's security features. This is an important process that will result in a new standard for government's sensitive, but unclassified, information and most likely will serve as the new worldwide standard for strong encryption similar to the Data Encryption Standard when it was introduced in the 1970's. Once the algorithm is selected, the PROTECT Act removes all export controls on encryption products using the 128-bit standard or its equivalent strength.

The PROTECT Act Should Be Further Improved To Reflect Market And Technological Realities

The PROTECT Act Does Not Provide Immediate Export Relief For Individual Consumers

The PROTECT Act does not go far enough to protect the millions and millions of consumers that are now engaging in electronic commerce. Foreign consumers still will not be able to obtain an American Internet browser with strong, non-recoverable encryption, making it impossible for them to securely purchase products from American companies. Also, an everyday foreign consumer who wants to protect an on-line diary, copies of health care records or a business proposal, may not easily obtain strong encryption to do so from American sources if any portion of the encryption used by the product is non-recoverable. Under the bill, all these individuals must wait until 2002.

The PROTECT Act Does Not Provide Immediate Export Relief For Small Businesses

We believe the PROTECT Act provides greater export relief for larger corporate customers. However, until 2002, small and privately-owned businesses face significant difficulty in easily obtaining U.S. encryption under any of the License Exceptions established by the PROTECT Act. So, for example, if two doctors in private practice together in Brazil or a restaurant owner in France or a small shopping market in Germany wants to purchase non-recoverable encryption, these small businesses probably would purchase a comparable foreign product as an American company could not easily export it to them.

Unfortunately, as companies install the security "plumbing" into their individual computers and company networks, it becomes increasingly difficult for American companies to replace the foreign software and hardware that already has been installed. Because the small business sector is, and most likely will continue to be, the fastest growing business sector, this puts American companies at a distinct disadvantage in selling encryption products at a later date.

The PROTECT Act's Export Relief For Mass Market Products And For Products Which Face Competition From Comparable Foreign Products Is Too Complicated And Creates An Unwieldy Bureaucracy

The PROTECT Act does recognize that mass market and publicly available encryption products, and encryption products for which comparable foreign products are available, should be treated differently under the U.S. export regime. The bill acknowledges the futility of trying to control a product that can be bought off of the Internet or easily purchased from commercial vendors such as CompUSA or from Circuit City by any individual in America regardless of nationality, or a comparable product can be easily purchased from similar stores in a foreign country. "Bad guys" certainly will have no problems obtaining the encryption products, and no concerns about "exporting" the products via telephone lines or the Internet or smuggled out on personally pressed CDs. The only impact of the export control will be to stop American companies from selling American products to legitimate users.

Unfortunately, the PROTECT Act establishes a complicated private/public board structure for deciding after-the-fact whether or not a product is a mass market product or whether comparable foreign products are available. The Secretary of Commerce has thirty days to approve or disapprove the Board determination, subject to judicial review, and the President may override any determination. Unfortunately, there is no guarantee of any consistency in the Board's decisions. Thus, while the Board procedure is an improvement, and the opportunity for judicial review provides a mechanism to ensure that exports are not denied in an arbitrary and capricious manner, it is not a predictable, clear process giving American companies certainty as to whether they can export their products. Such predictability is necessary so that American companies can have confidence designing and building security features into their products.

The PROTECT Act should, but does not, afford complete and immediate export relief for mass market encryption without any complicated oversight. The Act also does not recognize that if a comparable foreign product is available, *any* delay in exports provides a significant advantage to the foreign product.

The PROTECT Act's Relief For 128-Bit AES Products Is Too Little, Too Late

I want to make one final comment regarding the general exportability of mass market products. We support NIST's efforts to establish a new 128-bit Advanced Encryption Standard; however, under the bill, it will not be finalized until 2002. Because the PROTECT Act limits easy exportability of mass market products until the AES is adopted, general distribution of these products will have to wait almost three years. Considering the current speed of technological change, where Internet products are now on three-month product cycle times, and the fact that 128-bit comparable foreign encryption is currently available, this is an eternity in Internet time. Law enforcement and national security interests have known for a long time that ubiquitous use of strong encryption by consumers worldwide is just around the corner. They cannot hope to continue to delay the world from using strong encryption according to their timeframe.

The Time For Action Is Now

To keep American vendors on a level international playing field and American computer users adequately protected, U.S. export controls must be immediately updated to reflect technological and international market realities.

Thank you.

CHAPTER 35

TESTIMONY OF

DAVID AUCSMITH
CHIEF SECURITY ARCHITECT
INTEL CORPORATION

ON BEHALF OF THE
BUSINESS SOFTWARE ALLIANCE

THE NEED FOR FUNDAMENTAL REFORM OF AMERICA'S ENCRYPTION POLICY

BEFORE THE
COMMERCE, SCIENCE AND TRANSPORTATION COMMITTEE
OF THE
U.S. SENATE
WASHINGTON, D.C.

June 10, 1999

SUMMARY OF STATEMENT OF DAVID AUCSMITH
CHIEF SECURITY ARCHITECT FOR INTEL CORPORATION

Thank you Mr. Chairman for the opportunity to talk to you this morning about the need for fundamental reform of America's encryption policy. I am pleased to appear today on behalf of the Business Software Alliance which, together with ACP, has been in the forefront of efforts to persuade the U.S. government to adopt a new U.S. encryption policy. **We urge the Committee to pass the PROTECT Act with further amendments that would make the bill more fully comport with technological and market realities.**

This morning I would like to briefly make five points that we believe should underpin U.S. encryption policy.

First, encryption is essential to all business in an Internet economy. While private sector interest in encryption export reform is generally characterized in terms of the competitiveness of American encryption products in a worldwide market, it is becoming a much larger issue for all American business. The global economy, tied together with the Internet, is turning businesses into virtual enterprises, localized products into global products, and geographically limited networks into worldwide networks. In this environment, American businesses must be able to sell and support their products worldwide, must be able to securely coordinate with their business partners worldwide, and must be able to conduct safe electronic commerce worldwide.

Quite simply, cryptography has emerged as the only possible solution to many of the requirements of commercial security. It is the essential building block for building trust onto the open Internet. Without it, the hundreds of billions of dollars of e-commerce currently projected to occur by the year 2002 will not happen.

Second, encryption is vital to securing America's critical infrastructures. Much of the national economy is at risk from the decisions that are made today on the issues of infrastructure protection. Increasingly, these critical systems are driven by, and linked together with, computers making them vulnerable to disruption. The single best way, and sometimes the only way to affect effectively these critical networks and systems, is encryption. That's why the National Research Council found that encryption promotes the national security of the United States. However, the security of any network is only as good as its weakest link. America's infrastructures cannot be protected if they are networked with foreign infrastructures using weak encryption.

Third, the availability of encryption cannot be reasonably controlled. Cryptography is a branch of mathematics. Cryptographic technology can be reduced to mathematical formulas and protocols. Information about cryptography is available from many sources in many forms. It is the subject of numerous academic conferences. It is taught in universities worldwide. Moreover, while developing good algorithms is tough, implementing them is relatively easy.

Fourth, government promoted or required plaintext access will not work. While required plaintext access offers, at first glance, a solution to the government's problem: (1) it is not technically possible in most circumstances; (2) it does not let law enforcement verify compliance with access requirements; and (3) it does not give national security interests access to stored keys. There is simply no way that law enforcement can determine, in advance, that particular text had not been encrypted with more than one program or product. At the same time, targets of national security interests are unlikely to design or use a plaintext infrastructure which would allow the U.S. government to have secret access to plaintext.

Moreover, there is practically no commercial reason for storing communications keys—if the communication is disrupted or compromised a new session will be established. At the same time, the need for key recovery of stored data also is overstated—the frequent example is an employee hit by a bus. With the exception of personal notes, information is not solely possessed by an individual. In addition, most mission-critical data is held by the corporate data management system that has its own control and protection mechanism. Finally, most personal data has a time value and rapidly becomes obsolete.

If one factors in the additional costs and systemic vulnerabilities that result from building in access features, we conclude that there is no business or consumer need for key recovery or special plaintext access. To be blunt: Intel does not plan to implement a key recovery scheme for its own use.

Fifth, the government needs to find technological alternatives to meet its requirements for access to information. Intel agrees that access to data communications and stored data by law enforcement intelligence communities is both legitimate and extremely important. Clearly, Congress should adequately fund the technical efforts of these agencies so they can meet the challenges of the next century. Industry supports additional funding. Industry can also provide other assistance.

For example, ACP proposed last year the creation of a "NET center" to help law enforcement officials understand how to deal with encryption and other technological advances. ACP also has advocated that the U.S. government should work cooperatively with our nation's hardware and software manufacturers to develop the technical tools and know-how that they need. Technical innovation is predominantly centered in the private sector—only a government/industry cooperative effort can address effectively the challenge of continued technological change.

In conclusion, let me say that we strongly believe the *Protect Act* should be passed but with further improvements.

The *Protect Act* does begin to realize the realities of mass market products, eliminates reporting requirements for such products, and grants export control relief to products at all horizontal layers in the information technology sector. But the Act still does not grant widespread exportability for mass market and publicly available encryption products. There is a complicated, bureaucratic process which must be pursued. Not until 2002 will American industry be able to widely export products using the 128-bit Advanced Encryption Standard or its equivalent.

We believe it is in our national interest to permit such exportability now and urge the Committee to amend the bill accordingly.

Once again, many thanks for this opportunity to testify.

INTRODUCTION

My name is David Aucsmith, and as Chief Security Architect for the Intel Corporation I am responsible for research, development and deployment of data and communications security technologies and products, both hardware and software. Currently, my work is focusing on developing industry standard architectures for the application and interoperability of data security technologies for communications, electronic commerce, and content protection. I previously worked on security matters for two computer companies and as a Lieutenant Commander in Naval Intelligence.

Intel is the world's largest semiconductor manufacturer and a major supplier of information technology building blocks to the global computer and communications industries. We provide our customers with chips, printed circuit board assemblies and software that are the "ingredients" of PC's, servers and workstations. Our flagship business involves the mass production and sale of the Pentium® family of processors and other microprocessors, which are frequently described as the "brains" of a computer because they control the central processing of data in computers. In 1998, our sales exceeded $26 billion, and we employed more than 40,000 people in the United States.

Like most information technology companies, Intel's business model is global in scope. The bulk of our production takes place in the United States. Our products are sold worldwide to original equipment manufacturers of computer systems and peripherals, PC users who make purchases through various distribution channels including the Internet, and other manufacturers who produce a wide range of industrial and telecommunications equipment. Information security plays a prominent role in the conduct of our business.

Intel is headquartered in Santa Clara, California, and we have significant manufacturing facilities in a number of states, including Arizona, New Mexico, Oregon, California and Massachusetts.

Intel Corporation is a member of the Business Software Alliance ("BSA") and Americans for Computer Privacy ("ACP"). Both associations have been in the forefront of efforts to persuade the government to adopt a new encryption policy.

Since 1988, BSA has been the voice of the world's leading software developers before governments and with consumers in the international marketplace. BSA promotes the continued growth of the software industry through its international public policy, education and enforcement program in 65 countries throughout North America, Europe, Asia and Latin America. Its members represent the fastest growing industry in the world. BSA worldwide members include Adobe, Attachmate, Autodesk, Bentley Systems, Corel Corporation, Lotus Development, Macromedia, Microsoft,

Network Associates, Novell, Symantec and Visio. Additional members of BSA's Policy Council include Apple Computer, Compaq, Intuit, Sybase and my company Intel. BSA websites: www.bsa.org: www.nopiracy.com.

Intel Corporation takes, as a given, that access to data communications and stored data by the intelligence and law enforcement communities is both legitimate and extremely important. But, we also recognize that there is an inevitable tide of advancing technology that renders most conventional intercept methodologies obsolete. We also believe that all American businesses need access to strong cryptography to remain competitive in an ever increasing global economy.

We believe that these varied objectives can be met if only government does not seek to force solutions on industry that are incompatible with the development of technology and market demands. It is our view that, given the breathtaking pace at which information technology (including cryptography) is developing around the globe, the only way to achieve these goals is to adopt policies that will ensure American industry leadership in the area of information technology.

This morning I would like to discuss five points that we believe should underpin U.S. encryption policy:

1. Encryption is essential to conducting all business in an Internet economy;

2. Encryption is vital to securing America's critical infrastructures;

3. The availability of encryption cannot be reasonably controlled;

4. Government promoted or required plaintext access will not work; and

5. The government needs to find technological alternatives to meet its requirements for access to information.

1. ENCRYPTION IS ESSENTIAL TO CONDUCTING ALL BUSINESS IN AN INTERNET ECONOMY

While the private sector interest in encryption export reform is generally characterized in terms of the competitiveness of American encryption products in world markets, it is, in reality, a much larger issue for American businesses. In an Internet economy, all American businesses are affected by encryption export constraints.

The future of business is fundamentally changing. The Internet presents two distinctly different business opportunities.

- *Moving existing business to the Internet.* Taking our existing paper-based commerce models and moving them to the electronic world.

- *Creating new businesses because of the Internet.* The Internet provides a ubiquity, connectivity and speed that has never existed before. There are many hereto unimagined businesses that will arise to capitalize on these capabilities.

The global economy, tied together with the Internet, is turning businesses into virtual enterprises, localized products into global products, and geographically limited networks into worldwide networks. Taking place on a massive scale, this phenomenon rests on the following business principles:

- American businesses must be able to sell and support their products worldwide.
- American businesses must be able to securely communicate and coordinate with their foreign subsidiaries and business partners worldwide.
- American businesses must be able to conduct safe electronic commerce worldwide.

I will address each of these three principles in more detail. However, it should be obvious that they all depend on secure communications and financial infrastructures. Cryptography is an essential component of the security of these critical infrastructures, regardless of the nature of the company involved.

It is easy to underestimate the magnitude of the information technology industry in the U.S. and the importance of Internet driven electronic commerce. The Department of Commerce reported that:

> *Without information technology—and the electronic commerce it fosters—overall inflation would have hit 3.1% last year, more than a full percentage point higher than the 2% it was . . .*[1]

By the year 2002, Internet commerce is expected to be $327 billion[2] annually. By the year 2001, the U.S. information technology industry will be directly responsible for 5% of the GNP.[3]

American businesses must be able to sell their products worldwide

Much has been said about the need for American businesses to be able to sell their encryption products worldwide as will be discussed later in this testimony. What is not obvious is that encryption controls may make it difficult to sell non-encryption products on the world market as well. For example, a telecommunications application may need to have an integrated cryptographic component to meet an international standard.

---

1 Wall Street Journal, *Department of Commerce talks about Inflation,* 16 April 1998.

2 Forrester Research.

3 Dataquest.

American businesses must be able to securely communicate and coordinate with their foreign subsidiaries and business partners worldwide

Business practices demand tight coordination with both a companies overseas subsidiaries, their suppliers and their customers. It is essential that confidentiality and access control to business information be maintained. Frequently companies are suppliers or customers on one product and competitors on another. The tightly integrated networks required for coordination could rapidly become a source of competitive intelligence if not adequately protected. Only strong cryptography can offer the level of protection required.

American businesses must be able to conduct safe electronic commerce worldwide

In the near future, there will now longer be dedicated Internet companies—virtually every company will have to be an Internet company to survive. This requires that companies have the capability to securely sell products over the Internet to markets around the world. The ability to prevent fraud and protect intellectual property will depend heavily on the use of strong cryptography.

Importantly, corporate participation in electronic commerce includes both business-to-business and business-to-consumer transactions.

There is a need for commercial security

There has always been some level of need for data security in commercial environments. However, the Internet has enabled the connected PC and, with it, created both new business opportunities and new security vulnerabilities.

Both the value and volume of on-line information has sharply risen. This information includes organizational information such as financial data, manufacturing information, customer information, medical and legal records, and human resources data. Additionally, there is a growing amount of data which has intrinsic value, such as monetary instruments (e.g., credit cards, coupons, etc.) and intellectual property (e.g., movies, images, etc.).

In the past, such data was protected by physical and procedural controls. The connected PC largely negates those conventional controls and requires new security mechanisms, thus creating a need for commercial security technology.

After many years of false starts, commercial data security has become a viable business. The Internet has provided the driving force for this change. Physical barriers have all but disappeared, and security perimeters have become vague.

The Internet has created needs for security that were not present in isolated security domains. This has, in turn, created opportunities for vendors of security technologies and has also created a need for standards so those technologies can interoperate.

Cryptography is the only viable solution to most commercial security requirements

Cryptography has emerged as the only possible solution to many of the requirements of commercial security. It is the essential building block for projecting trust onto the open Internet.

The modern global commercial information infrastructure is characterized by more than 95 million Internet-connected computers,[4] most of which are in open environments with little or no physical control. They use a wide variety of hardware and software and implement no common security policy.

Only cryptographic technologies are capable of projecting security onto a completely open, arbitrary environment. Cryptography, by itself, does not guarantee any level of security. It is a necessary component but not a sufficient component.

Privacy, also known as confidentiality, is the characteristic that information is protected from being viewed in transit during communications and/or when stored in an information system. With cryptographically-provided confidentiality, encrypted information can fall into the hands of someone not authorized to view it without being compromised. It is almost entirely the confidentiality aspect of cryptography that has posed public policy dilemmas.

The commercial use of privacy (or confidentiality) encompasses not only the traditional view described above, but also the protection of intellectual property such as digital video and digital audio. The same technology used to keep communications private are required to ensure that a digital movie is not illegally copied.

2. ENCRYPTION IS VITAL TO SECURING AMERICA'S CRITICAL INFRASTRUCTURES

Governments also are recognizing that without encryption, the electronic networks that control such critical functions as airline flights, health care functions, electrical power and financial markets remain highly vulnerable. The U.S. General Accounting Office in its report issued in May of 1996 entitled *"Information Security: Computer Attacks at Department of defense Pose*

---

4 *Ibid.*, p. 8.

Increasing Risks" found that computer attacks are an increasing threat, particularly through connections on the Internet, such attacks are costly and damaging, and such attacks on Defense and other U.S. computer systems pose a serious threat to national security.

There is an awareness within the government of the vulnerability of the national information infrastructure to potential attack. The *Marsh Report*[5] highlighted the vulnerabilities very well. Much of the national economy is at risk from the decisions that are made today on the issues of infrastructure protection. Any action that degrades the security of Internet commerce or the vialibility of the industries involved must be viewed as a serious risk to the national security.

As the President said on January 22, 1999, before the National Academy of Sciences, "[w]e must be ready—ready if our adversaries try to use computers to disable power grids, banking, communications and transportation networks, police, fire and health services—or military assets. More and more, these critical system are driven by, and linked together with, computers, making them more vulnerability to disruption."

The President has been so concerned that he established a Commission on Critical Infrastructure Protection to provide him with guidance and issued two Presidential Directives based on the Commission's recommendations.

In the Report of the President's Commission on Critical Infrastructure Protection entitled *Critical Foundations: Protecting America's Infrastructures* (October 1997), the Commission emphasized that "Strong encryption is an essential element for the security of the information on which critical infrastructures depend." In fact "[p]rotection of the information our critical infrastructures are increasingly dependent upon is in the national interest and essential to their evolution and full use. A secure infrastructure requires the following:

- Secure and reliable telecommunications networks.
- Effective means for protecting the information systems attached to those networks.
- Effective means of protecting data against unauthorized use or disclosure.
- Well-trained users who understand how to protect their systems and data."

An earlier blue ribbon National Research Council (NRC) Committee similarly concluded in its (May 1996) CRISIS Report ("Cryptography's Role in

---

5 Marsh, R., Chairman, *Critical Foundations: Protecting America's Infrastructure,* The President's Commission on Critical Infrastructure Protection, October 1997.

Securing the Information Society") that encryption promotes the national security of the United States by protecting "nationally critical information systems and networks against unauthorized penetration."

Thus, the NRC Committee found that on balance the advantages of widespread encryption use outweighed the disadvantages and that the U.S. Government has "an important stake in assuring that its important and sensitive . . . information . . . is protected from foreign government or other parties whose interests are hostile to those of the United States."

In recognition of the risks and threats to information, on January 15, 1999, the National Institute of Standards and Technology (NIST) established a new draft Federal Information Processing Standard (FIPS 46-3) to require the use of stronger encryption in government systems. NIST stated that it "can no longer support the use of the DES for many applications" and that all new systems must use the significantly stronger Triple DES "to protect sensitive, unclassified data." Under the FIPS, all existing systems are now expected to develop a strategy to transition to Triple DES, with critical systems receiving a priority.

The vulnerability of national infrastructures has not been lost on other governments. Within the European Union, there is discussion on how to encourage companies to develop products to protect national infrastructures in their respective countries. Such mutual government encouragement will help to grow technical capabilities and fuel a viable world market.

Already the Swiss government is providing 128-bit encryption plug-ins for download off the Internet. The SecureNet system is required for use in accessing Telegiro, an Internet payment system. The plug-ins support SSL connections using IDEA encryption. Several Swiss banks are now using on-line banking systems compatible with the Telegiro cryptosystem.[6]

Information security is critical to the integrity, stability and health of individuals, corporations and governments. While cryptography is but one element of security, it is the keystone of secure, distributed systems. Frankly, there is no substitute for good, widespread, strong cryptography when attempting to prevent crime and sabotage through these networks. The security of any network, however, is only as good as its weakest link. America's infrastructures cannot be protected if they are networked with foreign infrastructures using weak encryption.

In the long-term, we believe it is in America's best interest to protect critical infrastructures and national security by relying on strong American encryption products. This will not happen if the U.S. Government limits the ability of U.S. companies to provide strong encryption to consumers. In-

---

6 *See* http://www.swisspost.ch/E/21.html

deed, the question is not whether critical infrastructures will be protected. Rather it is a question of who will protect them—U.S. or foreign companies. With individuals increasingly relying on critical infrastructures and governments increasingly desiring to safeguard these infrastructures, it is only a matter of time before strong encryption becomes a commodity feature of global networks and information systems.

U.S. encryption export controls hurt our national security

Our current export policy puts at risk America's global leadership in information security. U.S. export policy should, therefore, be changed so it no longer limits American participation in efforts to secure global e-commerce and related information infrastructures and no longer cedes the world market for encryption products to foreign competitors. Strong, high-quality encryption products already are widely available from foreign makers. Foreign producers of IT systems are finding that their ability to provide end-to-end systems incorporating stronger encryption than U.S. companies are permitted to export gives them a decided market advantage. We are concerned that as a result America will lose the critical encryption market to foreign companies. If that happens, it will be too late to change U.S. policy and too late to preserve U.S. leadership in this vital arena.

What will the loss of that U.S. leadership position mean? It will mean that the national security agencies will be confronting ubiquitous encryption made not by U.S. companies, but by foreign companies. Where then will the national security agencies go for technical help on encryption? It also will mean that the protection of our critical national infrastructure may depend on foreign-made systems incorporating foreign-made encryption-and that's unacceptable.

America must retain leadership in this vital technology if we are to meet our long-term national security objectives. That is why we must assess our encryption export policies from a long-term, not a short-term, perspective.

In the long run, U.S. national security objectives are best served by an IT world in which U.S. companies are market leaders in all aspects, especially encryption. U.S. export controls have had the effect of creating an encryption expertise outside the United States that is gathering momentum. Unfortunately, every time research and development of an encryption technique or product moves off-shore, U.S. law enforcement and national security agencies lose. We believe that continuing down this path will be ultimately more harmful to our national security and law enforcement efforts as American companies will no longer be the world leaders in creating and developing encryption products.

In fact, as long ago as 1996, the NRC Committee concluded that as demand for products with encryption capabilities grows worldwide, foreign competition could emerge at levels significant enough to damage the present

U.S. world leadership in information technology products. The Committee felt it was important to ensure the continued economic growth and leadership of key U.S. industries and businesses in an increasingly global economy, including American computer, software and communications companies. Correspondingly, the Committee called for immediate and easy exportability of products meeting general commercial requirements—which is currently 128-bit level encryption!

We recognize this is a difficult balance to strike, but we strongly believe that our long term national security objectives can only be achieved if the United States realistically acknowledges the inevitability of a world of ubiquitous, strong encryption. Trying to control the proliferation of encryption is like trying to control the proliferation of mathematics. For that is what we are talking about here. Encryption algorithms are nothing but sophisticated mathematics. And while the United States may realistically hope to remain the leader in such a field, it cannot realistically expect to monopolize it.

We are joined in this by the Center for Strategic and International Studies ("CSIS"). CSIS recently conducted a study of our nation's technical vulnerabilities; the study was chaired by William Webster, the former director of the FBI and Central Intelligence and former U.S. Circuit Judge. The subsequent report, entitled *Cybercrime. . . Cyberterrorism. . . Cyberwarfare. . . Averting an Electronic Waterloo,* calls for the "intelligence gathering communities—law enforcement and foreign intelligence—to examine the implications of the emerging environment and alter their traditional sources and means to address the SIW (strategic information warfare) needs of the twenty-first century. Continued reliance on limited availability of strong encryption without the development of alternative sources and means will seriously harm law enforcement and national security."

3. THE AVAILABILITY OF ENCRYPTION CANNOT BE REASONABLY CONTROLLED

Cryptography is a specialized branch of mathematics. Cryptographic technology can be reduced to mathematical formulas and protocols. Information about cryptography is available from many sources and in many forms. Implementation of cryptography is no more difficult than the implementation of any complicated mathematical technology such as digital video or digital signal processing.

Ease of implementation

Creation of good cryptographic algorithms that will withstand the test of time is amazingly difficult. Recent history is littered with failed attempts. Even so, many algorithms have survived and have become part of common usage. Inventing good cryptography is the mathematical equivalent of "rocket science." Implementing those algorithms is comparably "child's play."

Information security is such an important part of information technology that it is rare for a graduate level computer science student to graduate without having implemented a cryptographic algorithm or protocol. Many of these students become competent systems-level programmers who could easily fashion a production-quality cryptographic application. Many of these students are non-U.S. residents.

Open research

Cryptography and cryptanalysis are legitimate academic research topics. There is a growing, worldwide academic community specializing in the subject. Last year alone there were over 30 international conferences focusing on cryptography or related topics and over 100 books and journals. Many of these books include detailed specifications and source code of cryptography algorithms and protocols.[7] As an example, Bruce Schneier's popular cryptography text, *Applied Cryptography*, has sold over 100,000 copies world wide.[8]

Intangible software

The intangible nature of cryptographic software defies any physical controls. In an instant, software, cryptographic or otherwise, can be shipped virtually anywhere in the world. As an example, within hours of the U.S. release of PGP 5.0, it was available from sites in Western Europe.[9]

Cryptography exists in many uncontrollable forms, such as general knowledge, academic research, and network deliverable software.

Availability of strong encryption products abroad

Having export controls assumes that they are at least marginally effective. Cryptography is basically mathematics. The knowledge is inherently uncontrollable. This has led to the worldwide availability of strong encryption products and technologies.

One of the ironies of the U.S. cryptographic export regime is that it has fostered a growth in non-U.S. cryptographic technology providers who can sell strong cryptography worldwide without the constraints imposed by the U.S. government, while U.S. companies can not make the same claim.

---

7 Schneier, B., *Applied Cryptography*, John Wiley & Sons, Inc., New York, NY, 1996.

8 Schneier, B., Private correspondence, June 1998.

9 Hayward, D., *Europeans Break Encryption Barriers, TechWire*, 17 June 1997.

The belief that U.S. export regulations enable foreign cryptography businesses is held by the European Commission. The EC stated at the Copenhagen Hearing:

> *The current U.S. export regulations can provide a chance for European companies to enter the market for cryptographic products. Nevertheless this would require a concentrated effort of European industry and governments to prepare the basis for this market.*[10]

Some European companies and governments have turned this belief into practice. The following is quoted from a Siemens Nixdorf ad regarding a software product of theirs called *TrustedWeb:*

> *By simply downloading the TrustedWeb software from the Internet, you can create a highly secure Intranet infrastructure in a matter of days. The organization itself can decide on the level of security and adapt it in stages in line with needs—Ranging from simple password protection to authentication using cryptographic procedures (Public Key-Private Key) with full 128-bit key length. TrustedWeb is an independent European product and hence is not subject to the export restriction imposed by the US government in relation to encryption software.*[11]

Siemens Nixdorf runs similar ads covering their hardware products. Security products are available worldwide, in spite of, or perhaps because of, strong U.S. export controls.

Wide deployment of strong encryption is inevitable.

There are huge commercial incentives for the spread of cryptography. There is a legitimate need for the technology and a sharp increase in the amount of money being spent on security technology.[12] This has created a viable market for the technology, and there are many suppliers worldwide willing and able to meet the market demand.

The recognition of the importance of security to data communications has lead to the inclusion of security protocols within international standards. Examples of such standards include the Secure Sockets Layer (SSL) and the Internet Packet Security (IPSEC) protocols.

In most cases, the implementation of security components in international standards is optional. However, there is a strong trend to make many of

---

10 Ministry of Research and Information Technology Denmark for the European Commission Directorate-General XIII Telecommunications, Information Market and Exploitation of Research, *Report of Day 1 of the European Expert Hearing on Digital Signatures and Encryption (Copenhagen, April 23, 1998)*, Copenhagen, Denmark, 23-24 April 1998.

11 Siemens Nixdorf, Press Release, http://www.trustedwed.com/whats_new/ pressrelease.html, Hanover, Germany.

12 Burnahm, B., *The Electronic Commerce Report*, Piper Jaffray Research, p. 75, August 1997.

these features mandatory. Thus, compliance with international communications standards will promote the diffusion of security technologies.

4. GOVERNMENT PROMOTED OR REQUIRED PLAINTEXT ACCESS WILL NOT WORK

As the spread of strong cryptography threatens traditional intelligence methods, the government has used export control relief as an incentive for companies to build plaintext access capability into every product. There have also been attempts in Congress to mandate plaintext access capability in such products. The overall approach has revolved largely, though not exclusively, around key recovery requirements. This section primarily addresses specific concerns about key recovery issues, but it is applicable to all plaintext access solutions that may be promoted or mandated by the U.S. Government (hereinafter referred to as "required plaintext access"). The basic point is that non-market driven requirements to build *any* plaintext access mechanism into products will not work.

Key recovery, as a concept, now applies not only to the initial purpose of assuring law enforcement access to encrypted materials, but also to possible end-user or organizational requirements for a mechanism to protect against lost, corrupted, or unavailable keys. It can also mean that some process, such as authority to decrypt a header containing a session key, is escrowed with a trusted party, or it can mean that a corporation or individual is ready to cooperate with law enforcement to access encrypted materials. It may also mean that some technical mechanism must be put in place to bypass the use of the key entirely (strict "plaintext access").

While required plaintext access offers, at first glance, the promises of solving the technical problems of plaintext access, it is not technically possible for it to do so in most circumstances. It is unlikely to actually meet plaintext access requirements, and its deployment as a national strategy is fraught with technical challenges and dangers.

Required plaintext access systems will not satisfy government access requirements

Required plaintext access does not meet either law enforcement or national security requirements, but for slightly different reasons. Law enforcement can not verify compliance with key recovery requirements, and national security interests are unlikely to have access to stored keys.

Compliance can not be verified by law enforcement

Required plaintext access has a serious technical flaw in the area of *a priori* verification of compliance. Encryption, if applied, is likely to be applied at several different levels of the communications infrastructure. An example is having link-level encryption applied by IPSEC, having session-level encryption applied by SSL, and having application-level encryption applied by S/MIME.

Assuming one could construct a protocol to allow for the monitoring of IPSEC key recovery compliance, there is no physical way to verify that the other two levels have complied with the required plaintext access requirements unless one actually decrypts the IPSEC-data packet. If it requires probable cause to get a court order to obtain the IPSEC recovered key or mechanism, it would only be after law enforcement has probable cause of criminal activity that they would be able to verify whether or not the upper-level protocols have complied with the required plaintext access requirements.

Required plaintext access does not address national security requirements

While law enforcement may serve a warrant on a key recovery agent or other access mechanism provider to obtain encryption keys or the plaintext, national security interests are unlikely to have that opportunity. Required plaintext access does not provide any benefit to lawful access unless one is able to actually recover the plaintext. Targets of national security interests are unlikely to design a plaintext access infrastructure which would allow the U.S. government to have surreptitious access to stored keys or stored plaintext. This view has been born out by National Security Agency testimony before Congress.[13]

Required plaintext access systems are of limited commercial value

Product announcements of key recovery companies to the contrary, there is not a compelling market for commercial key recovery systems and no market for other plaintext access systems. There is no general reason to recover communications keys, and the use of key recovery for stored data ignores the fundamental properties of information.

A market for key recovery technology will emerge only when it is artificially created by government regulations. Prior to the current law enforcement push for key recovery, there were no widespread deployments of key recovery mechanisms even though the basic technology had been in existence for some time.

Not required for data communications

While key recovery may, debatably, be important in certain stored data systems, in communications cryptography there is little or no user demand for this feature. In particular, there is hardly ever a reason for an encryption user to want to recover the key used to protect a communication session

13 Crowell, W., Deputy Director National Security Agency, Testimony before Senate Commerce Committee, 1997.

such as a telephone call, FAX transmission, or Internet link. If such a key is lost, corrupted, or otherwise becomes unavailable, the problem can be detected immediately and a new key negotiated.[14] There is also no reason to trust another party with such a key.

Ignores the nature of stored data

Many of the proposed needs for key recovery of stored data operate under a false assumption about how data is actually stored and utilized. The frequent example is the assertion that a company will need to recover the encrypted files of an employee who has been hit by a bus.

There are three problems with this assertion. First, with the exception of personal notes, information is not solely possessed by an individual. Information is shared among a team of employees or partners in order to be of any benefit. Second, most mission-critical data is held by corporate data management systems (e.g., data bases) that have their own access control and protection mechanisms, which are administered by the corporation. Third, most personal data has a time value and rapidly becomes obsolete.

Given the observations above, we conclude that there is no business or consumer need for key recovery. Indeed, taking into account the observations and risks, Intel does not plan to implement a key recovery scheme.

Key recovery introduces additional vulnerabilities

Centralizing all of a user's secrets or access controls in a system with increased technological and procedural operational complexities can only increase the security vulnerabilities of the operation.

Centralized attack point

Regardless of the implementation, if key recovery systems must provide timely law enforcement access to a whole key or to plaintext, they present a new and fast path to the recovery of data that never existed before.

The key recovery access path is completely out of the control of the user. In fact, this path to lawful access is specifically designed to be concealed from the encryption user, removing one of the fundamental safeguards against the mistaken or fraudulent release of keys.

In contrast, non-recoverable systems can usually be designed securely without any alternative paths. Alternative paths to access are neither required for ordinary operation nor desirable in many applications for many users.[15]

---

14 Neumann, P., et.al., *The Risks of Key Recovery, Key Escrow, and Trusted Third Party Encryption*, Final Report of The Cryptographers' Working Group, 27 May 1997.

15 *Ibid.*

Complexity of implementation

Key recovery systems must be, in terms of functionally, a secure, distributed, open key management system. They have many of the properties of both large scale distributed databases and of command and control systems. Both types of systems have significant inherent complexity. As we have no practical experience, key recovery mechanisms represent a system of unknown and potentially daunting complexity.[16]

Commercial organizations would have to add the cost and risk of key recovery systems to their bottom line. Even government agencies participating in key recovery pilot programs have found the cost of centralized key recovery unacceptable.[17]

Key recovery mechanisms do not work in the horizontal information industry

The information technology industry is characterized by an open, international, horizontal architecture. Microprocessors are sold to OEMs who build motherboards, who then contract to have BIOSs and operating systems installed. The final product is then sold to an end user who adds whatever applications they wish,. New capabilities or requirements must have an active acceptance within each of the layers in order to be widely deployed. Key recovery discussion has focused only on the upper, application layer.

Low-level layers have no visibility into higher-level layers.

The nature of the information technology industry is that it is made-up of distinct horizontal architectural layers, from the microprocessor up through application programs. The components in each of these layers are supplied by different companies, having different economic models and different diffusion channels.

For valid security reasons, cryptography is migrating further "down" the layers toward the basic hardware. Key recovery, on the other hand, is a user-initiated protocol problem and can not be pushed down to the hardware. In short, cryptography implemented on hardware can not determine how it will ultimately be used.

Key recovery is under the end user's control and is performed by communications protocols or applications programs. The original microprocessor could have no knowledge of how its cryptography would be used any more than it could know how its multiplication instructions will be used.

---

16 *Ibid.*

17 Wayner, P., *administration Gest Sour Taste From Own Encryption Medicine,*New York Times, 1 July 1997.

Key recovery regulation is envisioned from the perspective of the end user. The end user "sees" a vertical single product, but the reality is that the PC is actually a collection of products from many different companies.

Horizontal interfaces are International standards

Within the horizontal architecture of the computer industry, the interfaces between horizontal layers are defined by established international industry standards. None of these interface standards currently support key recovery of keys stored in mass market hardware. To change these standards would be a slow and difficult process.

Key recovery does not work in an international setting.

The information technology industry is based on international standards. No U.S.-only solution is commercially feasible. Most U.S. information technology companies derive a large share of their revenue from non-U.S. sources. To restrict their products to only U.S. markets would be devastating.

Not all countries will adopt key recovery

Very few countries have embraced key recovery to the extent that the U.S. government has done. In particular, countries with strong privacy laws have generally regarded key recovery schemes as being in violation of those laws. As an example, Lotus Notes, which includes a key recovery feature, specifically lost a major sale to the Government of Sweden when the Swedish press discovered the key recovery feature.[18]

The European Commission has not endorsed key recovery as a solution to lawful access problems. It is therefore unlikely that a European-wide agreement can be reached. Indeed, the European Committee on Banking Standards (ECBS)—a powerful consortium of financial institutions—has filed a submission with the European Commission arguing against key recovery.[19]

Requires modification to existing standards

Data communications and architectural standards are internationally-negotiated standards. None of these standards include data recovery provisions. Products must be built to conform to these standards to become mass market products. Many of these standards are not controlled by any government, rather they are controlled by commercial or user communities (such as the IETF).

---

18 Laurin, F., and Froste, C., *Secret Swedish E-Mail Can Be Read by the U.S.A.*, Svenska Dagbladet, 18 Nov 1997.

19 Computing, *Banks Slam Snoops*, 26 March 1998.

Negotiating provisions for key recovery into these standards will require international agreement on the form and procedures of key recovery technology. Given the current international climate, it is unlikely that such negotiations would succeed.

Interoperability will require a non-recovery mode

If there is even one major country which prohibits key recovery, then all developed systems will have to have a "non-key recovery" mode to facilitate interoperability. There is little that one could do to ensure that the "non-key recovery" mode was not used in normal communications.

Mutual access to keys opens U.S. companies to industrial espionage

There is no way to guarantee that other countries will have the same level of constitutional safeguards on access to their key recovery agents as guaranteed in the U.S. U.S. corporations would be at high risk of international economic espionage if forced to deposit encryption keys with foreign key recovery agents.

According to the FBI, U.S. corporations are already targets of major industrial espionage efforts. The FBI says foreign spies have stepped up their attacks on American companies, and a new national survey estimates that intellectual property losses from foreign and domestic espionage may have exceeded $300 billion in 1997 alone.[20]

Governments of at least 23 countries, ranging from Germany to China, are targeting American companies, according to the FBI. More than 1,100 documented incidents of economic espionage and an additional 550 suspected incidents that could not be fully documented were reported last year by companies in a survey conducted by the American Society for Industrial Security.[21]

5. THE GOVERNMENT NEEDS TO FIND TECHNOLOGICAL ALTERNATIVES TO MEET ITS REQUIREMENTS FOR ACCESS TO INFORMATION

Given the global availability of strong, non-recoverable encryption and the fast pace of technological advancement, it is clear that current US policy is not working. An alternative means to gather lawful intelligence is needed by both national security and law enforcement interests.

---

20 Nelson, J.,*FBI: Commercial Spying Rises,* Los Angeles Times, 12 January 1998.

21 *Ibid.*

Clearly, Congress should adequately fund the technical efforts of our law enforcement and national security agencies so they can meet these challenges. And industry would support additional funding.

For example, ACP, for example, has advocated that the U.S. Government should work cooperatively with our nation's hardware and software manufacturers to develop the technical tools and know-how to achieve a policy that effectively responds to society's needs for law enforcement, national security, critical infrastructure protection, privacy preservation, and economic well-being.

NET center proposal

Last year, ACP proposed the creation of a National Center for Secure Network Communications ("NET Center"). The NET Center (now called "Tech Center") concept is aimed at helping law enforcement officials to understand how to deal with encryption and other technical advances when encountered in a criminal setting.

The Tech Center should be a public-private entity operating within a national laboratory for information technology to perform research and act as a forum for further discussions on technology trends and vulnerabilities. Clearly a Tech Center must operate within a legal framework that provides reasonable safeguards.

Attorney General Janet Reno announced plans for the Federal Bureau of Investigation to set up a new $64 million center to protect the nation's critical infrastructures, particularly computer networks, from both physical and cyber attack.

Industry cooperation

The national security is best secured by the American companies actively competing for and supplying the fundamental technologies of the national infrastructure. Only those companies directly involved in the research and development of information technology components can assess the security and vulnerabilities of the infrastructures created from those components. Technical innovation is predominantly centered in the private sector. Only a government/industry cooperation can effectively address the challenge of continued technological change.

CONCLUSION: THE PROTECT ACT SHOULD BE PASSED WITH FURTHER IMPROVEMENTS

The mass market model

Mass-market hardware manufacturers and software publishers sell products through multiple distribution channels such as OEMs (*i.e.*, hardware manufacturers that pre-load software onto computers), value-added resellers, retail stores and the emerging channel of on-line distribution. Thus, mass market products are available to the general public from a variety of sources.

The mass-market distribution model presupposes that hardware manufacturers and software publishers will take full advantage of these multiple channels to ship identical or substantially similar products worldwide (allowing only for differences resulting from localization) irrespective of specific customer location or characteristics. As mass market products are uncontrollable, Intel believes U.S. companies should be able to export the current market standard of 128-bit encryption. Unfortunately, the Administration only permits easy exports of 56-bit encryption even if foreign products exist in the marketplace. And the Administration continues to impose onerous controls on 56-bit toolkits and hardware encryption components, notably semiconductors.

The PROTECT Act grants export control relief to products at all horizontal levels

Intel believes that all distinct horizontal architectural layers, from the microprocessor up through application programs should be treated identically under any encryption export policy. However, contrary to the Administration's original announcement regarding export relief which included export relief for hardware, the new regulations still do not permit 56-bit encryption chips, integrated circuits, toolkits and executable or linkable modules to be easily exported except to subsidiaries of U.S. companies or otherwise relax export controls on stronger mass market hardware. We are pleased that the PROTECT Act remedies this problem and treats mass market hardware in the same manner as mass market software.

The PROTECT Act eliminates reporting requirements for mass market products

We are encouraged that the PROTECT Act recognizes the difficulties in complying with reporting requirements for mass market encryption products and eliminates such reporting requirements. It is virtually impossible for mass-market exporters to report the name and address of each end-user. Millions of these products are sold through multi-level distribution channels (*e.g.*, VAR's and chain stores). Moreover, as registration of mass market products is customarily voluntary. This is a vast improvement over the Administration's proposed regulations which effectively require companies to develop a system to obtain the names and addresses for each health and medical end-user of stronger encryption products and all foreign online merchants.

The PROTECT Act's export relief for mass market products and for products which face competition from comparable foreign products is too complicated and creates an unwieldy bureaucracy

We are pleased that the PROTECT Act does recognize that mass market and publicly available encryption products, and encryption products for which comparable foreign products are available, should be treated differently under the U.S. export regime. The bill acknowledges the futility of trying to control a product that can be bought off of the Internet or easily purchased from commercial vendors such as CompUSA or from Circuit City by any individual in America regardless of nationality, or a comparable product can be easily purchased from similar stores in a foreign country. "Bad guys" certainly will have no problems obtaining the encryption products, and no concerns about "exporting" the products via telephone lines or the Internet or smuggled out on personally pressed CDs. The only impact of the export controls will be to stop American companies from selling American products to legitimate users.

Unfortunately, the PROTECT Act establishes a complicated private/public board structure for deciding after-the-fact whether or not a product is a mass market product or whether comparable foreign products are available. The Secretary of Commerce has thirty days to approve or disapprove the Board determination, subject to judicial review, and the President may override any determination. There is no guarantee of any consistency in the Board's decisions. Thus, while the Board procedure is an improvement, and the opportunity for judicial review provides a mechanism to ensure that exports are not denied in an arbitrary and capricious manner, it is not a predictable, clear process giving American companies certainty as to whether they can export their products. Such predictability is necessary so that American companies can have confidence designing and building security features into their products.

The PROTECT Act should, but does not, afford complete and immediate export relief for mass market encryption without any complicated oversight. The Act also does not recognize that if a comparable foreign product is available, *any* delay in exports provides a significant advantage to the foreign product.

The PROTECT Act supports development of AES, but delays full export control relief until 2002

The PROTECT Act also provides Congressional support for, and sets a 5-year limit on the selection of, the 128-bit Advanced Encryption Standard which is being developed under the auspices of the National Institute of Standards and Technology. The 2002 deadline will provide impetus for NIST to finish developing the standard in a timely manner while providing NIST with sufficient time to study the final standard's security features.

This is an important process that will result in a new standard for government's sensitive, but unclassified, information and most likely will serve as the new worldwide standard for strong encryption similar to the Data Encryption Standard when it was introduced in the 1970's. Once the algorithm is selected, the PROTECT Act removes all export controls on encryption products using the 128-bit standard or its equivalent strength.

Unfortunately, because the PROTECT Act limits easy exportability of mass market products until the AES is adopted, general distribution of these products will have to wait almost three years. Considering the current speed of technological change, where Internet products are now on three-month product cycle times, and the fact that 128-bit comparable foreign encryption is currently available, this is an eternity in Internet time. Law enforcement and national security interests have known for a long time that ubiquitous use of strong encryption by consumers worldwide is just around the corner. They cannot hope to continue to delay the world from using strong encryption according to their timeframe.

A new approach

The preceding has made the argument that:

- Encryption is essential to conducting all business in an Internet economy;
- Encryption is vital to securing America's critical infrastructures;
- The availability of encryption cannot be reasonably controlled;
- Government promoted or required plaintext access will not work; and
- The government needs to find technological alternatives to meet its requirements for access to information.

If accepted, these arguments force one to the conclusion that a new approach to encryption policy is required.

CHAPTER 36

STATEMENT OF LANCE J. HOFFMAN PROFESSOR, THE GEORGE WASHINGTON UNIVERSITY BEFORE THE U.S. SENATE COMMITTEE ON COMMERCE, SCIENCE, AND TRANSPORTATION

June 10, 1999

My name is Lance J. Hoffman. I am a professor in the Department of Electrical Engineering and Computer Science at The George Washington University in Washington, D.C. I also am Director of the School of Engineering's Cyberspace Policy Institute and the author or editor of five books and numerous articles on computer security and privacy. My most recent book is a compendium of papers on the encryption policy problem entitled *Building in Big Brother* (Springer-Verlag, New York, 1995).

Currently, I am the principal investigator for a project entitled "Cryptography Products and Market Survey". As part of that project, we have recently produced a report entitled "Growing Development of Foreign Encryption Products in the Face of U.S. Export Regulations". I am leaving you copies of that report, which is also available from the Institute or on our Web site at http://www.seas.gwu.edu/seas/institutes/cpi/library/papers.html, where detailed tables and charts supporting this testimony are also available. We did this work in cooperation with NAI Labs, the Security Research Division of Network Associates, Inc., Glenwood, Md. The project manager for NAI Labs, Mr. David Balenson, is with me today. We were assisted in this project by three students.

In the project, we surveyed encryption products developed outside the United States and found that the **development of cryptographic products outside the United States is not only continuing but is expanding to additional countries; with rapid growth of the Internet, communications-related cryptography especially is experiencing high growth.**

As of June 8, 1999, **we identified 805 hardware and/or software products incorporating cryptography manufactured in 35 countries outside the United States.** As shown in Attachment 1, the greatest number of foreign cryptographic products are manufactured in the United Kingdom, followed by Germany, Canada, Australia, Switzerland, Sweden, the Netherlands, and Israel in that order. Other countries accounted for slightly more than a quarter of the world's total of encryption products.

These 805 foreign cryptographic products represent a 149-product increase (22%) over the most recent previous survey in December 1997. At

least 167 of them use strong encryption, the kind that one cannot export from the United States without applying for and receiving export license approval. The algorithms used in these are Triple DES, IDEA, BLOWFISH, CAST-128, or RC5.

Cryptography product manufacturers have appeared in six new countries since December 1997: Estonia, Iceland, Isle of Man, Romania, South Korea, and Turkey. **There has also been a large increase in the number of products produced by** certain countries. The **United Kingdom** jumped by 20 products from 119 to 139, and **Germany** jumped from 76 products to 104. Also notable was **Japan's** increase, from six products to 18, and **Mexico's,** from a single product to six.

There are now 512 foreign companies that either manufacture or distribute foreign cryptographic products in 70 countries outside the United States. Attachment 2 lists these countries.

On average, the quality of foreign and U. S. products is comparable. We have encountered poor products both within and outside the U.S., and we have encountered good products both within and outside the U.S. There are a number of very good foreign encryption products that are quite competitive in strength, standards compliance, and functionality.

A significant number of foreign competitors to U.S. manufacturers of software and hardware with encryption capabilities are developing products with strong encryption, and have as customers a number of large foreign or multinational corporations. The report gives thumbnail sketches of some of these companies and their offerings.

We found some examples of advertising used by non-U. S. companies that generally attempted to create the perception that purchasing American products may involve significant red tape and the encryption may not be strong due to export controls. As an example, we show in Attachment 3 material from Cybernetica's Web site in Estonia. We give several other examples of similar advertising in the report.

Companies want to sell encryption products that meet certain accepted worldwide standards. Encryption experts from all over the world have contributed to two important international standards efforts, IPsec and the Advanced Encryption Standard. In the case of IPsec, there are currently implementations (complete or in the works) from at least nine companies in five foreign countries. One effort, the KAME Project, is a joint effort of several Japanese companies (Fujitsu, Hitachi, IIJ Research Laboratory, NEC, Toshiba, and Yokogawa).

In 1997, the National Institute of Standards and Technology (NIST) solicited algorithms for the Advance Encryption Standard (AES) to replace the Data Encryption Standard (DES) as a U.S. government encryption standard. Individuals and companies from eleven different foreign countries proposed 10 out of the 15 candidate algorithms submitted to NIST. So it is very possible that the next U.S. government encryption standard will have been designed outside the United States. Details on who submitted what algorithm are given in Attachment 4.

Finally, our empirical product data could be combined with economic measures and economic theories to better explain why we are seeing the observed growth in the cryptography marketplace, and to examine the effects of Internet growth, e-commerce development, and regulatory actions on the international cryptographic market over time, thus getting better insights into the implications of various policy options. We should be able to combine previous work with studies already available on the information technology sector and the data in our study to better understand the changes we are seeing in the global marketplace, and thus be able to more easily adjust national laws for a global economy.

Attachment 1. Foreign Cryptographic Products by Country

Foreign Cryptographic Survey Results (as of May 1999)

The updated survey identified a total of 805 foreign cryptographic products from 35 countries:

Argentina	Australia	Austria
Belgium	Canada	Czech Republic
Denmark	Estonia	Finland
France	Germany	Greece
Hong Kong	Iceland	India
Iran	Ireland	Isle Of Man
Israel	Italy	Japan
Mexico	Netherlands	New Zealand
Norway	Poland	Romania
Russia	South Africa	South Korea
Spain	Sweden	Switzerland
Turkey	UK	

Country	Number of cryptographic products identified
Israel	31
Netherlands	32
Sweden	39
Switzerland	59
Australia	69
Canada	102
Germany	104
UK	139
Other	230

At least 167 of these foreign cryptographic products implement "strong" cryptographic algorithms, including Triple DES, IDEA, BLOWFISH, RC5, or CAST.

We identified 512 foreign cryptography companies (including distributors as well as manufacturers) in 70 countries.

Attachment 2. Foreign countries in which cryptography is manufactured or distributed

Argentina
Australia
Austria
Bahrain
Baltic Republics
Bangladesh
Belgium
Brazil
Brunei
Canada
Chile
Colombia
Cyprus
Czech Republic
Denmark
Estonia
Finland
France
Germany
Ghana
Greece
Hong Kong
Iceland
India
Indonesia
Iran
Ireland
Isle of Man
Israel
Italy
Ivory Coast
Japan
Kenya
Kuwait
Luxembourg
Madagascar
Malaysia
Malta
Mauritius
Mexico
Nepal
Netherlands
New Zealand
Nigeria
Norway
Oman
Philippines
Poland
Portugal
Qatar
Reunion
Romania
Russia
Saudi Arabia
Singapore
Slovak Republic
South Africa
South Korea
Spain
Sweden
Switzerland
Taiwan
Thailand
Turkey
United Arab Emirates
United Kingdom
Venezuela
West Indies
Yugoslavia
Zimbabwe

Attachment 3. Example of advertising used to create a perception that American products = red tape and weak encryption

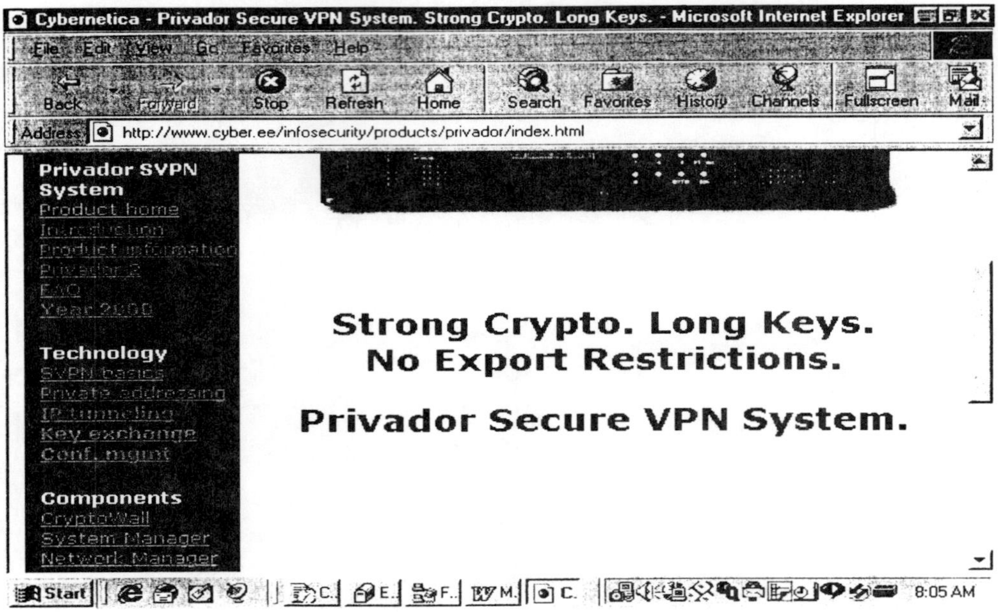

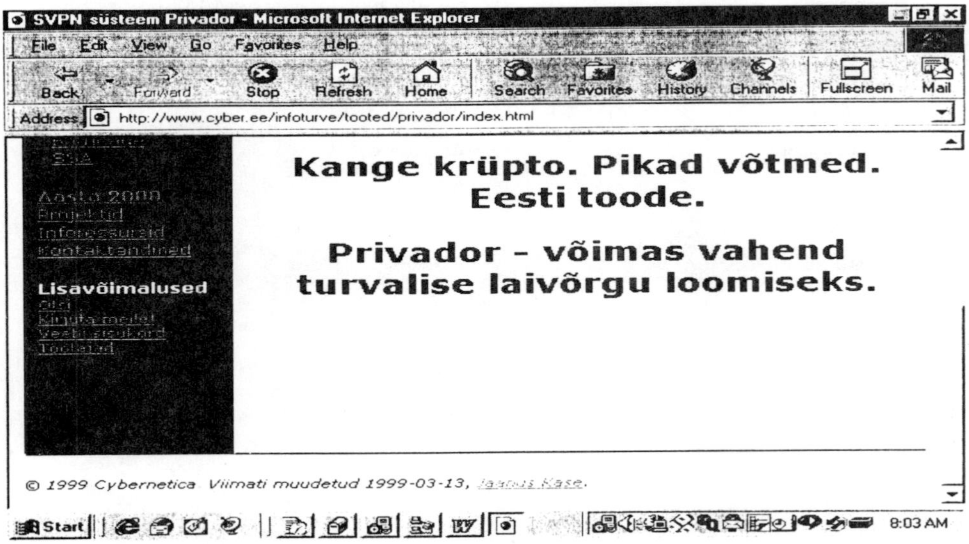

Attachment 4. Proposed Candidates for Advanced Encryption Standard

Country	Candidate Algorithm	Submittor(s)
Australia	LOK197	Lawrie Brown, Josef Pieprzyk, Jennifer Seberry
Belgium	RIJNDAEL	Joan Daemen, Vincent Rijmen
Canada	CAST-256	Entrust Technologies, Inc.
	DEAL	Outerbridge, Knudsen
Costa Rica	FROG	TecApro Internacional S.A.
France	DFC	Centre National pour la Recherche Scientifique (CNRS)
German	MAGENTA	Deutsche Telekom AG
Japan	E2	Nippon Telegraph and Telephone Corporation (NTT)
Korea	CRYPTON	Future Systems
USA	HPC	Rich Schroeppel
	MARS	IBM
	RC6	RSA Laboratories
	SAFER+	Cylink Corporation
	TWOFISH	Bruce Schneier, John Kelsey, Doug Whiting, David Wagner, Chris Hall, Niels Ferguson
UK/Israel/Norway	SERPENT	Ross Anderson, Eli Biham, Lars Knudsen

Smid, M., and M. Dworkin, Special Report on the First AES Conference, presented at Crypto '98 Conference, August 1998, http://csrc.nist.gov/encryption/aes/round1/crypto98.pdf.

CHAPTER 37

STATEMENT OF SENATOR LEAHY ON INTRODUCTION OF THE 'ELECTRONIC RIGHTS FOR THE 21ST CENTURY' OR 'E-RIGHTS' ACT

Mr. Leahy. Mr. President, concern over privacy is reaching an all time high. In 1978, 64 percent of Americans reported that they were 'very concerned' or 'somewhat concerned' about threats to their personal privacy. By 1998, this number had skyrocketed. According to the Center for Social and Legal Research, 88 percent of Americans reported being 'very' or 'somewhat concerned' about threats to their personal privacy. We in Congress must take this concern seriously, and in this regard I look forward to examining the privacy issues confronting us in hearings before the Senate Judiciary Committee.

Good privacy policies make good business policies. New technologies bring with them new opportunities, both for the businesses that develop and market them, and for consumers. It does not do anyone any good for consumers to hesitate to use any particular technology because they have concerns over privacy. That is why I believe that good privacy policies make good business policies.

Protecting privacy plays an important role in the exercise of First Amendment rights. Ensuring that we have adequate privacy laws has a more significant and important role in our democracy than just fostering hi-tech businesses, however. We also must defend our on-line free speech rights from heavy-handed content regulation. That was my purpose in voting against the unconstitutional Communications Decency Act that became law in 1996.

Stopping efforts to create government censors is critical to allow our First Amendment rights to flourish, but it is not enough. For people to feel comfortable in exercising their First Amendment rights—by speaking, traveling and associating freely online or in physical space—they must be able to keep their activities confidential and private. When Big Brother is watching, the exercise of First Amendment rights is chilled no less than the threat of a government censor.

It is therefore not surprising that our country has a long and honorable tradition of keeping our identities private when we exercise our First Amendment rights. The Federalist Papers, which is probably the most important political document ever written about our Constitution, was authored anonymously by James Madison, John Jay and Alexander Hamilton and published under a pseudonym.

Healthy advocacy and debate often rests on the ability of participants to keep their identities private and to act anonymously. Indeed, the Supreme Court has said, "Anonymity is a shield from the tyranny of the majority."

Healthy commerce also depends on satisfying consumers' desire to keep their business affairs private and secure. A report I released last month on Vermont Internet commerce is very telling on this point. The strongest obstacle among consumers from shopping and doing business online was their fear of the online security risks. This is why promoting the use of encryption is so important, so that businesses and consumers can use this technology to provide the privacy and security they want and best suits their needs.

The legislation I introduce today would help ensure that Americans' Fourth Amendment rights to be secure in their persons, houses, papers and effects against unreasonable government searches and seizures are given ample protection in a networked computer environment. In addition, several provisions address the concern Americans have about the use and handling of their personally identifiable records and information by businesses, satellite carriers, libraries and book sellers.

Industry self-regulation efforts should be encouraged. In contrast to a citizen's relationship with his or her government, consumers have a choice of whether they want to deal or interact with those in the private sector. In my view, this choice should be generally recognized in the law by allowing consumers and businesses in the marketplace to set the terms of their interaction. This is an area where the Congress should tread cautiously before regulating. Online businesses are engaging in serious efforts to make available to consumers information on privacy policies so that consumers are able to make more educated choices on whether they want to deal. I commend and applaud those efforts.

That being said, however, current laws do not apply privacy principles in an even-handed manner. Video rental stores and cable operators are subject to privacy laws to protect our right to keep our viewing habits private, but no protections exist for the books we borrow from the library or buy from a bookstore, or the shows we watch via satellite. This bill would provide more uniform privacy protection for both books and videos, no matter the medium of delivery.

Similarly, telephone companies and cable operators are subject to legal restrictions on how they may use personally identifiable information about their Internet subscribers, while other Internet and online service providers are not. The E-RIGHTS bill promotes a more level playing field in terms of the privacy protections available to Internet users, no matter whether they obtain their Internet access from AOL, their cable company or their local phone company.

This legislation addresses a broad range of emerging hi-tech privacy issues. For example:

When should the FBI be allowed to use cell phones to track a user's movements?

Should Kosovo human rights organizations that use a Web site to correct government misinformation be able to get a domain name without having their names publicly available on a database? Should we have the same ability to get an 'unlisted' domain name (or Internet address) as we are able to get an 'unlisted' phone number?

Should we allow other federal prosecutors to act like Special Prosecutor Kenneth Starr and go on fishing expeditions with subpoenas issued to bookstores to find out what we are reading? Should we protect our choices of reading and viewing materials the same way we protect our choice of videotapes that we rent from our local Blockbuster?

Should an Internet user who maintains a calendar on Yahoo! get the same privacy protection as people who keep their calendars on their desk or on their PCs' hard-drive? Will people avoid certain network services offered by Netscape or new Internet start-ups because they get less privacy protection for the information stored on the network than on their own PCs?

These are all important issues, and I have worked to propose solutions to each of these and to other questions, as well, in the E-RIGHTS bill. This bill has the following four titles:

TITLE I: Privacy Protection for Communications and Electronic Information. This title has ten sections that propose certain Fourth Amendment protections to guide the government's access to, or exercise of, law enforcement's enhanced surveillance capabilities due to new technologies. In addition, this title also contains sections that limit how domain name registrars and Internet/Online service providers may use information collected on Internet users.

Network Stored Information.—The bill would require that law enforcement give a subscriber notice of a subpoena or warrant before seizing electronic information stored on a network service. This is the same notice that the subscriber would get if the information were stored on his or her own computer.

Cell Phone Location Information.—Before law enforcement may use a person's cell phone as a tracking device, the bill would require a court order based on probable cause that the person is committing a crime.

A related provision that has already passed the House in February as part of the **"Wireless Communications and Public Safety Act of 1999," H.R. 438, would require wireless phone providers to inform a cell phone user's family and emergency services of their location in emergency situations, while requiring the prior customer consent before that location information may be used for any other purpose.**

Pen Registers.—The bill would authorize a judge to review information presented by a federal prosecutor to determine whether the pen register is likely to produce information relevant to an ongoing criminal investigation, since under current law the judge plays only a ministerial role and must approve any order upon presentation by a prosecutor. Current law compels judges to be only a rubber stamp.

Conference Calls.—The FBI has claimed that the Communications Assistance for Law Enforcement Act (CALEA) requires that they be given the capability to monitor conference calls which continue even after the target of a wiretap order has dropped out of the call. This provision would require that a court authorize such continued monitoring of conference calls in the absence of the target.

Roving Wiretaps.—A substantial change that provides easier access to roving wiretaps was inserted without debate or hearings into last year's Intelligence Authorization Act. With this change, the FBI is able to get a roving wiretap whenever a person's action could have the effect of thwarting interception. The bill would rectify this change to permit roving wiretaps only when the person actually changes phones in a way which has the effect of thwarting surveillance.

Domain Name Registrars.—Internet users or businesses who get an Internet address with a second level domain name must also provide information about contact names, physical and E-mail addresses, network location, and other information that is posted in a publicly available database called WHOIS. The bill would give users registering for a domain name/Internet address authority to prohibit disclosure of the information, and keep the information confidential. Of course, the registrar would be able to override the user's choice of confidentiality and to disclose the information as necessary to provide service or in response to a subpoena or court order.

Internet users who want an 'unlisted' Internet address just as they have the choice of getting an 'unlisted' telephone number will be able to do so.

Internet and Online Service Providers.—The 1986 Electronic Communications Privacy Act (ECPA) set up procedures for law enforcement to obtain records about subscribers from 'electronic communication service providers', but contained a blanket exemption allowing such providers to disclose a record or other information pertaining to a subscriber or customer to any non-governmental entity. Due to this exemption, ISPs and OSPs may sell their subscriber lists or track the online movements of their subscribers and sell that information—all without the subscribers' knowledge or consent.

The bill would cut back on this blanket exemption. The bill would require electronic communication service providers to give their subscribers an opportunity to prohibit disclosure of their personal information, and enumerates the situations in which the information may be used or disclosed without the subscriber's approval. These proposed rules are generally analogous to restrictions already in place for other providers of Internet services, including cable operators and phone companies, which are restricted in how they may use personally identifiable information about customers without the customers' approval.

No criminal penalties attach for violation. ECPA currently authorizes an aggrieved person to bring a civil action.

Title II: Promoting the Use of Encryption. This title contains three sections: (1) prohibiting domestic controls on encryption and government-compelled key escrow encryption; (2) requiring encryption products used by federal agencies to interoperate with commercial encryption products; and (3) adding a chapter to the federal criminal code detailing procedures to law enforcement and foreign government access to decryption assistance.

Specifically, the bill would require the release of decryption keys or assistance to law enforcement in response to a court order based upon a finding that the key or assistance is necessary to decrypt lawfully intercepted encrypted messages or data.

Title III: Privacy Protection for Library Loan and Book Sales Records. This title would extend the privacy protection in current law for video rental and sale records to library loan and book sale records.

Library.—The library provisions are a reprise of sections that were dropped from the Video Privacy Protection Act enacted in 1988. This provision would prohibit libraries from disclosing personally identifiable information about patrons without the written consent of the patron or in response to a court order to release the information to a law enforcement agency, with prior notice to the patron, if there is probable cause to believe a crime is being committed and the information sought is material to the investigation.

Booksellers.—The public outcry over Independent Counsel Kenneth Starr's subpoena in March 1988 to Kramerbooks & Afterwords for any books purchased by Monica Lewinsky, and the potential threat such government fishing expeditions pose to First Amendment rights, prompted examination of the privacy rules protecting the records maintained by bookstores. There are no rules barring book sellers from disclosing records about their customers.

This section would impose the same nondisclosure rules on booksellers—whether online or in physical spaces—that apply to video rental stores. Generally, book sellers would be barred from disclosing personally identifiable information concerning a book purchaser without that purchasers' written consent given at the time the disclosure is sought.

Title IV: Privacy Protection for Satellite Home Viewers.

In the 1984 Cable Act, Congress established a nationwide standard for the privacy protection of cable subscribers. Since the Cable Act was adopted, an entirely new form of access to television has emerged—home satellite viewing—which is especially popular in rural areas not served by cable. Yet there is no statutory privacy protection for information collected by home satellite viewing services about their customers or subscribers This title fills this gap by amending the privacy provisions of the Cable Act to cover home satellite viewing.

The amendments do not change the rules governing access to cable subscriber information. Instead, they merely add the words "satellite home viewing service" and "satellite carrier or distributor" where appropriate.

The amendment does not address another inconsistency in the law, which bears mentioning: should a cable company that provides Internet services to its customers be subject to the privacy safeguards in the Cable Act or in the Electronic Communications Privacy (ECPA), which normally applies to Internet service providers and contains obligations regarding the disclosure of personally identifiable information to both governmental and non-governmental entities different from those in the Cable Act? One court has described this as a "statutory riddle raised by the entrance of cable operators into the Internet services market."

New technologies and new uses for old technologies pose challenging 'riddles' for privacy, but they are solvable in ways that balance competing commerce, civil rights, and law enforcement interests. The E-RIGHTS bill proposes balanced solutions that protect our privacy rights. I invite others to share their ideas on these matters. There are few matters more important than privacy in maintaining our core democratic values, so I look forward to hearing their comments on ways to improve this legislation.

I ask unanimous consent that my full statement, the E-RIGHTS bill and the sectional analysis be printed in the RECORD.

CHAPTER 38

AS AMENDED OCT. 11, 1996

§ 1030. FRAUD AND RELATED ACTIVITY IN CONNECTION WITH COMPUTERS

(a) Whoever

(1) having knowingly accessed a computer without authorization or exceeding authorized access, and by means of such conduct having obtained information that has been determined by the United States Government pursuant to an Executive order or statute to require protection against unauthorized disclosure for reasons of national defense or foreign relations, or any restricted data, as defined in paragraph y. of section 11 of the Atomic Energy Act of 1954, with reason to believe that such information so obtained could be used to the injury of the United States, or to the advantage of any foreign nation willfully communicates, delivers, transmits, or causes to be communicated, delivered, or transmitted, or attempts to communicate, deliver, transmit or cause to be communicated, delivered, or transmitted the same to any person not entitled to receive it, or willfully retains the same and fails to deliver it to the officer or employee of the United States entitled to receive it;

(2) intentionally accesses a computer without authorization or exceeds authorized access, and thereby obtains—

(A) information contained in a financial record of a financial institution, or of a card issuer as defined in section 1602(n) of title 15, or contained in a file of a consumer reporting agency on a consumer, as such terms are defined in the Fair Credit Reporting Act (15 U.S.C. 1681 et seq.);

(B) information from any department or agency of the United States; or

(C) information from any protected computer if the conduct involved an interstate or foreign communication;

(3) intentionally, without authorization to access any nonpublic computer of a department or agency of the United States, accesses such a computer of that department or agency that is exclusively for the use of the Government of the United States or, in the case of a computer not exclusively for such use, is used by or for the Government of the United States and such conduct affects that use by or for the Government of the United States;

(4) knowingly and with intent to defraud, accesses a protected computer without authorization, or exceeds authorized access, and by means of such conduct furthers the intended fraud and obtains anything of value, unless the object of the fraud and the thing obtained consists only of the use of the computer and the value of such use is not more than $ 5,000 in any one-year period;

(5)

(A) knowingly causes the transmission of a program, information, code, or command, and as a result of such conduct, intentionally causes damage without authorization, to a protected computer;

(B) intentionally accesses a protected computer without authorization, and as a result of such conduct, recklessly causes damage; or

(C) intentionally accesses a protected computer without authorization, and as a result of such conduct, causes damage;

(6) knowingly and with intent to defraud traffics (as defined in section 1029) in any password or similar information through which a computer may be accessed without authorization, if

(A) such trafficking affects interstate or foreign commerce; or

(B) such computer is used by or for the Government of the United States;

(7) with intent to extort from any person, firm, association, educational institution, financial institution, government entity, or other legal entity, any money or other thing of value, transmits in interstate or foreign commerce any communication containing any threat to cause damage to a protected computer;

shall be punished as provided in subsection (c) of this section.

(b) Whoever attempts to commit an offense under subsection (a) of this section shall be punished as provided in subsection (c) of this section.

(c) The punishment for an offense under subsection (a) or (b) of this section is—

(1)

(A) a fine under this title or imprisonment for not more than ten years, or both, in the case of an offense under subsection (a)(1) of this section which does not occur after a conviction for another offense under this section, or an attempt to commit an offense punishable under this subparagraph; and

(B) a fine under this title or imprisonment for not more than twenty years, or both, in the case of an offense under subsection (a)(1) of this section which occurs after a conviction for another offense under this section, or an attempt to commit an offense punishable under this subparagraph; and

(2)

(A) a fine under this title or imprisonment for not more than one year, or both, in the case of an offense under subsection (a)(2), (a)(3), (a)(5)(C) or (a)(6) of this section which does not occur after a conviction for another offense under this section, or an attempt to commit an offense punishable under this subparagraph; and

(B) a fine under this title or imprisonment for not more than 5 years, or both, in the case of an offense under subsection (a)(2) if—

(i) the offense was committed for purposes of commercial advantage or private financial gain;

(ii) the offense was committed in furtherance of any criminal or tortious act in violation of the Constitution or laws of the United States or of any State; or

(iii) the value of the information obtained exceeds $5,000;

(C) a fine under this title or imprisonment for not more than ten years, or both, in the case of an offense under subsection (a)(2), (a)(3) or (a)(6) of this section which occurs after a conviction for another offense under this section, or an attempt to commit an offense punishable under this subparagraph; and

(3)

(A) a fine under this title or imprisonment for not more than five years, or both, in the case of an offense under subsection (a)(4), (a)(5)(A), (a)(5)(B), or (a)(7) of this section which does not occur after a conviction for another offense under this section, or an attempt to commit an offense punishable under this subparagraph; and

(B) a fine under this title or imprisonment for not more than ten years, or both, in the case of an offense under subsection (a)(4), (a)(5)(A), (a)(5)(B), (a)(5)(C), or (a)(7) of this section which occurs after a conviction for another offense under this section, or an attempt to commit an offense punishable under this subparagraph; and

(d) The United States Secret Service shall, in addition to any other agency having such authority, have the authority to investigate offenses under subsections (a)(2)(A), (a)(2)(B), (a)(3), (a)(4), (a)(5), and (a)(6) of this section. Such authority of the United States Secret Service shall be exercised in accordance with an agreement which shall be entered into by the Secretary of the Treasury and the Attorney General.

(e) As used in this section

(1) the term "computer" means an electronic, magnetic, optical, electrochemical, or other high speed data processing device performing logical, arithmetic, or storage functions, and includes any data storage facility or communications facility directly related to or operating in conjunction with such device, but such term does not include an automated typewriter or typesetter, a portable hand held calculator, or other similar device;

(2) the term "protected computer" means a computer

(A) exclusively for the use of a financial institution or the United States Government, or, in the case of a computer not exclusively for such use, used by or for a financial institution or the United States Government and the conduct constituting the offense affects that use by or for the financial institution or the Government; or

(B) which is used in interstate or foreign commerce or communications;

(3) the term "State" includes the District of Columbia, the Commonwealth of Puerto Rico, and any other commonwealth, possession or territory of the United States;

(4) the term "financial institution" means

(A) an institution with deposits insured by the Federal Deposit Insurance Corporation;

(B) the Federal Reserve or a member of the Federal Reserve including any Federal Reserve Bank;

(C) a credit union with accounts insured by the National Credit Union Administration;

(D) a member of the Federal home loan bank system and any home loan bank;

(E) any institution of the Farm Credit System under the Farm Credit Act of 1971;

(F) a brokerdealer registered with the Securities and Exchange Commission pursuant to section 15 of the Securities Exchange Act of 1934;

(G) the Securities Investor Protection Corporation;

(H) a branch or agency of a foreign bank (as such terms are defined in paragraphs (1) and (3) of section 1(b) of the International Banking Act of 1978); and

(I) an organization operating under section 25 or section 25(a) of the Federal Reserve Act.

(5) the term "financial record" means information derived from any record held by a financial institution pertaining to a customer's relationship with the financial institution;

(6) the term "exceeds authorized access" means to access a computer with authorization and to use such access to obtain or alter information in the computer that the accesser is not entitled so to obtain or alter;

(7) the term "department of the United States" means the legislative or judicial branch of the Government or one of the executive departments enumerated in section 101 of title 5; and

(8) the term 'damage' means any impairment to the integrity or availability of data, a program, a system, or information, that

(A) causes loss aggregating at least $5,000 in value during any 1year period to one or more individuals;

(B) modifies or impairs, or potentially modifies or impairs, the medical examination, diagnosis, treatment, or care of one or more individuals;

(C) causes physical injury to any person; or

(D) threatens public health or safety; and

(9) the term 'government entity' includes the Government of the United States, any State or political subdivision of the United States, any foreign country, and any state, province, municipality, or other political subdivision of a foreign country.

(f) This section does not prohibit any lawfully authorized investigative, protective, or intelligence activity of a law enforcement agency of the United States, a State, or a political subdivision of a State, or of an intelligence agency of the United States.

(g) Any person who suffers damage or loss by reason of a violation of the section may maintain a civil action against the violator to obtain compensatory damages and injunctive relief or other equitable relief. Damages for violations involving damage as defined in subsection (e)(8)(A) are limited to economic damages. No action may be brought under this subsection unless such action is begun within 2 years of the date of the act complained of or the date of the discovery of the damage.

(h) The Attorney General and the Secretary of the Treasury shall report to the Congress annually, during the first 3 years following the date of the enactment of this subsection, concerning investigations and prosecutions under section 1030(a)(5) of title 18, United States Code.

CHAPTER 39

OFFICE OF THE CIRCUIT EXECUTIVE
U.S. COURT OF APPEALS FOR THE NINTH CIRCUIT

Case Name: BERNSTEIN V USDOJ
Case Number: 97-16686
Date Filed: 05/06/99

DANIEL J. BERNSTEIN,
Plaintiff-Appellee,
v.
UNITED STATES DEPARTMENT OF JUSTICE;
UNITED STATES DEPARTMENT OF COMMERCE;
DEPARTMENT OF STATE; UNITED STATES DEPARTMENT OF DEFENSE;
UNITED STATES ARMS CONTROL AND DISARMAMENT AGENCY;
NATIONAL SECURITY AGENCY;
UNITED STATES DEPARTMENT OF ENERGY;
CENTRAL INTELLIGENCE AGENCY;
MADELINE E. ALBRIGHT, United States Secretary of State;
WILLIAM M. DALEY, United States Secretary of Commerce;
WILLIAM COHEN, United States Secretary of Defense;
KENNETH A. MINIHAN, Director, United States National Security Agency;
JOHN B. HOLUM, Director, United States Arms Control and Disarmament Agency;
WILLIAM G. ROBINSON; GARY M. ONCALE; AMBASSADOR MICHAEL NEWLIN; CHARLES RAY; MARK KORO; GREG STARK; DOES 1-100,
Defendants-Appellants.

Appeal from the United States District Court
for the Northern District of California

Marilyn Hall Patel, District Judge, Presiding

Argued and Submitted
December 8, 1997—San Francisco, California

Filed May 6, 1999

Before: Myron H. Bright, * Betty B. Fletcher, and Thomas G. Nelson, Circuit Judges.

Opinion by Judge B. Fletcher; Concurrence by Judge Bright; Dissent by Judge T.G. Nelson

COUNSEL

Scott R. McIntosh (argued), Douglas N. Letter, United States Department of Justice, Washington, D.C., for the defendants-appellants.

Cindy A. Cohn (argued), McGlashan & Sarrail, San Mateo, California, and Lee Tien, Berkeley, California, for the plaintiff-appellee.

Ivan K. Fong, Covington & Burling, Washington, D.C., for amicus curiae Electronic Privacy Information Center; American Civil Liberties Union; American Civil Liberties Union of Northern California; Center For Democracy and Technology; Computer Professionals for Social Responsibility; Economic Strategy Institute; Free Congress Research and Education Foundation; Human Rights Watch; Independence Institute; International Information System Security Certification Consortium; Internet Mail Consortium; Internet Society; National Association of Manufacturers; Privacy International; U.S. Public Policy Committee of the Association for Computing; Dr. Whitfield Diffie; Dr. Peter Neumann; and Dr. Ronald Rivest.

Garrett Epps, University of Oregon School of Law, Eugene, Oregon, for amicus curiae Silicon Valley Software Industry Coalition; Professor Keith Aoki; Professor Margreth Barrett; Professor James Boyle; Professor Garrett Epps; Professor Peter Jaszi; Professor David Lange; and Professor Eugene Volokh.

Brian Conboy, Wilkie Farr & Gallagher, Washington, D.C., for amicus curiae Maynard Anderson; D. James Bidzos; National Computer Security Association; Mark Rasch; RSA Data Security, Inc.; Dr. Eugene Spafford; and Dr. Ross Stapleton-Gray.

J. Joshua Wheeler, Charlottesville, Virginia, for amicus curiae Thomas Jefferson Center for the Protection of Free Expression.

Richard D. Marks, Vinson & Elkins, Washington, D.C., for amicus curiae Association for the Advancement of Science.

OPINION

B. FLETCHER, Circuit Judge:

The government defendants appeal the grant of summary judgment to the plaintiff, Professor Daniel J. Bernstein ("Bernstein"), enjoining the enforcement of certain Export Administration Regulations ("EAR") that limit Bernstein's ability to distribute encryption software. We find that the EAR regulations (1) operate as a prepublication licensing scheme that burdens scientific expression, (2) vest boundless discretion in government officials, and (3) lack adequate procedural safeguards. Consequently, we hold that the challenged regulations constitute a prior restraint on speech that offends the First Amendment. Although we employ a somewhat narrower rationale than did the district court, its judgment is accordingly affirmed.

BACKGROUND

A. Facts and Procedural History

Bernstein is currently a professor in the Department of Mathematics, Statistics, and Computer Science at the University of Illinois at Chicago. As a doctoral candidate at the University of California, Berkeley, he developed an encryption method—"a zero-delay private-key stream encryptor based upon a one-way hash function"[1]—that he dubbed "Snuffle." Bernstein described his method in two ways: in a paper containing analysis and mathematical equations (the "Paper") and in two computer programs written in "C," a high-level computer programming language ("Source Code"). Bernstein later wrote a set of instructions in English (the "Instructions") explaining how to program a computer to encrypt and decrypt data utilizing a one-way hash function, essentially translating verbatim his Source Code into prose form.

Seeking to present his work on Snuffle within the academic and scientific communities, Bernstein asked the State Department whether he needed a license to publish Snuffle in any of its various forms. The State Department responded that Snuffle was a munition under the International Traffic in Arms Regulations ("ITAR"), and that Bernstein would need a license to "export" the Paper, the Source Code, or the Instructions.[2] There followed a protracted and unproductive series of letter communications between Bernstein and the government, wherein Bernstein unsuccessfully attempted to determine the scope and application of the export regulations to Snuffle.[3]

---

1 The term "hash function" describes a function that transforms an input into a unique output of fixed (and usually smaller) size that is dependent on the input. For some purposes (*e.g.* error checking, digital signatures), it is desirable that it be impossible to derive the input data given only the hash function's output—this type of function is known as a "one-way hash function." Hash functions have many uses in cryptography and computer science, and numerous one-way hash functions are widely known. "Zero delay" means that Snuffle can be used for interactive communications because it encrypts and decrypts on a character-by-character basis—the users need not complete an entire message before encrypting and sending.

2 In June 1995, after Bernstein initiated this suit, the State Department clarified its earlier determination, explaining that while ITAR did restrict the Source Code and the Instructions, it did not restrict the Paper.

3 Bernstein notes that his difficulties with the State Department are by no means unique. Declarations provided by Bernstein demonstrate ongoing suppression of academic publication by the State Department under ITAR. *See* Demberger Decl. (found in violation of ITAR for posting encryption program on the internet); Junger Decl. (stated that ITAR caused him to censor publication of his work for fear of violating

Bernstein ultimately filed this action, challenging the constitutionality of the ITAR regulations. The district court found that the Source Code was speech protected by the First Amendment, see Bernstein v. Department of State, 922 F. Supp. 1426 (N.D. Cal. 1996) ("Bernstein I"), and subsequently granted summary judgment to Bernstein on his First Amendment claims, holding the challenged ITAR regulations facially invalid as a prior restraint on speech, see Bernstein v. Department of State, 945 F. Supp. 1279 (N. D. Cal. 1996) ("Bernstein II").

In December 1996, President Clinton shifted licensing authority for non-military encryption commodities and technologies from the State Department to the Department of Commerce. *See* Exec. Order No. 13,026, 61 Fed. Reg. 58,767 (1996). The Department of Commerce then promulgated regulations under the EAR to govern the export of encryption technology, regulations administered by the Bureau of Export Administration ("BXA"). *See* 61 Fed. Reg. 68, 572 (1996) (Codified at 15 C.F.R. Pts. 730-74). Bernstein subsequently amended his complaint to add the Department of Commerce as a defendant, advancing the same constitutional objections as he had against the State Department. The district court, following the rationale of its earlier Bernstein opinions, once again granted summary judgment in favor of Bernstein, finding the new EAR regulations facially invalid as a prior restraint on speech. *See* Bernstein v. Department of State, 974 F. Supp. 1288 (N.D. Cal. 1997) ("Bernstein III"). The district court enjoined the Commerce Department from future enforcement of the invalidated provisions, an injunction that has been stayed pending this appeal.

B. Overview of Cryptography

Cryptography is the science of secret writing, a science that has roots stretching back hundreds, and perhaps thousands, of years. *See generally* DAVID KHAN, THE CODEBREAKERS (2d ed. 1996). For much of its history, cryptography has been the jealously guarded province of governments and militaries. In the past twenty years, however, the science has blossomed in the civilian sphere, driven on the one hand by dramatic theoretical innovations within the field, and on the other by the needs of modern communication and information technologies. As a result, cryptography has become a dynamic academic discipline within applied mathematics. It is the cryptographer's primary task to find secure methods to encrypt messages, making them unintelligible to all except the intended recipients:

> Encryption basically involves running a readable message known as "plain-text" through a computer program that translates the message according to

---

the regulations); Zimmerman Decl. (target of a criminal investigation for publishing encryption software on the Internet).

> an equation or algorithm into unreadable "ciphertext." Decryption is the translation back to plaintext when the message is received by someone with an appropriate "key".

Bernstein III, 974 F. Supp. at 1292. The applications of encryption, however, are not limited to ensuring secrecy; encryption can also be employed to ensure data integrity, authenticate users, and facilitate nonrepudiation (*e.g.*, linking a specific message to a specific sender). *See id.*

It is, of course, encryption's secrecy applications that concern the government. The interception and deciphering of foreign communications has long played an important part in our nation's national security efforts. In the words of a high-ranking State Department official:

> Policies concerning the export control of cryptographic products are based on the fact that the proliferation of such products will make it easier for foreign intelligence targets to deny the United States Government access to information vital to national security interests. Cryptographic products and software have military and intelligence applications. As demonstrated throughout history, encryption has been used to conceal foreign military communications, on the battlefield, aboard ships and submarines, or in other military settings. Encryption is also used to conceal other foreign communications that have foreign policy and national security significance for the United States. For example, encryption can be used to conceal communications of terrorists, drug smugglers, or others intent on taking hostile action against U.S. facilities, personnel, or security interests.

Lowell Decl. at 4 (reproduced in Appellant's Excerpts of Record at 97). As increasingly sophisticated and secure encryption methods are developed, the government's interest in halting or slowing the proliferation of such methods has grown keen. The EAR regulations at issue in this appeal evidence this interest.

C. The EAR regulations[4]

The EAR contain specific regulations to control the export of encryption software, expressly including computer source code. Encryption software is treated differently from other software in a number of significant ways. First, the term "export" is specifically broadened[5] with respect to encryption software to preclude the use of the Internet and other global mediums if such publication would allow passive or active access by a foreign

4 Because the district court capably detailed the ITAR and EAR regulatory regimes, see Bernstein III, 974 F. Supp. at 1292-96, we present only an overview of the relevant provisions here.

5 "Export", even as applied to software generally, is defined quite broadly to include any release, including oral exchanges of information and visual inspections, in a foreign country or to a foreign national within the United States. 15 C.F.R. S 734.2(b)(2) & (3).

national within the United States or anyone outside the United States. 15 C.F.R. S 734.2(b)(9)(B)(ii).[6] Second, the regulations governing the export of nonencryption software provide for several exceptions that are not applicable to encryption software.[7] In addition, although printed materials containing encryption source code are not subject to EAR regulation , the same materials made available on machine-readable media, such as floppy disk or CD-ROM, are covered. 15 C.F.R. S. 734. 3(b), Note to Paragraphs (b)(2) & (b)(3). The government, moreover, has reserved the right to restrict source code in printed form that may be easily "scanned", thus creating some ambiguity as to whether printed publications are necessarily exempt from licensing. *See* 61 Fed. Reg. 68,575 (1996).

If encryption software falls within the ambit of the relevant EAR provisions, the "export" of such software requires a prepublication license. When a prepublication license is requested, the relevant agencies undertake a "case-by-case" analysis to determine if the export is "consistent with U.S. national security and foreign policy interests". 15 C.F.R. S 742.15(b). All applications must be "resolved or referred to the President no later than 90 days" from the date an application is entered into the BXA's electronic license processing system. 15 C.F.R. S 750.4(a). There is no time limit, however, that applies once an application is referred to the President. Although the regulations do provide for an internal administrative appeal procedure, such appeals are governed only by the exhortation that they be completed "within a reasonable time." 15 C.F.R. S 756.2(c)(1). Final administrative decisions are not subject to judicial review. 15 C.F.R. S 756.2(c)(2).

DISCUSSION

I. Prior Restraint

The parties and amici urge a number of theories on us. We limit our attention here, for the most part, to only one: whether the EAR restrictions

---

6 Specifically, 15 C.F.R. S 734.2(b)(9)(B)(ii) provides that "export" includes:
downloading or causing the downloading of, such software to locations (including electronic bulletin boards, Internet file transfer protocol, and World Wide Web sites) outside the U.S., or making such software available for transfer outside the United States, over wire, cable, radio, electromagnetic, photo-optical, photoelectric or other comparable communications facilities accessible to persons outside the United States, including transfers from electronic bulletin boards, Internet file transfer protocol and World Wide Web sites, unless the person making the software available takes precautions adequate to prevent unauthorized transfer of such code outside the United States.

7 These exceptions allow for export of software that is publicly available, 15 C.F.R. S 734.7(c); results from fundamental research or is educational, 15 C.F.R. SS 734.3(b)(3), 734.8, 734.9; is already available from foreign sources, 15 C.F.R. S 768. 1(b); or contains only a de minimis quantity of domestically-derived content, 15 C.F.R. S 734. 4(b)(2).

on the export of encryption software in source code form constitute a prior restraint in violation of the First Amendment. We review de novo the district court's affirmative answer to this question. *See* Roulette v. Seattle, 97 F.3d 300, 302 (9th Cir. 1996).

[1] It is axiomatic that "prior restraints on speech and publication are the most serious and least tolerable infringement on First Amendment rights". Nebraska Press Ass'n v. Stuart, 427 U.S. 539, 559 (1976). Indeed, the Supreme Court has opined that "it is the chief purpose of the [First Amendment] guaranty to prevent previous restraints upon publication". Near v. Minnesota, 283 U.S. 697, 713 (1931). Accordingly, "[a]ny prior restraint on expression comes . . . with a heavy presumption' against its constitutional validity." Organization for a Better Austin v. Keefe, 402 U.S. 415, 419 (1971). At the same time, the Supreme Court has cautioned that" [t]he phrase 'prior restraint' is not a self-wielding sword. Nor can it serve as a talismanic test". Kingsley Books, Inc. v. Brown, 354 U.S. 436, 441 (1957). We accordingly turn from "[t]he generalization that prior restraint is particularly obnoxious" to a "more particularistic analysis." *Id.* at 442.

[2] The Supreme Court has treated licensing schemes that act as prior restraints on speech with suspicion because such restraints run the twin risks of encouraging self-censorship and concealing illegitimate abuses of censorial power. *See* Lakewood v. Plain Dealer Publishing Co., 486 U.S. 750, 759 (1988). As a result, "even if the government may constitutionally impose content-neutral prohibitions on a particular manner of speech, it may not condition that speech on obtaining a license or permit from a government official in that official's boundless discretion." *Id.* at 764 (emphasis in original). We follow the lead of the Supreme Court and divide the appropriate analysis into two parts. The threshold question is whether Bernstein is entitled to bring a facial challenge against the EAR regulations. *See id.* at 755. If he is so entitled, we proceed to the second question: whether the regulations constitute an impermissible prior restraint on speech. *See id.* at 769.

A. Is Bernstein entitled to bring a facial attack?

[3] A licensing regime is always subject to facial challenge[8] as a prior restraint where it "gives a government official or agency substantial power

8 In using the term "facial challenge" in the prior restraint context, the Supreme Court has meant two distinct things. First, if entitled to bring a facial challenge, a plaintiff need not apply for a license before challenging the licensing regime. *See* Lakewood, 380 U.S. at 755-56. This is a question of standing. Second, a litigant challenging an enactment on its face champions the rights of those not before the court and thus may attack the statute "whether or not his conduct could be proscribed by a properly drawn statute." Freedman v. Maryland, 380 U.S. 51, 56 (1965); see also Secretary of State of Md. v. J. H. Munson Co., 467 U.S. 947, 957 (1984); Roulette, 97 F.3d at 303 n.3. This goes to the scope of the constitutional challenge.

to discriminate based on the content or viewpoint of speech by suppressing disfavored speech or disliked speakers", and has "a close enough nexus to expression, or to conduct commonly associated with expression, to pose a real and substantial threat of . . . censorship risks". *Id.* at 759.

[4] The EAR regulations at issue plainly satisfy the first requirement—"the determination of who may speak and who may not is left to the unbridled discretion of a government official". *Id.* at 763. BXA administrators are empowered to deny licenses whenever export might be inconsistent with "U.S. national security and foreign policy interests." 15 C.F.R. S 742.15 (b). No more specific guidance is provided. Obviously, this constraint on official discretion is little better than no constraint at all. *See* Lakewood, 486 U.S. at 769-70 (a standard requiring that license denial be in the "public interest" is an "illusory" standard that "renders the guarantee against censorship little more than a high-sounding ideal."). The government's assurances that BXA administrators will not, in fact, discriminate on the basis of content are beside the point. *See id.* at 770 (presumption that official will act in good faith "is the very presumption that the doctrine forbidding unbridled discretion disallows."). After all, "the mere existence of the licensor's unfettered discretion, coupled with the power of prior restraint, intimidates parties into censoring their own speech, even if the discretion and power are never actually abused." *Id.* at 757.

The more difficult issue arises in relation to the second requirement—that the challenged regulations exhibit "a close enough nexus to expression". We are called on to determine whether encryption source code is expression for First Amendment purposes.[9]

We begin by explaining what source code is.[10] "Source code," at least as currently understood by computer programmers, refers to the text of a

---

9 As an initial matter, we note that the fact that the regulations reach only "exports" does not reduce the burden on Bernstein's First Amendment rights. It is Bernstein's right to speak, not the rights of foreign listeners to hear, that we are concerned with here. The government does not argue, nor could it, that being cut off from a foreign audience, as distinguished from a domestic one, does not implicate First Amendment concerns. *See* Bullfrog Films, Inc. v. Wick, 847 F.2d 502, 509 n.9 (9th Cir. 1988). In addition, because the regulations define "export" to include the use of Internet fora that may be accessible by foreign nationals, as well as domestic communications with foreign nationals, we think it plain that the regulations potentially limit Bernstein's freedom of speech in a variety of both domestic and foreign contexts. *See* Reno v. American Civ. Lib. Union, 117 S. Ct. 2329, 2348-49 (1997) (rejecting government argument that restriction of expression on the Internet is justified because ample alternative channels of communication exist).

10 In undertaking this task, we are mindful that computer technology, and the lexicon of terms that accompanies it, is changing rapidly. Nevertheless, because the regulations speak in terms of "source code," we premise our discussion on the meaning commonly ascribed to this term by the programming community.

program written in a "high-level" programming language, such as "PASCAL" OR "C". The distinguishing feature of source code is that it is meant to be read and understood by humans and that it can be used to express an idea or a method. A computer, in fact, can make no direct use of source code until it has been translated ("compiled") into a "low-level" or "machine" language, resulting in computer-executable "object code". That source code is meant for human eyes and understanding, however, does not mean that an untutored layperson can understand it. Because source code is destined for the maw of an automated, ruth-lessly literal translator—the compiler—a programmer must follow stringent grammatical, syntactical, formatting, and punctuation conventions. As a result, only those trained in programming can easily understand source code.[11]

Also important for our purposes is an understanding of how source code is used in the field of cryptography. Bernstein has submitted numerous declarations from cryptographers and computer programmers explaining that cryptographic ideas and algorithms are conveniently expressed in source code.[12]

---

11 It must be emphasized, however, that source code is merely text, albeit text that conforms to stringent formatting and punctuation requirements. For example, the following is an excerpt from Bernstein's Snuffle source code:

```
for (; ;)
(
uch = gtchr();
if (! (n & 31))
(
for (i = 0; i64; i++)
1 [ ctr[i] ] = k[i] + h[n -64 + i]
Hash512 (wm, wl, level, 8);
)
```

As source code goes, Snuffle is quite compact; the entirety of the Snuffle source code occupies fewer than four printed pages.

12 Source code's power to convey algorithmic information is illustrated by the declaration of MIT Professor Harold Abelson:

The square root of a number X is the number Y such that Y times Y equals X. This is declarative knowledge. It tells us something about square roots. But it doesn't tell us how to find a square root.

In contrast, consider the following ancient algorithm, attributed to Heron of Alexandria, for approximating square roots:

To approximate the square root of a positive number X,

—Make a guess for the square root of X.

—Compute an improved guess as the average of the guess and X divided by the guess.

—Keep improving the guess until it is good enough.

Heron's method doesn't say anything about what square roots are, but it does say how to approximate them. This is a piece of imperative "how to" knowledge.

Computer science is in the business of formalizing imperative knowledge—developing formal notations and ways to reason and talk about methodology. Here is Heron's method formalized as a procedure in the notation of the Lisp computer

That this should be so is, on reflection, not surprising. As noted earlier, the chief task for cryptographers is the development of secure methods of encryption. While the articulation of such a system in layman's English or in general mathematical terms may be useful, the devil is, at least for cryptographers, often in the algorithmic details. By utilizing source code, a cryptographer can express algorithmic ideas with precision and methodological rigor that is otherwise difficult to achieve. This has the added benefit of facilitating peer review—by compiling the source code, a cryptographer can create a working model subject to rigorous security tests. The need for precisely articulated hypotheses and formal empirical testing, of course, is not unique to the science of cryptography; it appears, however, that in this field, source code is the preferred means to these ends.

[5] Thus, cryptographers use source code to express their scientific ideas in much the same way that mathematicians use equations or economists use graphs. Of course, both mathematical equations and graphs are used in other fields for many purposes, not all of which are expressive. But mathematicians and economists have adopted these modes of expression in order to facilitate the precise and rigorous expression of complex scientific ideas.[13] Similarly, the undisputed record here makes it clear that cryptographers utilize source code in the same fashion.[14]

---

language:

```
(define (sqrtx)
(define (good-enough? guess)
((abs (-(square guess) x)) tolerance))
(define (improve guess)
(average guess (/ x guess)))
(define (try guess)
(if (good-enough? guess)
guess
(try (improve guess))))
(try 1))
```

13 We are reminded of at least one occasion in which a judicial thinker resorted to a mathematical equation to express a legal principle. *See* United States v. Carroll Towing Co., 159 F.2d 169, 173 (2d Cir. 1947) (Judge Hand's famous BPL formula to determine "when the absence of a bargee or other attendant will make the owner of the barge liable for injuries to other vessels if she breaks away from her moorings.").

14 Bernstein's Snuffle, in fact, provides an illustration of this point. By developing Snuffle, Bernstein was attempting to demonstrate that a one-way hash function could be employed as the heart of an encryption method. The Snuffle source code, as submitted by Bernstein to the State Department, was meant as an expression of how this might be accomplished. The Source Code was plainly not intended as a completed encryption product, as demonstrated by the fact that it was incomplete and not in a form suitable for final compiling. The Source Code, in fact, omits the hash function entirely—until combined with such a function and compiled, Snuffle is incapable of performing encryption functions at all.
Snuffle was also intended, in part, as political expression. Bernstein discovered that the ITAR regulations controlled encryption exports, but not one-way hash functions.

[6] In light of these considerations, we conclude that encryption software, in its source code form[15] and as employed by those in the field of cryptography, must be viewed as expressive for First Amendment purposes, and thus is entitled to the protections of the prior restraint doctrine. If the government required that mathematicians obtain a prepublication license prior to publishing material that included mathematical equations, we have no doubt that such a regime would be subject to scrutiny as a prior restraint. The availability of alternate means of expression, moreover, does not diminish the censorial power of such a restraint—that Adam Smith wrote Wealth of Nations without resorting to equations or graphs surely would not justify governmental prepublication review of economics literature that contain these modes of expression.

The government, in fact, does not seriously dispute that source code is used by cryptographers for expressive purposes. Rather, the government maintains that source code is different from other forms of expression (such as blueprints, recipes, and "how-to" manuals) because it can be used to control directly the operation of a computer without conveying information to the user. In the government's view, by targeting this unique functional aspect of source code, rather than the content of the ideas that may be expressed therein, the export regulations manage to skirt entirely the concerns of the First Amendment. This argument is flawed for at least two reasons.

[7] First, it is not at all obvious that the government's view reflects a proper understanding of source code. As noted earlier, the distinguishing feature of source code is that it is meant to be read and understood by humans, and that it cannot be used to control directly the functioning of a computer. While source code, when properly prepared, can be easily compiled into object code by a user, ignoring the distinction between source and object code obscures the important fact that source code is not meant solely for the computer, but is rather written in a language intended also for human analysis and understanding.

[8] Second, and more importantly, the government's argument, distilled to its essence, suggests that even one drop of "direct functionality" overwhelms any constitutional protections that expression might otherwise

---

Because he believed that an encryption system could easily be fashioned from any of a number of publicly-available one-way has functions, he viewed the distinction made by the ITAR regulations as absurd. To illustrate his point, Bernstein developed Snuffle, which is an encryption system built around a one-way hash function.

15 We express no opinion regarding whether object code manifests a "close enough nexus to expression" to warrant application of the prior restraint doctrine. Bernstein's Snuffle did not involve object code, nor does the record contain any information regarding expressive uses of object code in the field of cryptography.

enjoy. This cannot be so.[16] The distinction urged on us by the government would prove too much in this era of rapidly evolving computer capabilities. The fact that computers will soon be able to respond directly to spoken commands, for example, should not confer on the government the unfettered power to impose prior restraints on speech in an effort to control its "functional" aspects. The First Amendment is concerned with expression, and we reject the notion that the admixture of functionality necessarily puts expression beyond the protections of the Constitution.

[9] The government also contends that the challenged regulations are immune from prior restraint analysis because they are "laws of general application" rather than being "directed narrowly and specifically at expression". Lakewood, 486 U.S. at 760-61. We cannot agree. Because we conclude that source code is utilized by those in the cryptography field as a means of expression, and because the regulations apply to encryption source code, it necessarily follows that the regulations burden a particular form of expression directly.

[10] The Supreme Court in Lakewood explored what it means to be a "law of general application" for prior restraint purposes. In that case, the Court cited a law requiring building permits as a "law of general application" that would not be subject to a facial attack as a prior restraint, reasoning that such a law carried "little danger of censorship, "even if it could be used to retaliate against a disfavored newspaper seeking to build a printing plant. *Id.* at 761. In the Court's view, "such laws provide too blunt a censorship instrument to warrant judicial intervention prior to an allegation of actual misuse." *Id.* Unlike a building permit ordinance, which would afford government officials only intermittent and unpredictable opportunities to exercise unrestrained discretion over expression, the challenged EAR regulations explicitly apply to expression and place scientific expression under the censor's eye on a regular basis. In fact, there is ample evidence in the record establishing that some in the cryptography field have already begun censoring themselves, for fear that their statements might influence the disposition of future licensing applications. *See, e.g.,* NATIONAL RESEARCH COUNCIL, CRYPTOGRAPHY'S ROLE IN SECURING THE INFORMATION SOCIETY 158 (1996) ("Vendors contended that since they are effectively at the mercy of the export control regulators, they have considerable incentive to suppress any public expression of dissatisfaction with the current process."). In these circumstances, we cannot conclude that the export

16 If it were, we would have expected the Supreme Court to start and end its analysis of David Paul O'Brien's burning of his draft card with an inquiry into whether he was kept warm by the ensuing flames. *See* United States v. O'Brien, 391 U.S. 367 (1968).

control regime at issue is a "law of general application" immune from prior restraint analysis.[17]

[11] Because the prepublication licensing scheme challenged here vests unbridled discretion in government officials, and because it directly jeopardizes scientific expression, we are satisfied that Bernstein may properly bring a facial challenge against the regulations.[18] We accordingly turn to the merits.

First, the dissent conceives of the exchange of source code among scientists as "conduct". We disagree. The source code at issue here is text intended for human understanding, albeit in a specialized language. To say that the "export:" of this text is "conduct" for First Amendment purposes, rather than straightforward scientific "expression", is to call into question all distribution and circulation of scientific texts that communicate ideas by using specialized languages. Of course, source code may be functional as well as expressive. We are not persuaded, however, that that fact transmogrifies the distribution of scientific texts from "expression" into "conduct" deserving of diminished First Amendment protection.

Having cast the question as one relating to "conduct", the dissent then takes a second step. Drawing from Lakeside, the dissent asks whether the "conduct"—the exchange of cryptographic source code—is "commonly

---

17 The government also argues that the EAR regulations are "laws of general application" because they are not purposefully aimed at suppressing any particular ideas that may be expressed in source code. With respect to this contention, the panel (including the dissenter) agree that the purpose of the regulations is irrelevant to prior restraint analysis. It is clear that a prior restraint analysis applies equally to content-neutral or content-based enactments. *See* FW/PBS, Inc. v. Dallas, 493 U.S. 215, 223 (1990) (plurality opinion of O'Connor, J.) ("Because we conclude that the city's licensing scheme lacks adequate procedural safeguards, we do not reach . . . whether the ordinance is properly viewed as a content-neutral time, place, and manner restriction. . . ."); Lakewood, 486 U.S. at 764 ("[E]ven if the government may constitutionally impose content-neutral prohibitions on a particular manner of speech, it may not condition that speech on obtaining a license or permit from a government official in that official's boundless discretion.") (emphasis in original). Indeed, where unbridled discretion is vested in a governmental official, it is difficult to know whether a licensing regime is content-based or content-neutral. Accordingly, the government's purpose in censoring encryption source code is, at this stage of our First Amendment inquiry, beside the point. In other words, a prepublication licensing regime that has a chilling and censorial effect on expression is properly subject to facial attack as a prior restraint, whatever the purpose behind its enactment. *See* Lakewood, 486 U.S. at 759 (upholding facial attack against newsrack ordinance because of censorial effects, without discussing governmental purpose for enacting the ordinance).

18 It is at this juncture that we part ways with the dissent. The dissent concedes that source code can be expressive. Nevertheless, the dissent contends that Bernstein is not entitled to bring a facial attack against the EAR regulation. This argument, it seems to us, is based on two foundations.

associated with expression". This question the dissent answers in the negative; in other words, the dissent concludes that source code is not used expressively often enough. We find this conclusion somewhat perplexing, as there is nothing in the record to support it. Bernstein has introduced extensive expert evidence to support his contention that source code is frequently used for expressive purposes. The government, however, has failed to introduce anything into the record to rebut this evidence. In fact, the government has made it clear that it means to control the export of source code no matter how commonly associated it may be with expression: "Whatever ideas may be reflected in the software, or the intent of the exporter to convey ideas, the NSA recommends that encryption software be controlled for export solely on the basis of what it does. . . ." Second Lowell Decl., Appellant's Excerpts of Record at 104.

B. Are the regulations an impermissible prior restraint?

[12] "[T]he protection even as to previous restraint is not absolutely unlimited." Near, 283 U.S. at 716. The Supreme Court has suggested that the "heavy presumption" against prior restraints may be overcome where official discretion is bounded by stringent procedural safeguards. *See* FW/PBS, 493 U.S. at 227 (plurality opinion of O'Connor, J.); Freedman v. Maryland, 380 U.S. 51, 58-59 (1965); Kingsley Books, 354 U.S. at 442-43; 11126 Baltimore Blvd. v. Prince George's County, 58 F. 3d 988, 995 (4th Cir. 1995) (en banc). As our analysis above suggests, the challenged regulations do not qualify for this First Amendment safe harbor.[19] In Freedman v. Maryland, the Supreme Court set out three factors for determining the validity of licensing schemes that impose a prior restraint on speech: (1) any restraint must be for a specified brief period of time; (2) there must be expeditious judicial review; and (3) the censor must bear the burden of going to court to suppress the speech in question and must bear the burden

---

19 The Supreme Court has also suggested that the presumption against prior restraints may be overcome where publication would directly and imminently imperil national security. *See* New York Times Co. v. United States, 403 U.S. 713, 730 (1971) (Stewart, J., joined by White, J., concurring); Near, 283 U.S. at 716; see also United States v. The Progressive, Inc., 467 F. Supp. 990, 992 (W.D. Wisc. 1979). In order to justify a prior restraint on national security grounds, the government must prove the publication would "surely result in direct, immediate, and irreparable damage to our Nation or its people." New York Times, 403 U.S. at 730 (Stewart, J., joined by White, J., concurring); see also id. at 726-27 (Brennan, J., concurring) (finding that national security is a sufficient interest only where there is "governmental allegation and proof that publication must inevitably, directly, and immediately cause the occurrence of an event kindred to imperiling the safety of a transport already at sea"); Burch v. Baker, 861 F.2d 1149, 1155 (9th Cir. 1988) ("Prior restraints are permissible in only the rarest of circumstances, such as imminent threat to national security.").
The government does not argue that the prior restraint at issue here falls within the extremely narrow class of cases where publication would directly and immediately imperil national security.

of proof.[20] *See* 380 U.S. at 58-60. The district court found that the procedural protections provided by the EAR regulations are "woefully inadequate" when measured against these requirements. Bernstein III, 974 F. Supp. at 1308. We agree.

[13] Although the regulations require that license applications be resolved or referred to the President within 90 days, see 15 C.F.R. S 750.4(a), there is no time limit once an application is referred to the President. Thus, the 90-day limit can be rendered meaningless by referral. Moreover, if the license application is denied, no firm time limit governs the internal appeals process. *See* 15 C.F.R. S 756.2(c)(1) (Under Secretary "shall decide an appeal within a reasonable time after receipt of the appeal."). Accordingly, the EAR regulations do not satisfy the first Freedman requirement that a licensing decision be made within a reasonably short, specified period of time. *See* FW/PBS, 493 U.S. at 226 (finding that "a prior restraint that fails to place time limits on the time within which the decisionmaker must issue the license is impermissible"); Riley v. National Fed. of the Blind, 487 U.S. 781, 802 (1988) (licensing scheme that permits "delay without limit" is impermissible); Vance v. Universal Amusement Co., 445 U.S. 308, 315-17 (1980) (prior restraint of indefinite duration is impermissible). The EAR regulatory regime further offends Freedman's procedural requirements insofar as it denies a disappointed applicant the opportunity for judicial review.[21] *See* 15 C.F.R. S 756.2(c)(2); FW/PBS, 493 U.S. at 229 (plurality opinion of O'Connor, J.) (finding failure to provide "prompt" judicial review violates Freedman); Freedman, 380 U.S. at 59 (licensing procedure must assure a prompt final judicial decision).

[14] We conclude that the challenged regulations allow the government to restrain speech indefinitely with no clear criteria for review. As a result, Bernstein and other scientists have been effectively chilled from engaging

---

20 Whether all three Freedman factors apply to all prior restraints is the subject of dispute. Compare FW/PBS, 493 U.S. at 229-30 (plurality opinion of O'Connor, J.) (finding the government does not bear the burden of going to court to defend its licensing requirement where restrained speakers are likely to challenge the restraint in court) with id. at 239 (Brennan, J., concurring in judgment) ("We have never suggested that our insistence on Freedman procedures might vary with the particular facts of the prior restraint before us."). Because we conclude that the EAR regulations fail Freedman's first two procedural requirements, we need not reach the issue of whether the third Freedman factor applies in this case.

21 As noted earlier, the BXA enjoys essentially unbounded discretion under the EAR regulations in administering the license process. Accordingly, even if the challenged regulations provided for judicial review, the lack of explicit limits on the decisionmaker's discretion would likely make such review meaningless. In this sense, the presence of unbounded discretion itself may be considered fatal for purposes of prior restraint review. *See* Lakewood, 486 U.S. at 769-70 (striking down a licensing scheme where the mayor could merely claim that the license" is not in the public interest' when denying a permit application").

in valuable scientific expression. Bernstein's experience itself demonstrates the enormous uncertainty that exists over the scope of the regulations and the potential for the chilling of scientific expression. In short, because the challenged regulations grant boundless discretion to government officials, and because they lack the required procedural protections set forth in Freedman, we find that they operate as an unconstitutional prior restraint on speech.[22] *See* Lakewood, 486 U.S. at 769-772 (holding that newsrack licensing ordinance was an impermissible prior restraint because it conferred unbounded discretion and lacked adequate procedural safeguards).

C. Concluding comments

We emphasize the narrowness of our First Amendment holding. We do not hold that all software is expressive. Much of it surely is not. Nor need we resolve whether the challenged regulations constitute content-based restrictions, subject to the strictest constitutional scrutiny, or whether they are, instead, content-neutral restrictions meriting less exacting scrutiny. We hold merely that because the prepublication licensing regime challenged here applies directly to scientific expression, vests boundless discretion in government officials, and lacks adequate procedural safeguards, it constitutes an impermissible prior restraint on speech.

We will, however, comment on two issues that are entwined with the underlying merits of Bernstein's constitutional claims. First, we note that insofar as the EAR regulations on encryption software were intended to slow the spread of secure encryption methods to foreign nations, the government is intentionally retarding the progress of the flourishing science of cryptography. To the extent the government's efforts are aimed at interdicting the flow of scientific ideas (whether expressed in source code or otherwise), as distinguished from encryption products, these efforts would appear to strike deep into the heartland of the First Amendment. In this regard, the EAR regulations are very different from content-neutral time, place and manner restrictions that may have an incidental effect on expression while aiming at secondary effects.

Second, we note that the government's efforts to regulate and control the spread of knowledge relating to encryption may implicate more than the First Amendment rights of cryptographers. In this increasingly electronic age, we are all required in our everyday lives to rely on modern technology to communicate with one another. This reliance on electronic communica-

---

22 Our conclusion relating to the Source Code also resolves the status of the regulations as applied to the Instructions. Because the Instructions are essentially a translation of the Source Code into English, they are, if anything, nearer the heartland of the First Amendment. Consequently, to the extent the challenged regulations are unconstitutional as applied to the Source Code, they necessarily are unconstitutional as applied to the Instructions.

tion, however, has brought with it a dramatic diminution in our ability to communicate privately. Cellular phones are subject to monitoring, email is easily intercepted, and transactions over the Internet are often less than secure. Something as commonplace as furnishing our credit card number, social security number, or bank account number puts each of us at risk. Moreover, when we employ electronic methods of communication, we often leave electronic "fingerprints" behind, fingerprints that can be traced back to us. Whether we are surveilled by our government, by criminals, or by our neighbors, it is fair to say that never has our ability to shield our affairs from prying eyes been at such a low ebb. The availability and use of secure encryption may offer an opportunity to reclaim some portion of the privacy we have lost. Government efforts to control encryption thus may well implicate not only the First Amendment rights of cryptographers intent on pushing the boundaries of their science, but also the constitutional rights of each of us as potential recipients of encryption's bounty. Viewed from this perspective, the government's efforts to retard progress in cryptography may implicate the Fourth Amendment, as well as the right to speak anonymously, see McIntyre v. Ohio Elections Comm'n, 115 S. Ct. 1511, 1524 (1995), the right against compelled speech, see Wooley v. Maynard, 430 U.S. 705, 714 (1977), and the right to informational privacy, see Whalen v. Roe, 429 U.S. 589, 599-600 (1977). While we leave for another day the resolution of these difficult issues, it is important to point out that Bernstein's is a suit not merely concerning a small group of scientists laboring in an esoteric field, but also touches on the public interest broadly defined.

II. Scope of Declaratory Relief

The government also challenges the scope of the declaratory relief granted by the district court. The government argues that the relief provided is invalid in two respects: (1) that the relief extends to encryption object code and encryption commodities; (2) that the relief extends to encryption technology. The district held that

> the Export Administration Regulations, 15 C.F.R. pt. 730 et seq. (1997) and all rules, policies and practices promulgated or pursued thereunder insofar as they apply to or require licensing for encryption and decryption software and related devices and technology are in violation of the First Amendment on the grounds of prior restraint and are, therefore, unconstitutional as discussed above, and shall not be applied to plaintiff's publishing of such items, including scientific papers, algorithms or computer programs.

Bernstein III, 974 F. Supp. at 1310. We review the district court's grant of declaratory relief de novo. *See* Crawford v. Lungren, 96 F.3d 380, 384 (9th Cir. 1996); Ablang v. Reno, 52 F.3d 801, 803 (9th Cir. 1995).

This inquiry leads us into the uncertain jurisprudence of "severability" *See generally* John Copeland Nagle, Severability, 72 N.C. L. REV. 203 (1993). The general principle is clear: "[A] court should refrain from invalidating more of [a] statute than is necessary. . . . '[W]henever an act of Congress contains unobjectionable provisions separable from those found to be unconstitutional, it is the duty of this court to so declare, and to maintain the act in so far as it is valid'." Alaska Airlines, Inc. v. Brock, 480 U.S. 678, 684 (1987) (quoting Regan v. Time, Inc., 468 U.S. 641, 652 (1984)); see also National Collegiate Athletic Ass'n v. Miller , 10 F.3d 633, 640 (9th Cir. 1993). The applicable legal standard has also been oft repeated: "[u]nless it is evident that the Legislature would not have enacted those provisions which are within its power, independently of that which is not, the invalid part may be dropped if what is left is fully operative as a law." Buckley v. Valeo, 424 U.S. 1, 108 (1976) (per curiam); accord NCAA v. Miller, 10 F.3d at 640. Thus, in the general case, severability analysis properly focuses on legislative intent. *See* Alaska Airlines, Inc., 480 U.S. at 685.

This case, however, is not the general case. First, the challenged enactment here is a regulation, rather than a statute. As a result, we cannot look to the usual public sources to determine the intentions of the drafters. Nevertheless, we agree with the government that the EAR regulations can be conceptually severed into component parts governing commodities, software, and technology. We also assume that the Department of Commerce, even if barred from imposing prepublication licensing on encryption source code, would have enacted regulations controlling the export of encryption commodities, object code, and technology.

But while the district court may have erred in treating software and commodities as the same item, the integrated structure of the regulations does not permit us to sever the various provisions in the manner requested by the government. To sever the unconstitutional portion of the regulations, we would have to line edit individual sections, deleting or modifying the definition of "software" while retaining "commodities" and "technology." We would then have to redefine general terms such as "items" which refer collectively to commodities, software, and technology. We have neither the power nor the capacity to engage in line by line revisions of the challenged regulations or to redefine terms within the regulations. *See* Hill v. Wallace, 259 U.S. 44, 70-71 (1922); American Booksellers Ass'n v. Hudnut, 771 F.2d 323, 332-33 (7th Cir. 1985). To do so would be to improperly invade the province reserved to the Executive. Accordingly, we affirm the district court's grant of declaratory relief.

CONCLUSION

Because the prepublication licensing regime challenged by Bernstein applies directly to scientific expression, vests boundless discretion in government officials, and lacks adequate procedural safeguards, we hold that it constitutes an impermissible prior restraint on speech. We decline the invitation to line edit the regulations in an attempt to rescue them from constitutional infirmity, and thus endorse the declaratory relief granted by the district court.

AFFIRMED.

BRIGHT, Circuit Judge, separately concurring.

I join Judge Fletcher's opinion. I do so because the speech aspects of encryption source code represent communication between computer programmers. I do, however, recognize the validity of Judge Nelson's view that encryption source code also has the functional purpose of controlling computers and in that regard does not command protection under the First Amendment. The importance of this case suggests that it may be appropriate for review by the United States Supreme Court.

T.G. NELSON, Circuit Judge, Dissenting:

Bernstein was not entitled to bring a facial First Amendment challenge to the EAR, and the district court improperly granted an injunction on the basis of a facial challenge. I therefore respectfully dissent.

The basic error which sets the majority and the district court adrift is the failure to fully recognize that the basic function of encryption source code is to act as a method of controlling computers. As defined in the EAR regulations, encryption source code is "[a] precise set of operating instructions to a computer, that when compiled, allows for the execution of an encryption function on a computer". 15 C.F.R. pt. 722. Software engineers generally do not create software in object code—the series of binary digits (1's and 0's)—which tells a computer what to do because it would be enormously difficult, cumbersome and time-consuming. Instead, software engineers use high-level computer programming languages such as "C" or "Basic" to create source code as a shorthand method for telling the computer to perform a desired function. In this respect, lines of source code are the building blocks or the tools used to create an encryption machine. *See e.g.*, Patrick Ian Ross, Bernstein v. United States Department of State, 13 Berkeley Tech. L.J. 405, 410-11 (1998) ("[E]lectronic source code that is ready to compile merely needs a few keystrokes to generate object code—the equivalent of flipping an on' switch. Code used for this purpose can fairly easily be characterized as essentially functional.' "); Pamela Samuelson et al., A Manifesto Concerning Legal Protection of Computer Programs, 94 Colum. L. Rev. 2308, 2315-30 (1994) ("[P]rograms are, in fact, machines (entities that bring about useful results, i.e., behavior) that have been constructed in the medium of text (source code and object code)."). Encryption source code, once compiled, works to make computer communication and transactions secret; it creates a lockbox of sorts around a message that can only be unlocked by someone with a key. It is the function or task that encryption source code performs which creates its value in most cases. This functional aspect of encryption source code contains no expression; it is merely the tool used to build the encryption machine.

This is not to say that this very same source code is not used expressively in some cases. Academics, such as Bernstein, seek to convey and discuss their ideas concerning computer encryption. As noted by the majority, Bernstein must actually use his source code textually in order to discuss or teach cryptology. In such circumstances, source code serves to express Bernstein's scientific methods and ideas.

While it is conceptually difficult to categorize encryption source code under our First Amendment framework, I am still inevitably led to conclude that encryption source code is more like conduct than speech. Encryption source code is a building tool. Academics and computer programmers can convey this source code to each other in order to reveal the encryption machine they have built. But, the ultimate purpose of encryption code is, as its name suggests, to perform the function of encrypting messages. Thus, while encryption source code may occasionally be used in an expressive manner, it is inherently a functional device.

We are not the first to examine the nature of encryption source code in terms of First Amendment protection. Judge Gwin of the United States District Court for the Northern District of Ohio also explored the function versus expression conundrum of encryption source code at some length in Junger v. Daley, 8 F. Supp. 2d 708 (N.D. Ohio 1998). Junger, like Bernstein, is a professor, albeit a law professor, who wished to publish in various forms his work on computers, including a textbook, Computers and the Law. The book was determined by the Government to be subject to export without a license, but his software programs were determined to come within the licensing provisions of the EAR. In the course of rejecting Junger's claims, the court said:

> Like much computer software, encryption source code is inherently functional; it is designed to enable a computer to do a designated task. Encryption source code does not merely explain a cryptographic theory or describe how the software functions. More than describing encryption, the software carries out the function of encryption. The software is essential to carry out the function of encryption. In doing this function, the encryption software is indistinguishable from dedicated computer hardware that does encryption.
>
> In the overwhelming majority of circumstances, encryption source code is exported to transfer functions, not to communicate ideas. In exporting functioning capability, encryption source code is like other encryption de-

> vices. For the broad majority of persons receiving such source code, the value comes from the function the source code does.

Id. at 716. The Junger decision thus adds considerable support for the propositions that encryption source code cannot be categorized as pure speech and that the functional aspects of encryption source code cannot be easily ignored or put aside.

Both the district court and the majority hold that because source code can be used expressively in some circumstances, Bernstein was entitled to bring a facial challenge to the EAR. Such an approach ignores the basic tenet that facial challenges are inappropriate "unless, at a minimum, the challenged statute 'is directed narrowly and specifically at expression or conduct commonly associated with expression'." Roulette v. City of Seattle, 97 F.3d 300, 305 (9th Cir. 1996) (quoting City of Lakewood v. Plain Dealer Publishing Co., 486 U.S. 750, 760 (1988)). That encryption source code may on occasion be used expressively does not mean that its export is "conduct commonly associated with expression" or that the EAR regulations are directed at expressive conduct. *See id.* at 303 ("The fact that sitting can possibly be expressive, however, isn't enough to sustain plaintiffs' facial challenge."); see also Junger, 8 F. Supp. 2d at 718 ("[T]he prior restraint doctrine is not implicated simply because an activity may on occasion be expressive.").

The activity or conduct at issue here is the export of encryption source code. As I noted above, the basic nature of encryption source code lies in its functional capacity as a method to build an encryption device. Export of encryption source code is not conduct commonly associated with expression. Rather, it is conduct that is normally associated with providing other persons with the means to make their computer messages secret. The overwhelming majority of people do not want to talk about the source code and are not interested in any recondite message that may be contained in encryption source code. Only a few people can actually understand what a line of source code would direct a computer to do. Most people simply want to use the encryption source code to protect their computer communications. Export of encryption source code simply does not fall within the bounds of conduct commonly associated with expression such as picketing or handbilling. *See* Roulette, 97 F.3d at 303-04.

Further, the EAR regulates the export of encryption technology generally, whether it is software or hardware. *See* 15 C.F.R. S 742.15; Junger, 8 F. Supp. 2d at 718 ("The Export Regulations do not single out encryption software."). These regulations are directed at preventing the functional capacity of any encryption device, including its source code, from being exported without a government license. The EAR is not specifically directed towards stifling the expressive nature of source code or Bernstein's academic discussions about cryptography. This is demonstrated by the fact that the regulations do not object to publication in printed form of learned articles containing source code. *See* 15 C.F.R. S 734.3. Thus, the EAR is generally directed at non-expressive conduct—the export of source code as a tool to make messages secret and impervious to government eavesdropping capabilities.

Because this is a law of general application focused at conduct, Bernstein is not entitled to bring a facial challenge. The district court's injunction based upon the finding of a facial prior restraint is thus impermissible. This is not to say that Bernstein's activities would not be entitled to First Amendment protection, but that the legal path chosen to get that protection must be the correct one. We should be careful to "entertain[]facial freedom-of-expression challenges only against statutes that, 'by their terms,' sought to regulate 'spoken words,' or patently 'expressive or communicative conduct.' " Roulette, 97 F.3d at 303 (citing Broadrick v. Oklahoma, 413 U.S. 601, 612-13(1973)). Bernstein may very well have a claim under an as-applied First Amendment analysis; however, such a claim must be left to the district court's determination in the first instance. Here, the district court did not rule on Bernstein's as-applied claims. I would therefore vacate the district court's injunction and remand for consideration of Bernstein's as-"applied challenges to the EAR. Accordingly, I respectfully dissent.

CHAPTER 40

BUREAU OF EXPORT ADMINISTRATION U.S. DEPARTMENT OF COMMERCE

BXA STATEMENT ON ENCRYPTION CASE

May 6 Court Decision in Bernstein Encryption Case

You may have read about a recent court decision regarding encryption exports. **Please be advised that this decision does not mean that encryption products may be exported without regard to the Export Administration Regulations (EAR).** Regardless of how the decision might be interpreted, the decision is subject to a stay. This stay is in effect for at least 45 days. (See *Department of Justice press release.*)

On May 6, the U.S. Court of Appeals for the Ninth Circuit rendered a decision in *Bernstein v. The United States Department of Justice.* Professor Daniel Bernstein filed suit against the U.S. Government after he was notified by the State Department that his "Snuffle" encryption program was subject to the International Traffic in Arms Regulations (ITAR) and would require an export license to post the source code on the Internet. Bernstein subsequently amended his petition to challenge the controls on encryption products maintained under the EAR after President Clinton placed encryption exports under the Commerce Department's jurisdiction in 1996. In a 2-1 decision, the Ninth Circuit court upheld the district court's decision that the regulation of Bernstein's export of his "Snuffle" program "constitute[s] an impermissible prior restraint on speech."

Exporters should be aware that the decision **does not** affect the applicability of the EAR to exports and reexports of encryption hardware and software products or encryption technology. **This includes controls on the export of encryption software in source code.** The EAR remain in effect for these items. The Commerce Department will apprise exporters of any changes to the encryption controls.